W9-CLO-199

ATOMIC HISTORIES

Masters of Modern Physics

Advisory Board

Published Volumes

ATOMIC HISTORIES

Rudolf E. Peierls

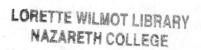

Springer

Library of Congress Cataloging-in-Publication Data
Peierls, Rudolf Ernst, Sir, 1907–1995
  Atomic histories / Rudolf E. Peierls.
    p. cm. — (Master of modern physics)
  Includes index.
  ISBN 1-56396-243-8 (hardcover)
  1. Physics. 2. Nuclear physics. I. Title. II. Series.
QC71.P38 1996   96-22298
530´.09´04—dc20   CIP

Printed on acid-free paper.

Printed and bound by United Book Press, Inc., Baltimore, MD.
Printed in the United States of America.

9 8 7 6 5 4 3 2

ISBN 1-56396-243-8 Springer-Verlag New York Berlin Heidelberg SPIN 10652590

Contents

About the Series

Masters of Modern Physics introduces the work and thought of some of the most celebrated physicists of our day. These collected essays offer a panoramic tour of the way science works, how it affects our lives, and what it means to those who practice it. Authors report from the horizons of modern research, provide engaging sketches of friends and colleagues, and reflect on the social, economic, and political consequences of the scientific and technical enterprise.

Authors have been selected for their contributions to science and for their keen ability to communicate to the general reader—often with wit, frequently in fine literary style. All have been honored by their peers and most have been prominent in shaping debates in science, technology, and public policy. Some have achieved distinction in social and cultural spheres outside the laboratory.

Many essays are drawn from popular and scientific magazines, newspapers, and journals. Still others—written for the series or drawn from notes for other occasions—appear for the first time. Authors have provided introductions and, where appropriate, annotations. Once selected for inclusion, the essays are carefully edited and updated so that each volume emerges as a finely shaped work.

Masters of Modern Physics is overseen by an advisory panel of distinguished physicists. Sponsored by the American Institute of Physics, a consortium of major physics societies, the series serves as an authoritative survey of the people and ideas that have shaped twentieth-century science and society.

Editor's Note

Sir Rudolf Peierls passed away in September 1995. At the time of his death, the preparation of this collection was very nearly, but not entirely, completed. We would like to acknowledge the kind assistance of those who helped to conclude the editorial process.

Our thanks to Professor R. H. Dalitz of Oxford University, who aided Professor Peierls during an earlier editorial phase, and who later provided further assistance in coordinating information, and to Dr. Normand Mousseau of the Université de Montréal, who helped to clarify Professor Peierls's final comments on the manuscript.

We also gratefully acknowledge the assistance of Mrs. Jo Hookway, Professor Peierls's daughter, in organizing her father's effects and providing important material therefrom.

Lastly, we are particularly grateful to Professor Sir Roger Elliott of Oxford University, who gave generously of both time and effort to resolve a number of editorial details.

According to the author's wishes and due to the historical value of Rudolf Peierls's writings, we have retained the British spelling in some of the chapters.

Preface

This is a collection of my non-technical essays. It includes a number of book reviews, as such reviews often provide a good opportunity to express a point of view. Many of the titles were chosen by editors and do not necessarily represent my choice. In these cases I have substituted my own titles; the published titles are also given. In each subject group the essays are ordered by date of publication, except where several deal with the same topic or person, when they are grouped together.

Part I, "A Physicist's Portrait Gallery," consists of profiles and obituaries. Included here are three biographical memoirs written for the Royal Society. The earliest is of Pauli, one of my teachers, greatly respected for his high standards, and feared for his sharp tongue. Heisenberg, the great originator of quantum mechanics, became a controversial figure because of his work on atomic energy during the war; the biographical memoir was written jointly with Sir Nevill Mott. "Heisenberg's Recollections" is a review of his auto-biographical book. "A Heisenberg Biography" reviews a biography of Heisenberg by David Cassidy, which seems to me good on personal history, but weak on physics. "The Bomb that Never Was" disagrees with a book that tries to show that Heisenberg deliberately prevented a German bomb being made.

"A Physicist who Enjoys it" is a review of the autobiography of O. R. Frisch, an original and inventive experimentalist; "Otto Robert Frisch" is his biographical memoir.

"Paul Adrien Maurice Dirac" is a brief obituary of this great physicist. There is also an informal address at a memorial meeting for Dirac, and an article in a collective volume about him.

Herbert Skinner was a personal friend and a great personality in the physics community. Of the great Niels Bohr there is a brief appreciation and a light-hearted address, "Truth and Clarity" at the centenary meeting.

"Rutherford and Bohr" is an account of the relation between two great scientists, of very different personalities, yet very close friends. "J. Robert Oppenheimer" is the entry in the Dictionary of Scientific Biography. "The Growing Pains of J. Robert Oppenheimer" reviews a book about his problems as a young man, based on his letters.

"Two Mathematicians" reviews a book about John von Neumann and Norbert Wiener, which seems to me too negative about the first, and perhaps a little too positive about the other. "Physics and Homi Bhabha" reviews the biography of a distinguished theoretical physicist who later became important in Indian science policy. "Physicist Extraordinary" reviews a biography of Oliphant, a great and original personality in physics. "Landau in the 1930s" was a talk at a Landau Memorial conference in Tel Aviv and Copenhagen.

"William George Penney" is an obituary of Lord Penney, who was responsible for the design and development of British nuclear weapons. "Conservative Revolutionary" reviews a book about Max Planck, who started quantum theory, and thereby revolutionized physics, to his own surprise.

"Recollections of James Chadwick" was a talk at a Chadwick centenary meeting in Liverpool. "Bell's Early Work," a contribution to a collection of articles to commemorate John Bell, talks about his brilliance as a young man at Birmingham and Harwell.

Part II is concerned with problems of atomic energy. The first paper is the memorandum which O. R. Frisch and I wrote in March 1940, when we first recognized the feasibility and the danger of the atom bomb. After 1945 there was a great demand for simple expositions of the basic facts. "Atomic Energy—Threat and Promise" is an example of such an attempt at a simple explanation. "Britain in the Atomic Age" sketches the situation for an American audience. Two essays are written to argue against mistaken ideas: "Defense Against the Atom Bomb" deals with the idea that we could all move underground and thus be safe from nuclear attack; "Limited Nuclear War?" protests against the claim that escalation can be avoided by the use of tactical nuclear weapons. "Agonizing Misappraisal" (the title mocks the then-current phrase "agonizing reappraisal") is a review of Herman Kahn's book in which he claims that a country could reasonably survive a full-scale nuclear war. "Counting Weapons" argues that an arsenal above a minimum deterrent confers no advantage. "The Case for the Defense" reviews a collection of essays by Edward Teller, and disagrees both with his views on the Strategic Defense Initiative and on his recollection about Los Alamos.

Other essays are retrospective. "Forty Years into the Atomic Age" reviews the celebrations of the fortieth anniversary of the first chain reaction. "Memories of the Secret City" reviews a collection of reminiscences of Los Alamos. "Nuclear Weapons: How Did We Get There and Where Are We Going?" is a personal account of the development and the prospects. "Reflections of a

British Participant" is a similar account. "The Making of the Atom Bomb" is a review of the excellent book by Richard Rhodes. The essay "Atomic History" reviews the book by Morton Szasz about the British contribution to the Manhattan project, particularly at Los Alamos, and Robert Serber's "Los Alamos Primer", the book used to bring newcomers up to date. "Energy from Heaven and Earth" reviews and, on the whole, agrees with Edward Teller's views on energy policy.

Two essays are related to the misuse of security. "In the Matter of J. Robert Oppnheimer" describes the unfair and prejudiced treatment of Oppenheimer in the famous trial. "'Security' Troubles" is a chapter of my autobiography recounting my experience in this field.

Part III is a collection of miscellaneous items. "The Concept of the Positron" reviews a book by Norwood R. Hanson, who approaches the problem with the attitude of a philosopher; the review stresses the difference in outlook between physicist and philosopher.

"The Scientist in Public Affairs" discusses the role of scientists at a time when their views on policy receive some attention. "Born-Einstein Correspondence" reviews the relations between two scientists who share many liberal views but ultimately disagree over Born's decision to return to Germany, and over Einstein's support for an American bomb during the war. "Is There a Crisis in Science?" is a Wunsch Lecture at the Technion, Haifa, which tries to separate facts from prejudice.

"The Jew in Twentieth-Century Physics" asks how far the contribution by Jews to modern physics is greater than corresponds to their numbers. It concludes that this is true at some periods and in some countries, but there are great variations.

"The Physicists" reviews a book with that title by C. P. Snow. "Fact and Fancy in Physics" is a review of two books; one about Robert Millikan, an important physicist and organizer, the other about an imaginary physicist at the time of the development of quantum theory.

"What Einstein Did" reviews the book by Abraham Pais about Einstein's work, and attempts to summarize his work in simple language.

"Reminiscences of Cambridge in the Thirties" is a contribution to a collection of stories about the period.

"Struggling with Quantum Mechanics" is a review of a book by John Gribbin, which tries to explain the basis of quantum mechanics in layman's language. The book is lively but contained serious errors. I note that a second edition corrected these errors satisfactorily. "Kapitza Detained" reviews a book about Kapitza not being allowed to return to his work in Cambridge from a vacation in Moscow and the efforts of Rutherford and others to get this decision reversed.

"Does Physics Ever Come to an End?" is an account of a talk given at a Nishina Memorial Conference in Tokyo. It concludes that the answer can be "yes" or "no" depending on what precisely is meant by "physics."

"Microwave Cooking for 'Foxes'" is a review of a book by the German nuclear physicist Heinz Maier-Leibnitz and a friend about microwave cookery. The "foxes" are people with an intellectual approach to the art. "First Encounter with English Food" is a contribution to a volume, edited by Nicholas Kurti, of essays by fellows of the Royal Society on thoughts about food.

Rudolf Peierls

Rudi Peierls—An Appreciation

Sir Rudolf Peierls (as he became) was one of the great physicists of his generation who had an enormous influence over the development of the subject and indirectly over world affairs. His long and varied career falls into three periods of almost equal length. He was born in Berlin in 1907 into an assimilated partly Jewish family. It is significant that he never subsequently professed any religion but maintained a humane and rational approach to all issues, including the dilemmas which beset those who had worked on nuclear weapons. He originally wanted to become an engineer like his father but was dissuaded by his family and eventually studied physics at the Universities of Berlin, Munich (with Sommerfeld) and Leipzig (with Heisenberg), and he then became a research assistant of Pauli in Zurich. Pauli was renowned as a trenchant critic of other's work with a sharp tongue. Peierls inherited the former talent but he combined it with a courtesy and creativeness which meant that in his later career his criticism was avidly sought by his colleagues rather than feared.

There followed a period of brief sojourns in all the great European centers of physics which at that time were bursting with ideas and developments as the impact of the new quantum mechanics was digested. This included some time in Russia where, at a conference in Odessa, he met his future wife Genia. His main contributions during this period involved the application of quantum mechanics to the properties of electrons in solids and the phenomena of electrical and thermal conductivity which could now be better understood using the new ideas. But his work also stretched into the properties of nuclei as well as into the structure of materials. He belonged to the last generation where it seemed possible for one man to encompass the knowledge to make detailed contributions across the whole range of physics.

Toward the end of this period, which is splendidly described in his autobiography "Bird of Passage," the family came to England to temporary appointments in Cambridge and in Manchester. His reflections on how he and his wife coped with the unfamiliar culture is entertaining reading. But the Peierls came to love Britain for the reasonable and civilized behavior they encountered in society at all levels, especially when they compared it with the problems they had seen in Europe. They settled in Birmingham where he was appointed Professor in 1937 and where he was to remain for the middle, and most creative, period of his life. Here he built a "school" of Theoretical Physics quite unlike any other in Britain, which produced a stream of talented scientists many of whom went on to fill important academic positions in Britain and the U.S. Its reputation grew to a point where in addition to large numbers of students (some 200 passed through during his tenure) there were frequent visits from active scientists from all over the world. The department was often compared to a family with Rudi taking a fatherly interest in the scientific development of everyone under his charge while Genia was a motherly figure who kept open house and was always ready to assist with any personal problems. In his funeral speech Ronnie Peierls said that there often seemed to be no difference in the home between himself and his sisters and the various students and visitors who were continually present.

Genia used to divide physicists into two classes, "golfers" who single–mindedly pursued a lonely path to a clear goal (Dirac was the archetypical example) and "tennis players" who made progress by batting ideas backwards and forwards. In this classification Rudi was the tennis player par excellence who not only played the game brilliantly but always brought out the best in his opponents.

It was during the period at Birmingham that he wrote in 1940 his famous memorandum with Otto Frisch (reproduced in this volume) that showed clearly for the first time how an atomic (super) bomb was feasible, how it might be constructed, and what sort of effect it would produce. This led directly to the beginning of a British program which was eventually merged with the Manhattan Project with Peierls as the leader of the British team at Los Alamos. The many issues raised by the bomb played a major part in his future contribution to public activities, and he became a leading member of the Atomic Scientists Association and subsequently of the Pugwash Group.

Rudi Peierls came to Oxford as Wykeham Professor of Physics in 1964 and was to remain there for the rest of his life. The decision to move from Birmingham was a difficult one but in the end he decided that he would like, after twenty years, to accept the broader horizons and challenges of that ancient university. He realized from the beginning that it would not be possible to recreate the highly centralized department which had grown up around him in Birmingham because of the very different structure of the university. But he was able to augment the existing department with several new members of staff and give it a new coherence by creating a separate geographical entity. He continued to emphasize that a broad training was necessary for theoretical physicists through a pattern of courses for graduate students which were more formalized and wide ranging than was usual at that time. He took an interest in all the research programs which his catholic knowledge of the subject made uniquely possible. He once said that one of the most important lessons which he had learned was that in physics, as in life, the hardest thing was to ask the right questions, more so than finding the right answers.

Although the department could never be the family affair which existed in Birmingham he and his wife worked hard socially to create a sense of community. Their parties and hospitality helped the many visitors enjoy their time in Oxford and they are fondly remembered by all who worked with them. His scientific interest did not wane with retirement in 1974 and he continued to travel extensively until shortly before his death. His discussions on contemporary problems were always widely welcomed, while at the same time he had become a living historical resource as one of our few remaining contacts with those who had worked in the early developments of quantum mechanics and atomic weapons. Genia often accompanied him on his travels and her death in 1986 was a great blow, but he coped splendidly in his calm and rational way with this tragedy as he had with all the other crises of his life.

In the epilogue to his autobiography Rudi describes himself as an optimist, with the definition that an optimist is a person who believes the future is uncertain. He goes on to say, "our fathers' generation took it for granted that changes would all be improvements and that progress must lead to a better quality of life. We have learned to question this, we believe today that progress provides opportunities for improvement but also opportunities for disaster and for new kinds of cruelty and suffering. The world will not become a better place unless we try hard to make it so. I hope that I have not only added a few small bricks to the growing edifice of science but also contributed a little to the fight against its misuse." Those who were privileged to work with him know from personal experience that he fulfilled both these hopes to an overflowing measure. With his death we have lost not only one of the last direct links with the pioneers of quantum theory but also a quiet voice of reason and moderation in the wider world, while those who knew him well have lost a warm, generous and courteous colleague.

Roger Elliott

I. A PHYSICIST'S PORTRAIT GALLERY

Wolfgang Ernst Pauli, 1900–1958

W olfgang Pauli was born on April 25, 1900 in Vienna. His father, Wolfgang Joseph Pauli, was distinguished as a biochemist and was a Professor in the University of Vienna; previously he had practiced as a doctor, and his patients included many prominent figures in Vienna society. The mother, Bertha, *née* Schütz, was a writer and had many contacts in the world of the theatre and the press. It is probable that this background and the acquaintance with the leading authorities in many fields had a profound effect in creating the high standards and the impatience with anything but the best of its kind, which later became important characteristics of Pauli.

The young Wolfgang showed early signs of exceptional ability. He was outstanding at school in scientific subjects and particularly in mathematics, and first-rate in all other subjects, except in languages in which his performance was good but not exceptional. He was probably a 'bookish' child and was not interested in games, though he was, and remained throughout his life, fond of walking, particularly in the mountains, and of swimming. As a small child he disliked fairy tales, which seemed to worry him. As a growing boy he was fascinated by books like those of Jules Verne, and he later developed a special interest in astronomy. His younger sister, Hertha (now Mrs. Ashton), had to serve as an audience for his current discoveries in astronomy, and he was most impatient when the audience did not respond adequately to the instruction.

In 1918 he became a student at the University of Munich under A. Sommerfeld, and he obtained his Ph.D. there after three years, the shortest period allowed by the University regulations, and exceptionally short for a subject like theoretical physics.

Sommerfeld's group in Munich provided a stimulating environment in which Pauli found it easy to develop an understanding of the whole of current theoretical physics and of the mathematical methods required for its study. Sommerfeld was one of the best teachers, with his physical intuition and his command of mathematical tools; he was particularly successful in getting his pupils to work on problems which challenged their ability and initiative, yet were certain to lead to concrete and quantitative answers. At the time he had a collection of brilliant young men working with him, but he recognized at once the exceptional promise and maturity of the young Pauli, and he arranged for Pauli to be entrusted with the writing of the article on relativity for the *Encyclopaedia of Mathematical Sciences.*

As a result, Pauli completed during the three years at Munich not merely his Ph.D. dissertation on the hydrogen molecule ion, probably the most ambitious application of the old Bohr-Sommerfeld quantum theory that has ever been attempted, but also his article on relativity, which has remained to this day unrivalled as a survey of this field. An English translation with a brief review by Pauli of the developments since 1921 has recently been published.

Pauli always remained conscious of how much he owed to Sommerfeld. In later years it was surprising, when Sommerfeld visited him, to watch the respect and awe in his attitude to his former teacher, particularly striking in a man who was not normally inclined to be diffident. He would himself laughingly refer to his "Schüler complex."

Working on his Ph.D. thesis in the daytime and on his relativity article at night did not leave him much time for the more mundane problems of life. He held a small grant from the University, but his father, who was anxious that the son should not be in need, sent some money to supplement the award. However, the early 1920s saw the beginning of the German inflation and, by the time young Pauli remembered to go and claim the money from the bank, it had become practically worthless.

He spent the academic year 1921–22 in the University of Göttingen, another center of modern theoretical physics, and the following year in Niels Bohr's Institute in Copenhagen. This was the beginning of a very close and warm friendship with Bohr, renewed in many later visits.

In 1923 he was appointed to a post in the University of Hamburg where he became Privatdozent, and he remained there until 1928. He always spoke of his period in Hamburg with particular pleasure. It must have been a period when he began to sense his power and to realize that his work in physics came up to his own high standards. At the same time, largely under the influence of Otto Stern, he began to take more interest in practical matters and to organize his way of life to suit his interests.

He was appointed in 1928 to the Chair of physics in the Eidgenössische

Technische Hochschule (Federal Institute of Technology), Zurich, where he remained, with an interruption during the war, until his death.

He spent the war years 1940–45 at the Institute for Advanced Study, Princeton, and after that remained a permanent member of the Institute, which he visited on several later occasions.

After a short first marriage, which broke up almost immediately, he married Francisca Bertram in 1934; his wife's help and understanding became an invaluable factor in keeping him at peace with the world in his later years.

Pauli's first publications were concerned with the theory of relativity; he was attracted to this subject no doubt as a result of his early interest in astronomy. These papers already show the incisive clarity which was to become characteristic of all his later papers and which found its early expression in the review article on relativity. But soon his main interest changed to atomic theory and, in particular, to the quantum theory.

At that time the ideas of Planck, Bohr and others had become firmly established; it had become clear that the old Newtonian mechanics was not adequate for dealing with problems on the atomic scale, that the rules involving the quantum constant h played an essential part in understanding atomic phe-

Wolfgang Pauli. (Photograph published by permission from CERN and courtesy of AIP Emilio Segrè Visual Archives.)

nomena, and that in most cases in which these rules led to unambiguous predictions, they were confirmed by observation. However, the physics of the time was still unsatisfactory in two major ways. One was that the quantum rules developed by Bohr and their refinement by Sommerfeld did not form a complete logical system, and in many physical problems different ways of applying them appeared to give different answers. Important examples of such difficulties were the behavior of atoms in external electric and magnetic fields and the many-body problem.

The other trouble was that one was still using the dynamical concepts of Newtonian physics and relied on the orbits of particles calculated from the Newtonian equations of motion, selecting amongst these possible orbits certain sets which satisfied the quantum rules. This required ideas like the "quantum jumps" from one orbit to another under the effect of radiation, which meant that the dynamical equations were applied during the motion on a steady orbit but not to the transition from one orbit to another. It also meant that one had to select in an external field orbits oriented in certain special ways relative to this field, however weak it might be; this created a major logical difficulty, if one considered a situation in which a field in one direction was reduced to zero and then a field in a different direction applied gradually. It was difficult to see how the orbits could suddenly change from those quantized in relation to the initial field direction into those appropriate for the new orientation.

Also, since the new rules were to be applied by first finding complete solutions of the equations of motion and then selecting the appropriate orbits, their application was difficult in problems which, because of the number of degrees of freedom or because of lack of symmetry, did not lead to soluble mathematical problems.

Thus, the problems with which the early Pauli papers on quantum theory were concerned, i.e. the effect of magnetic and electric fields the two-centre problem, and the problem of perturbation theory, which expresses the solution of a complicated problem approximately in terms of the properties of a similar simplified problem, related to the vital unsolved problems of the time.

One of these problems, the Zeeman effect of complex atoms, had been analyzed to a large extent. It was realized that, for example, in alkali atoms the main features of the spectrum could be understood in terms of the orbit of the last electron, regarding the inner electrons or "core" as forming a stable inert gas configuration, which remained unchanged except at very high excitation. But applying the known quantum rules to the motion of the last electron, one predicted only half as many states as were in fact found in a magnetic field. The first suggestion to explain this was to attribute the additional degree of freedom to the core, but this view was not easily reconciled with

Hilda Levi, Wolfgang Pauli and Sir Rudolf E. Peierls at the 1934 Copenhagen Conference, Bohr Institute. (Photograph by Paul Ehrenfest, Jr. and published courtesy of AIP Emilio Segrè Visual Archives, Weisskopf Collection.)

the idea that the core was in a definite physical state and almost unaffected by the presence of the last electron and by the field. In his discussion of this problem, Pauli demonstrated that the evidence favored clearly the alternative view that the outer electron itself possessed an additional degree of freedom, which allowed it two possible states for each orbit permitted by the Bohr–Sommerfeld rules, though the physical meaning of this additional degree of freedom was not immediately clear.

This view also had important consequences for the structure of complex atoms. The guiding principle in the understanding of atomic spectra and of the structure of the periodic system had been that one could classify atomic states by assigning to each electron in a complex atom one of the orbits of the Bohr theory of the hydrogen atom, i.e. one possible orbit of one electron moving in a central field of force. This picture had implied the existence of shells in the atom, a shell being made up of all the electrons in orbits of a given principal quantum number, n. But this had made it necessary to dispose rather arbitrarily of the number of electrons in each shell. For example, the K-shell with $n=1$ allowed only one choice of the Bohr–Sommerfeld quantum numbers, yet to be in line with the behavior of atoms, one required that two, but no more than two, electrons could be found in the K-shell.

Pauli pointed out that with the idea of a new degree of freedom, there would just be two different states of an electron in the K-shell and that the correct answer could be obtained by postulating that no two electrons could be found in equivalent orbits, i.e. in orbits of the same atom with the same set of quantum numbers, including that related to the new degree of freedom.

Similarly in the L-shell one had now to allow two different states for each of the four Bohr orbits, so that with the new principle the L-shell could contain up to 8 electrons. This fitted in with the L-shell being filled in the neon atom.

Pauli realized clearly that the new principle now known as the exclusion principle, or Pauli principle, could not be expected to be a consequence of the mechanics of the motion or of the quantum rules as previously formulated, but that one was dealing here with a drastic new restriction which had to be added to the basic postulates.

The conclusions of this paper found immediate acceptance because they belonged to the type of argument whose simplicity and agreement with observation made it immediately convincing as soon as it is pointed out.

If he had made no other contribution to physics, this one paper would have assured him an important place in the history of physics and would have justified the Nobel Physics Prize, which he was awarded in 1945.

Shortly afterwards physics made a major advance through the new mechanics, first in the form of the matrix mechanics of Heisenberg and others, and almost simultaneously the wave equation of Schrödinger.

Pauli was immediately attracted by the possibilities of using these new ideas to put quantum theory on a more consistent footing. He succeeded in solving the hydrogen problem in terms of the new mechanics (20), which was a considerable achievement. The Heisenberg formulation of quantum mechanics is exceedingly easy to apply to problems with linear equations of motion, such as that of the harmonic oscillator, but the presence of the inverse square law in the attractive force in the hydrogen problem leads to mathematical problems of considerable complexity. This paper therefore added important support to the new mechanics. In his discussion of the hydrogen atom, he of course included the new degree of freedom to which he had previously drawn attention and referred to the new hypothesis of Uhlenbeck and Goudsmit, which identified this degree of freedom with an internal rotation or "spin" of the electron, a hypothesis which by this time after some initial controversy was beginning to find acceptance.

Meanwhile, Fermi had investigated the consequences of applying Pauli's exclusion principle not only to the electrons inside the atom but to the molecules of a gas, and Dirac had shown how to formulate the principle in the new wave mechanics in terms of the symmetry of the wave function describing many identical particles, so that it was possible to apply it even to situations when the interaction between the particles is so important that one may not assign individual orbits or quantum states to each particle.

Pauli applied Fermi's generalization of the exclusion principle to the conduction electrons of a metal, and in particular to their behavior in external magnetic fields. The magnetic behavior of metals was then one of the puzzles in physics. If each electron carried a spin, as was by now established, a magnetic field would tend to orient the spins in one direction; the distribution of spin directions should obey Boltzmann's law. This would lead to a paramagnetic susceptibility following Curie's law, giving a susceptibility inversely proportional to the absolute temperature. Normal metals did not behave in this way, and Pauli pointed out that this was explained by the operation of the exclusion principle or Fermi statistics, since the electrons were normally distributed over a wide range of states of orbital motion, each containing two electrons of opposite spin. The exclusion principle does not permit the spin of an electron to change unless it is also moved to another orbital state not already occupied by an electron with the new spin direction. Hence, spin reversal must be accompanied by a considerable increase in kinetic energy, except for those electrons which could find an unoccupied orbit of about the same energy as they had before, and this can apply only in an energy region of an extent of kT, near the highest occupied state.

This result explained both the order of magnitude and the absence of a strong temperature dependence of the paramagnetic susceptibility of normal metals, and stimulated therefore further research into the electron theory of metals, which developed with impressive speed over the following years.

Up to this point the spin of the electron had remained outside the quantum mechanical description, except insofar as it allowed a doubling of the states counted for the exclusion principle and in allowing for the spin moment and its interaction with an external field of given direction. Pauli showed how to incorporate the spin dynamics into the Schrödinger theory by extending the wave equation to a pair of simultaneous equations for one electron, the two wave functions representing the amplitudes for the two spin orientations. This paved the way to applications involving variable fields and interactions between spinning electrons. This equation was later shown to represent the nonrelativistic limit of the wave equation of Dirac, in which spin was shown to occur naturally in at least one form of the relativistic quantum theory of a single particle.

Pauli was always interested in the relation between matter and radiation. The similarity between light beams and beams of particles, which had been vital both in the early work on quantum theory and in the de Broglie–Schrödinger formulation of waves, had not yet found full expression in the formalism of quantum mechanics. In an important paper with Jordan, Pauli showed how to formulate the connection between field and particle aspects for electromagnetic radiation in the absence of charged particles, and later in collaboration with Heisenberg he made the first attempt to give a full quantum treatment of the interaction of radiation with matter. These papers are among the earliest contributions to what is now known as quantum field theory. The classical theory by Lorentz of the interaction of a particle with its own electromagnetic field had led to difficulties because the field energy of a point charge was infinite and the concept of an electron of finite extent, which was favored by Lorentz, was not easily reconciled with relativity. There seemed hope that the new concepts of space and time, which were used in quantum mechanics, would alter this situation. But the conclusion of Heisenberg and Pauli was that the self energy remained infinite. This difficulty remained a serious obstacle until the post-war years when the ideas of "renormalization" showed at least how to obtain unambiguous answers from electrodynamics in spite of the infinities. It remains to be seen whether further drastic modifications of the basic concepts are necessary before we have a completely consistent theory.

Apart from drawing attention to this difficulty, the paper by Heisenberg and Pauli raised a number of points that have remained of importance in quantum field theory, particularly the part played in electromagnetic problems by gauge invariance.

In the early 1930s quantum mechanics had become an established discipline, which led to definite answers in almost all problems of atomic structure and properties, and an extensive literature developed rapidly in which the consequences of the new mechanics for a large variety of practical problems were studied. While Pauli was interested in these applications, he remained concerned with the basic structure of quantum mechanics; his article on the principles of wave mechanics, published in 1933, was invaluable to many of the new recruits to quantum theory, who were anxious to get a clear view of the fundamental principles. This article again shows the characteristic attitude of Pauli not to be satisfied with superficial plausibility but to think out clearly all the connections between the different ingredients of the theory and to face the logical difficulties. Like his earlier article on relativity, it has stood the test of time, so that it could be taken over with minor changes into the reedition in 1958 of the *Handbuch der Physik*.

Another major contribution of the pre-war period was the paper with Weisskopf on the quantum theory of particles without spin or with an integral spin. The success of Dirac's work on the relativistic wave equation had left the impression that only a particle of spin 1/2 (or possibly more generally of a half integral spin) allowed a consistent relativistic quantum theory; the main point being that the Klein–Gordon equation for spinless particles had no room for a particle density which was capable of only positive values. Pauli and Weisskopf tackled this problem afresh, starting from the fact that a relativistic quantum theory of photons was perfectly possible and that these were particles of integral spin. They showed that the same approach could be extended to charged particles, provided one gave up the one-particle description and incorporated in the theory processes in which the particle number could change. In that way one avoided the need for defining a particle density in space. For the interaction with the electromagnetic field it was essential to introduce a charge density, but this need not necessarily be positive if particles of both positive and negative charge were contained in the field. The absence of a positive particle density thus was seen to be the counterpart of the absence in Dirac's theory of a positive energy density. There the reinterpretation of negative energy states in terms of antiparticles ('hole theory') also led to a formalism in which a consistent relativistic description required the inclusion of pair creation and annihilation. These conclusions, which at the time appeared academic, were an essential requirement for dealing with the theory of mesons, which are today known to possess zero spin.

At about this time Pauli made a remark which must rank as one of his most important contributions to physics, even though it was not at the time published as a paper. This is the suggestion of the neutrino hypothesis. At that time one of the difficulties of nuclear physics was the existence of beta decay, with the emission of electrons with a continuous spectrum. In each individual event therefore the electron energy could not equal the energy difference between the initial and the final nucleus, and careful work had shown that this energy difference was equal not to the average but to the maximum of the electron energy, so that in each event an unpredictable amount of energy appeared to be lost. At the same time observations on spins of nuclei had made it clear that the nuclear spin was always consistent with the nucleus consisting of neutrons and protons, so that the spin was integral for even mass number and half integral for odd mass number. Since in beta decay the mass number does not change, the half integral spin of the electron cannot be compensated for by a spin change of the nucleus, and it can of course not be balanced by the orbital moment of the electron, which is always integral. The study of symmetry effects in band spectra of molecules had also shown that the statistics or symmetry of the wave functions supported the empirical rule that Fermi–Dirac statistics (i.e. the exclusion principle) applied to particles or systems of half integral and Bose–Einstein statistics to the case of integral spin. The violation of spin conservation thus also corresponded with a change in the statistics of the whole system.

Pauli postulated the existence of a neutral particle of low mass but with spin 1/2, for which, to distinguish it from the heavier neutron, Fermi coined the name neutrino, using the Italian diminutive. It was evident that this postulate could restore the conservation of energy, spin and statistics. Later, Fermi showed how to formulate a theory of this process in quantitative detail and that it led to predictions for the energy distribution of the electrons and for the decay constants of beta emitters in terms of one single fundamental constant, which have been well confirmed by experiment. The idea also encouraged experiments in which one looked for the recoil of the decaying nucleus. These showed that the amount of momentum, apparently lost in the process, was connected with the loss of energy in the right way for interpreting both as the momentum and energy of an undetected particle of negligible rest mass. More recently, the ingenious experiments of Cowan and Reines, which led to the detection of inverse beta decay, have given a final proof of Pauli's hypothesis.

Pauli remained concerned about the part played by the exclusion principle as an apparently *ad hoc* addition to the rules of quantum theory and about the empirical rule which connected the statistics with spin. He showed that, while an attempt to incorporate in theory particles with integral spin obeying the exclusion principle, or particles of half integral spin obeying Bose–Einstein statistics, did not lead to mathematical inconsistencies, it would lead to physically unreasonable consequences. The reason is that the elimination of negative particle densities or negative energies, which arose in the wave equation for integral and half integral spin respectively, could only be carried out by the procedures of the Pauli–Weisskopf theory or the "hole theory" respectively if the spin and statistics were related according to the empirical rule.

In the years from 1940–45 at Princeton, Pauli's main interest was the problem of meson theory. He followed the ideas of the "strong coupling" hypothesis. This formed the extreme opposite to the picture of weak coupling, which had in general terms been successful in electrodynamics, but had led to serious difficulties in meson theory. This culminated in the paper with Dancoff, which solved the strong-coupling problem for the case of charged pseudoscalar mesons, leading to an isobaric state, i.e. a bound state of a meson and nucleon, of spin 3/2 and isotopic spin 3/2. We know today that the coupling between nucleons and mesons is not strong enough to justify this extreme limit, but nevertheless the result has acquired special significance through the discovery of a "virtual" or resonance state of spin and isotopic spin 3/2 in the meson-nucleon interaction. How far this justifies the use, at least in part, of the strong-coupling picture remains to be seen.

He followed with great interest the development of quantum electrodynamics and, for example, helped to clarify the difficulties connected with the "infrared catastrophe", which gives rise to apparent infinities in the theory at the long wavelength end. This is now understood, in part as a result of Pauli's work, as representing simply the fact that any collision involving a charged particle always results in the emission of an infinite number of photons of very low energy, though the total energy of all these photons is small. The natural approach of the ordinary theory in which one would calculate the probability of a collision in which no photon or only a small number of photons is emitted is therefore inappropriate. The more relevant question is to find the probability of the particle being scattered into a certain direction with its energy lying within a small interval. To this question the theory gives a finite and sensible answer.

Later the work of Tomonaga, Schwinger and others showed how to reinterpret the divergent results of quantum electrodynamics, so as to obtain reasonable answers. Pauli helped to clarify the structure of the new theory by careful analysis of the definition and structure of the various propagators which appeared in the new theory, and also (with Villars) by introducing the concept of "regularization". He realized the inadequacies of the new formalism and examined attempts to get over them by using non-local theories and by widening the interpretation of the theory so as to allow negative probabilities, as in the particular model proposed by Lee.

He was as excited as any physicist about the discovery of violation of conservation of parity, and helped in the elucidation of the two-component theory of the neutrino. This is equivalent with the hypothesis that a neutrino had always a definite "helicity" i.e. that the component of its spin in its direction of motion could have only one sign. This hypothesis led to an equation which in its mathematical form had already been postulated many years ago by Weyl, in an attempt to avoid the trouble with negative energies. Pauli had analyzed this equation in his 1933 article and had rejected it because, while not eliminating the negative energies, it violated the principle of the conservation of parity, which at that time one saw no reason to renounce.

This brief survey of Pauli's major contributions has, of course, omitted reference to many papers concerned with more detailed points and with less topical subjects. But in addition we must refer to another aspect of Pauli's role in modern physics, through his participation in discussions and, above all, in correspondence. The neutrino hypothesis, which was put forward in private discussions and in letters, has already been mentioned, but it would be impossible to list all the ideas, constructive or critical, by which he has influenced the work of pupils and colleagues in innumerable letters. Some of these letters are written in reply to requests for advice. Others were spontaneous and written either by way of comment on somebody else's work or when he had arrived at some new thought himself and just sat down to put these thoughts to somebody he knew would be interested to hear of them. All of his pupils and friends are familiar with these letters, invariably written by hand, invariably relating to problems of crucial importance at the time, pungent in criticism. In these letters, as well as in conversation, he would often discuss conjectures and intuitive judgments, which went far beyond anything he would regard as worthy of publication, but he would draw a clear distinction between knowledge and conjecture.

To quote a very charming remark from one of these letters, he had expressed the opinion that the experiment on parity violation would give a negative answer, and had offered to bet substantial odds on this. Shortly after this he heard that the experiment had indeed established parity violation, and in a letter to V. F. Weisskopf he remarked that he was lucky that the bet had not been taken, because he would have lost some money, which he could not afford, whereas as it was he had only lost some of his reputation, which he thought he could afford.

These letters supplemented the profound influence he exerted personally on his pupils and collaborators and on many others who came to him for advice. To discuss some unfinished work or some new and speculative ideas with Pauli was a great experience, because of the clarity of his understanding and of his high standard of intellectual honesty, which would never let a slipshod or superficial argument get by. At critical times in physics, when it was not clear whether some new idea should be taken seriously, one tended naturally to ask "What does Pauli say about it?"

On some occasions he was hard to convince when the advocate of a new idea had not yet examined all its aspects and did not have answers ready for all possible objections, but usually he was quick to see the important truth through some unimportant inaccuracies of detail and would use his sharp criticism not to discourage but to force one into facing the objections and finding an answer to them.

He made on his own work the same, or higher, demands of soundness as on the work of others. He would never be satisfied with an argument, let alone publish it, before he had achieved perfect clarity about its basis. As a result his published work is not only remarkably free from errors (I am aware of only one incorrect published statement under his name, and even that occurs in the summary of a short talk given at a meeting), but they became extremely lucid in presentation because of the amount of thought that had gone into clarifying the basic ideas.

While these qualities made him a very powerful teacher at the research level, he was not specially distinguished as a lecturer to undergraduates or to more general audiences. His elementary lectures were well thought out, but he did not have too much sympathy with the difficulties of the weaker students or the patience to go back to first principles in talking to a general audience. Probably the oversimplification and glossing over of complications, which is often essential to explain a subject briefly to a non-specialist audience, were against his nature. In lectures to groups of experimental physicists his mathematical ability would soon outrun the capacity of the listeners.

While so far only his thoughts on physics have been mentioned, it should not be thought that his interests were confined to his own subject. He was keenly appreciative of art, of music, and of the theater, and he had a discriminating taste which would be as impatient of the second-rate in the arts as in physics. He had many close friends in the world of arts with whom he corresponded or talked about their fields as seriously as he could debate his own subject.

He was specially interested in philosophy, to which he was led both from the border-line where it impinges on physics and which was familiar to him through his thoughts on the basis of quantum theory, and from a very personal interest in psychology, and much of this interest was stimulated by (or perhaps responsible for) his friendship with the psychologist Jung.

In appearance he was a most unusual figure, a round face and a well-rounded body, with rather awkward movements and a very characteristic habit of rocking his body, particularly when in thought ('Wenn er mit Problemen kämpft, ist die Schwingung ungedämpft' read an irreverent verse about him at one international conference). He was not good with his hands, or with mechanisms, which he tackled with a careful and suspicious deliberation. He passed his driving test after 100 lessons, but did not drive much afterwards.

One can see here the probable origin of the legend of the "Pauli effect." This was a kind of spell he was supposed to cast on people or objects in his neighborhood, particularly in physics laboratories, causing accidents of all sorts. Machines would stop running when he arrived in a laboratory, a glass apparatus would suddenly break, a leak would appear in a vacuum system, but none of these accidents would ever hurt or inconvenience Pauli himself.

At one reception this Pauli effect was to be parodied, and a chandelier had been suspended carefully by a rope which was to be released when Pauli entered, causing the chandelier to crash down. But when Pauli came, the rope became wedged on a pulley and nothing happened—a typical example of the Pauli effect!

Anyone who met Pauli for the first time would have been tempted to class him as an absent-minded professor, but such a judgment would have been wide of the mark. He had in fact learned to arrange his affairs and his way of life so as to interfere least with his work and with the things to which he was inclined to devote his time. He had no patience for conventions, unless he saw some sense in them. His ways of going about practical things were often unorthodox, but they generally achieved the object of saving time and effort and concentrating on what he regarded as essential. One aspect of this was that he was selective in the type of problem in which he was prepared to take an interest; he could be very discouraging to colleagues who wanted to arouse his interest in problems which he regarded as too technical or too complicated or just too uninteresting. This would be done with a disarming frankness. He generally avoided meetings or discussions on subjects that bored him, but, if he did find himself at such a meeting, he was apt to announce cheerfully that he did not know anything about the subject and thought it dull.

But this is not to say that his interest was restricted to fundamental problems. He knew well that big things are often made up of many points of detail, and he could be extremely patient in discussing minor points once he regarded them as relevant. Similarly, his interest in people was by no means confined to the great authorities. He was extremely interested in young people and their problems, and, if people were in difficulty with their personal problems, he could be most patient and helpful in listening and advising.

On one occasion he was talking with a group of young people, who were young enough to think they knew everything. Sensing that they were not inclined to listen to him he said, "I have more experience than you. I have once been young, but you have never yet been old."

No account of Pauli and his attitude to people would be complete without mention of his critical remarks, for which he was known and sometimes feared throughout the world of physics. He not merely did not spare the other person's feelings but he often deliberately selected the sensitive spot.

No doubt many of the stories of this kind circulated about him are apocryphal, but the examples below come from reliable sources or from conversations at which the writer was present. The oldest of the famous remarks dates back to the Munich days, when Pauli was a brilliant but unknown research student, and at a crowded colloquium meeting Einstein, who was visiting, made a comment in the discussion. Young Pauli rose at the back of the hall and said. "You know, what Mr. Einstein said is not so stupid," a remark characteristic for his lack of respect for authority but not yet of the bite which came later with his greater assurance.

N. Kemmer reports a more characteristic remark. "I do not mind, Mr. X, if you think slowly, but I do object when you publish more quickly than you think."

When L. Landau, after a long argument in Zurich, pleaded for an admission that not everything he had said was complete nonsense, Pauli replied, "Oh, no. Far from it. What you said was so confused that one could not tell whether it was nonsense or not."

When a charming colleague whose papers had not impressed Pauli had given him directions how to find a certain place in a strange town and inquired the next day whether Pauli had found the place, he said, "Oh, yes. You express yourself quite intelligibly when you don't talk about physics."

Quite recently, a friend showed him the paper of a young physicist which he suspected was not of great value but on which he wanted Pauli's views. Pauli remarked sadly, "It is not even wrong." People have tried to attribute these sharp remarks to Pauli's impatience with slipshod reasoning and wishful thinking. There is no doubt that he was using them as a tool to drive home valid and often constructive criticism, but equally often they were so remote from any specific point in the argument that it is doubtful whether this is the full story. He himself once said to the writer, "Many people have sensitive corns and the only possible way of living with them is to step on these corns often enough until they get used to it," but that remark, too, probably oversimplifies the problem.

The remarkable thing is that, although the victims often felt hurt at the time, none of then ever bore a grudge for long. It is a tribute to his greatness as a physicist and as a man, and to his understanding of other people, that all who knew him, who all must at one time or another have been exposed to remarks of this sort, had as much affection for him as they had respect for his knowledge, his judgment and his integrity.

His death was quite unexpected. Few people knew that he was unwell and he himself did not realize the seriousness of his condition. Until almost the last day he would not let discomfort and severe pain interrupt his routine of work and of thought. But the feeling of loss caused amongst the family of physicists by his death was so severe, not only because of the suddenness of the event but because of the unique position he held in physics as a colorful personality, as a thinker of unusual clarity, as a powerful critic and adviser, and as representing through his example and through his criticism what has been called the "conscience of physics."

The writer would like to acknowledge his indebtedness in writing these notes to many colleagues and in particular to Mrs. Pauli and to Mrs. Hertha Pauli-Ashton.

Professor H. W. B. Skinner, F.R.S., 1900–1960

H erbert Wakefield Banks Skinner was born on October 7, 1900. He was educated at Rugby and Trinity College, Cambridge, where he remained for research at the Cavendish Laboratory from 1922–1927. He joined the staff of the H. H. Wills Physical Laboratory of the University of Bristol in 1927 and remained at Bristol until 1939, with the exception of the year 1932–1933, which he spent at M.I.T. on a Rockefeller Fellowship. On the outbreak of war he joined the radar team at Dundee which transferred to Swanage and later to Malvern. In 1944 he moved to Berkeley to help in the development of the electromagnetic isotope separation plant in the atomic energy project.

At the end of 1945 he returned to England as one of the first senior staff members of the new Atomic Energy Research Establishment at Harwell, where he directed the "general physics division" until his appointment to the Lyon Jones professorship in physics in the University of Liverpool, as a successor to Sir James Chadwick.

Skinner's early work in the Cavendish Laboratory followed the current interest in β-ray spectra and radiation, and early papers show him as a skilled experimenter with a taste for accuracy and clarity.

His spectroscopic work led him to take a particular interest in the borderland between short-wave ultraviolet and soft X-rays, a field not yet adequately explored at the time because of the experimental difficulties. It is probable that the challenge of just these difficulties was for him one of the attractions to this subject.

This was during his Bristol period, where contact with N. G. Mott and H. Jones, who were then developing the now generally accepted ideas of the electron theory of metals, led to a powerful combination of an experimental and theoretical study.

This showed in particular that the energy spectra of the electron bands in metals, which were fundamental to the theory, could show up in the X-ray spectra and that Skinner's techniques for handling the soft X-ray region were capable of revealing an impressive amount of detail of these electron bands. This possibility of seeing the dynamical details of the electron motion rather than only the integral features which are shown up in the magnetic and electric properties played a great part in establishing our knowledge of metals on sound and realistic foundations.

The papers of the Bristol period represent Skinner's main direct contribution to pure physics, and this work rightly earned him a name as an experimenter of outstanding skill, judgment, and reasoning power. But an assessment of his role in physics based only on his own publications would ignore his great contribution of later years, surviving in the record only through the publications of others.

His war work on radar and on atomic energy was, of necessity, applied physics. It required for its execution the best and most experienced of the modern physicists, and Skinner's contribution was very much that of a physicist. Yet it was, in purpose and motivation, more engineering than physics. Skinner was outstandingly successful in adapting himself to this situation, and on his return to Harwell he became one of the key figures in building up an institution which combined strength in its practical outlook and its handling of large-scale equipment with a strong position in fundamental physics.

His own general physics division became a good model of this duality. Its cyclotron was a sound piece of engineering which ever since its construction has been working smoothly with as little trouble as the smaller laboratory instruments of earlier generations of physicists, allowing the research workers to concentrate on their results and techniques without concerning themselves with the problems of machine operational performance. In spite of its unfashionable energy range the machine is even today making important contributions to fundamental nuclear problems.

His influence at Harwell far transcended his own division and even after his move to Liverpool, his advice, his enthusiasm and his pungent criticism continued to be available for many of the projects arising at the Atomic Energy Establishment.

He was clearly the right man to take over the work started by Chadwick in building up high-energy research at Liverpool, and to look after the completion of the cyclotron then in construction. He increased the team at Liverpool by attracting more outstanding physicists and by training others, and the group of people he had brought together became one of the important centers for work in the several 100 MeV range, in particular for pion physics, in spite of the fact that they had entered this field rather later than many groups elsewhere. This is not the occasion to write at length of his personal qualities but no account of Skinner's work could omit some mention of the part he played amongst his colleagues in his University, in his subject and in his country. In all discussions to form policy he was liable to express strong views, and he would stand up for what he judged to be right regardless of whether his views were fashionable or conformed with conventions. He would listen to argument but was not impressed by authority.

His outspoken comments could irritate his seniors but his judgment and his integrity were always respected and his counsel was in great demand. The number of committees and discussions in which he was asked to take part was such that only a man of his sense of duty could agree to serve, and only a man of his energy and capacity for work could combine them with the work of running his own department. In the last few years, these outside commitments involved many contacts with CERN, the progress of which was close to his heart, and work on the development of a National Bubble Chamber, by a combined effort of several groups of physicists in Britain.

In looking after his collaborators no trouble was ever too much and anyone in difficulty or with problems would find Skinner's house open and could count on a patient hearing, active interest and ready help.

The suddenness of his unexpected death in Geneva deepened the sense of loss felt by all his colleagues and by all those who had come to rely on his judgment and help and who were inspired by the example he had set in standing up for what he believed to be right.

An Appreciation of Niels Bohr

With the death on November 18, 1962 of Niels Bohr, the world lost a truly great scientist, and physicists lost their most respected and most beloved figure.

Niels Henrik Bohr was born in Copenhagen on October 7, 1885, the son of Professor Christian Bohr and Ellen, *née* Adler. He grew up in a home of culture and of warm personal relationships, even by the standards of education and of friendliness of Copenhagen. As he went through the thorough and slow process of Scandinavian education the outstanding quality of his mind became evident very early, without preventing his enjoying less intellectual pursuits. He was a first-rate football player, though not quite the equal of his brother Harald, who later became the distinguished mathematician.

He graduated in physics in 1909, having gained the Gold Medal of the Royal Danish Academy of Sciences in 1907. After some work on surface tension, he turned to the problems of the structure of matter. His dissertation for the Dr. Phil. was a study of the electron theory of metals, a field in which the successes, and the difficulties, of the new picture of matter in terms of electrons were then particularly evident. This dissertation was written in 1911, and about that time he must have made the first of his many visits to England, where he came into contact with J. J. Thomson and Ernest Rutherford.

He spent some time in Manchester, where Rutherford had just made his great discovery which had solved the problem of the structure of the atom, and had shown the existence of the nucleus. His discussions with Rutherford soon became of immense value to both, and from them grew a close friendship that was to last until Rutherford's death.

Bohr's first paper written at Manchester dealt with the slowing-down of charged particles in their passage through matter. It contains a clear explanation of the response of an atomic electron to the electric pulse of a passing particle, and its connection with the response to radiation. This theory re-

Niels H. D. Bohr. (Photograph published courtesy of AIP Emilio Segrè Visual Archives, Wheeler Collection.)

mains today the basis of our understanding of the process. He always retained a very special interest in this problem, and returned to it on many occasions.

The work at Manchester was of absorbing interest, and Bohr could spare only a short time to return home to marry Margrethe Nørlund.

In 1913 he published three papers, "On the constitution of atoms and molecules," which laid the foundation of his quantum theory of the atom. This work is so well known to every physicist today that it is not necessary to describe it here. On reading these papers today one notices how clearly he appreciated from the very beginning the revolutionary nature of his theory in its departure from classical mechanics.

This is in great contrast to the attitude of Planck, on whose quantum hypothesis Bohr's theory was built. Planck long remained reluctant to accept the drastic break with classical physics to which his hypothesis seemed to lead him, and made many attempts to reconcile his results with the classical picture, whereas Bohr realized, and accepted, from the very beginning the fact that Planck's ideas and his own were bringing in new concepts quite foreign to the older physics. He did not take this step lightly, but only on the strength of the evidence that classical mechanics had failed to give a correct account of atomic phenomena, and that the very existence of stable matter forces on us something entirely new. The immediate success of his theory in accounting for the spectrum of hydrogen, and of hydrogen-like atoms, left no doubt that the particular departure from classical ideas which he proposed was in the right direction.

In 1913 he was appointed docent in the University of Copenhagen, but a year later he became a lecturer in the University of Manchester, where he stayed two years. In 1916 he accepted the Chair of Theoretical Physics in Copenhagen, which he occupied until his retirement, in 1956.

The years from 1913 until about 1923 yielded a large number of papers in which the hypothesis first put forward in 1913 was developed and extended. In the course of this work the "Bohr theory of the atom" reached the form in which we know it today.

In this theory the electrons are taken to follow orbits in accordance with Newtonian mechanics; but of all possible such orbits only those may occur for which the action integrals are integral multiples of Planck's constant. While the motion in the orbit is described by classical theory, the transition from one orbit to another in the process of emission or absorption of radiation, or in atomic collisions, occurs in "quantum jumps" which cannot be understood in classical terms.

Nevertheless, it is essential to refer to classical radiation theory for many of the features of such transitions, in particular for understanding selection rules and the polarization of the radiation, and generally for predicting the relative intensities of different lines. The way in which the nonclassical radiation processes were dominated by the features of the analogous problems in classical theory was a point on which Bohr laid great stress. He later used the term "correspondence principle" for this connection.

The importance of Bohr's work was recognized quickly. One measure of the value that scientific opinion attached to it is the series of honors he received, including amongst the earlier ones the Hughes Medal of the Royal Society in 1921, and the Nobel Prize in Physics in 1922. But a far truer measure is found in its effect on physics. Within a few years large numbers of theoretical physicists had taken up the study of the atom following Bohr's lead, and a rapidly growing body of experimental work was being analyzed in terms of the new ideas.

In 1920, Bohr became Director of the Institute for Theoretical Physics of the University of Copenhagen, and this soon became the world's center of atomic physics, to which physicists from everywhere came for short visits or for longer stays to find inspiration and understanding.

During the next few years the foundations of quantum theory were consolidated and extended, both by Bohr himself and by others. Sommerfeld sharpened and extended the quantum conditions which select the stationary orbits for electrons in atoms. He also studied the relativistic corrections on the hydrogen orbits and related them to the observed fine structure of the hydrogen lines. Pauli showed that the arrangement of electrons over the various orbits, which Bohr had deduced from the periodic system and from atomic spectra, required the postulate of the exclusion principle. Goudsmit and Uhlenbeck introduced the idea of spin to explain the last of the quantum numbers required to classify atomic states.

New experimental techniques, such as the molecular beam method of Stern and Gerlach, made it possible to test predictions of the quantum theory other than those relating to energy levels and spectra.

By the mid–1920s, quantum theory had become a firmly established part of physics. Yet it was also clear that it was not yet complete. There remained questions to which quantum theory provided no answers, such as the problems relating to nonperiodic orbits. There were questions to which the answers appeared inconsistent, such as the behavior of atomic orbits when an infinitely weak external field changed its direction. Above all, the distinction between the use of classical mechanics in parts of the problem, as in the determination of the quantized orbits, and the failure of classical concepts, as in the quantum jumps between such orbits, seemed arbitrary and unsatisfactory.

Bohr was very much concerned with the problem of clarifying the fundamental concepts. He had realized from the beginning that quantum theory required the abandonment, or modification, of many ideas that had been taken for granted in classical physics. This was symbolized by the appearance of Planck's constant h, which dimensionally defined a distinction between large and small-scale phenomena for which there was no room in classical physics. He was therefore not afraid of using concepts which appeared to contradict our intuition, which is derived from experience on the macroscopic scale. But if the old ideas are to be replaced by new ones for the atomic phenomena, the new ones must form a clear and consistent scheme.

In the study of the basic ideas an attempt was even made, in the paper by Bohr, Kramers and Slater, to abandon the conservation of energy in individual atomic processes.

The next step came, almost simultaneously, through Heisenberg's quantum mechanics, which started from a mathematical formulation of Bohr's correspondence principle, and through Schrödinger's wave mechanics, which built on de Broglie's ideas about the association of waves with particles. The two forms of the new mechanics were found to be completely equivalent; their relation to classical concepts was clarified by Born's interpretation of the wave function as determining the probability of the position of the particle, and a little later by Heisenberg's uncertainty principle.

Bohr at once took an active part in the discussion of these new developments. Now quantum theory, or at least nonrelativistic quantum theory, had become a complete and consistent scheme, capable of giving a unique answer to any question relating to actual or possible observations. It was even further removed than the old form from any description in terms of intuitive pictures, but the limitations on the use of classical concepts were stated quantitatively, and the reasons for their failure beyond these limits, which in many aspects had been foreseen by Bohr, could be formulated.

Bohr had been the first to accept a complete break with the classical ideas, and to have confidence that a new and consistent scheme was possible. He was now concerned with the task of showing that the new scheme was consistent, and of explaining the nature of the new way of thinking. The key to the understanding of the new approach lay in the uncertainty principle, which Bohr clarified substantially, and which led him to the wider concept of complementarity. Many of his papers, lectures, and books deal with these problems. They contain as much epistemology as physics, and in this work Bohr could be called a philosopher. But since the approach was based on an intimate familiarity with the nature of the physical laws, his kind of philosophy required a philosopher who was also a physicist of Bohr's stature.

Relativistic quantum theory, and the related theory of fields, were still beset with difficulties. In spite of great progress, fundamental difficulties remain today. Irked by a paper by Landau and Peierls which tried to establish in the relativistic domain limitations on the process of measurement not included in the uncertainty principle, Bohr set out to prove that without the framework of quantum theory it is possible to measure electromagnetic fields to precisely that accuracy which is permitted by the uncertainty principle, provided one assumes the existence of any suitable measuring device consistent with quantum theory. This involves ignoring the limitations arising from the fact that matter, and hence the physicist's tools, must be made of the elementary particles provided by nature; any such limitation could be considered only in a future scheme which incorporated the reasons for the existence of only certain types of particles. This program was carried out in collaboration with Rosenfeld and resulted in a paper which throws much new light on the nature of the measuring process in quantum theory.

One of the physicists who was reluctant to accept the principles of quantum theory as satisfactory was Einstein. He had himself revolutionized our ideas of space and time, and had been the first to show that ideas which had been regarded as self-evident could be prejudices derived from everyday experience. He had also been the first to suggest that a light wave might be regarded as a guiding field, which determined the probability of the presence of light quanta, an idea which was used by Born in his interpretation of the relation between waves and particles. Nevertheless Einstein was suspicious of the ultimate validity of the uncertainty principle and, many times over the years, produced counterexamples which appeared to violate it. On each occasion Bohr was able to disprove the objection, usually by showing that Einstein's example had failed to allow for some physical principle which would result in a source of error sufficient to restore the limit imposed by the uncertainty principle. So in the end the controversy with Einstein, which Bohr thoroughly enjoyed, resulted in an increased confidence in the consistency of quantum theory combined with the other important principles of physics.

One of the consequences of quantum theory was that the idea of causality took on an entirely new form, and this tempted some writers to draw rather facile conclusions on the problem of determinism, free will, and the nature of life. This was too superficial for Bohr, and in a number of articles he showed what lessons there are in quantum theory for the study of biological problems.

In 1936 one might well have expected Bohr to continue only with his work on epistemology and the clarification on the basic principles, without contributing further to the solution of specific, practical problems of physics. But he still produced two ideas of outstanding importance. The first related to the problem of collisions of neutrons with nuclei. At that time the neutron had been used by Fermi and others for the study of nuclei, and this work had shown many surprising features.

It was then usual to think of neutrons and protons in a nucleus as moving largely independently of each other, each being subject to a field of force which was the average result of the motion of all the others. This followed the way in which Bohr had so successfully described the behavior of electrons in the atom. These ideas suggested that the effectiveness of neutrons in collisions should not depend sensitively on their energy, and that radiative capture of neutrons should be a rather rare event. The experiments showed many extremely sharp resonances in the cross-section for neutron collisions as a function of energy, and at reasonably low energies radiative capture was the dominant process. Bohr pointed out that a nucleus must be regarded as a system of many particles all strongly coupled together, like the molecules in a drop of water.

The numerous resonances were then at once understood as the large number of quantum levels of such a system at a high excitation energy; their sharpness reflected the long life of such a state. Although there was enough energy to eject a neutron (since the state had been formed by neutron impact) this energy was shared between many degrees of freedom, and it took a long time before, by chance, the whole energy was again concentrated on one neutron. During this time some of this energy could easily be lost as radiation.

This picture of the "compound nucleus" had a profound influence on the theory of nuclear reactions. It was, for a time, misinterpreted as ruling out the description in terms of independent particles, even in the normal state of a nucleus or in states of low excitation. We later learned that the two pictures are perfectly compatible, and that even at the high energies to which the compound-nucleus picture applies one can also recognize single-particle features if one looks at energy averages (which by the uncertainty principle relate to the short-time behavior of the process) rather than at well-defined energies.

Yet even the misunderstanding was fruitful since it led to speculations about a liquid-drop model for a nucleus, of which a direct descendant, the collective model, became, in the hands of Aage Bohr and Mottelson and their collaborators, an extremely useful tool in the understanding of nuclei.

Bohr's other contribution of that period was the analysis of the fission process. He had taken a great interest in the discovery of fission, and his ideas about the compound nucleus provided just the appropriate approach for the study of its mechanism. His paper with Wheeler in 1939 showed, in quantitative detail, how to understand the competition between the decay of the compound nucleus by fission, radiation and neutron emission, as well as the different behavior of fast and slow neutrons and the parts played by the uranium isotopes.

This paper was published in 1939, and soon WWII started. Before long Denmark was occupied. Bohr could easily have left the country, but he felt he belonged there, with his friends and his pupils. He stayed and did much to help where help was needed, until in 1943 he was warned that as a result of a new tougher policy of the occupying forces he was about to be arrested. He escaped to Sweden with his family in a fishing boat, and he and his son, Aage, were taken to England and later to the United States.

Here the project of applying nuclear fission to the making of a weapon of war had become a gigantic enterprise, and Bohr was enthusiastically welcomed at Los Alamos, the laboratory charged with the development of the bomb. He was interested in the scientific problems of the project, and took part in many discussions about its progress. But his real concern was not the solution of the remaining technical problems (which by this time were largely in the nature of technological details), but the problem of the effect of this new weapon on humanity.

He saw that the existence of a weapon with such unprecedented power created a new situation in the world, and that this provided the opportunity of making a fresh start on international relations. He saw the possibility of danger and destruction, but also the possibility of using this situation to persuade nations to adopt a new attitude of openness and mutual confidence. He used every opportunity to impress his views on the statesmen.

When he returned to Copenhagen after the war, these problems remained his main interest. In 1950 he published his ideas on the "Open World" in an open letter to the United Nations.

The demands on his time for various important administrative and political activities continued to increase. The institute was expanding, and for a time it housed a theoretical physics section of CERN, the European organization for nuclear research. Later a joint Scandinavian research institute in theoretical atomic physics, NORDITA, was founded in association with Bohr's institute. He became chairman of the Danish Atomic Energy Commission, and in many other ways, official and unofficial, his help and advice was sought.

But he still found time for physics, time to follow the post-war developments of the subject, and to add contributions to his favorite problems, the foundations of quantum theory, the problem of measurement, and the passage of charged particles through matter.

His last publication, it seems, was the Rutherford Memorial Lecture, in the *Proceedings of the Physical Society*, which tells of his reminiscences of the Manchester days and brings out not only a very vivid picture of Rutherford, but also of Niels Bohr himself.

Bohr's influence on modern physics should not be viewed merely through his own contributions, great as they are; an important part of it consisted of the inspiration, encouragement and guidance he gave to others.

Since the early days of quantum theory the institute in Copenhagen was a place to which theoretical physicists came for ideas, for advice and for criticism. They all took away with them a little of the "Copenhagen spirit," which was really Bohr's way of looking at physics. To take part in discussions with Bohr was a unique experience. He loved discussion and argument, both because he was fond of people and enjoyed helping them to understand and to bring out what was of value in their ideas, and also because he liked expressing his own thoughts and perfecting their formulation in debate. In such discussions his extreme courtesy and kindness of manner tended to struggle with the strength of his convictions and his insistence on getting things absolutely right.

There were never any silences on these occasions, because Bohr always had something to say; others had to assert themselves if they wanted to be heard. His energy seemed inexhaustible. At the end of a long day the liveliness of his ideas and his comments would be undiminished, while his younger collaborators might be worn out, though full of wonderful new insights.

The dialectic approach also applied to the process of writing papers, which invariably went through many stages of drafting, many versions in typescript, and usually many sets of proofs, all of which were tried on some of the collaborators or visitors, and many nuances of wording discussed. The papers did not always become easier to read in this process, since getting a true statement of the argument mattered more than superficial simplicity.

His work was often concerned with rather abstract ideas, but he never forgot that physics is concerned with experimental facts. It is not an accident that, in spite of its name, the Institute for Theoretical Physics contains important experimental research teams and facilities, in which he took great pride. The discovery of hafnium is probably the best-known single discovery resulting from this work, but it belongs to a large body of important experimental work in atomic, and later in nuclear, physics that has come from the institute.

When it came to helping any of the people in the institute, any physicists, or indeed sometimes strangers in need of help, no trouble was too much for Bohr. Particularly during the exodus of scientists from Nazi Germany, and later during the occupation of Denmark, anyone in need could count on sympathy and advice and, wherever possible, on active help.

Bohr's kind interest in people made the personal relations in his institute very much like those in a family. This was, in fact, an extension of his own family in whose warm and intimate circle the affairs and the members of the institute were allowed to intrude at any time.

In his relations with people, as well as in all other problems of his rich and fruitful life, he derived great strength from the support of Mrs. Margrethe Bohr, whose gentleness charmed all those who had the good fortune of joining the large family of the institute.

Truth and Clarity

Niels Bohr, revered for his many contributions to science and to other areas of serious thought is also fondly remembered as a human being, full of charm and with many amusing traits.

For example, his great kindness and reluctance to hurt anyone's feelings, coupled with his inability to let any inexact or wrong statement pass, led to the famous comment: "I am not saying this in order to criticize, but this is sheer nonsense!"

But while he was intolerant of nonsense, he was interested in simple problems and simple people. He could take a genuine interest in anyone's views and talk with them without condescension. I remember an occasion when he had a serious conversation with my son, then age four, with obvious interest.

On his attitude to the truth, we have been reminded of his saying that truth and clarity were complementary. This came out strongly in his papers, in which he tended to give all possible weight to the truth. As a result his papers were usually not easy to read. It helped if one was able to see an early draft, in which often the clarity had not yet been sacrificed to the truth. Papers always went through innumerable drafts, followed sometimes by 12 sets of proofs, and in the course of this many changes made the paper more true, but not often clearer.

He seems to have had the same attitude to other matters, to judge by the story of his visit to the site of a new extension to the Bohr Institute, when the old foreman, who knew him well, said: "Professor Bohr, do you see that wall? If you want to move it again, you must be quick, because in three hours the concrete will have set!"

I experienced some of the problems of drafting in trying to write a paper jointly with Bohr and George Placzek. There were many drafts, but it never got published. It is probably the most frequently cited unpublished paper in the literature.

As we know, he thought deeply about problems outside of physics, and he used to defend the right of scientists to take part in political and other general debates. He said:

We are no wiser and no less biased than other people. But as a physicist, or a biologist, you are certain to have gone through the experience of making a confident assertion, and then being proved wrong. A philosopher or a sociologist might never have had this wholesome lesson.

He had a fund of stories to illustrate his views. He was opposed to any form of nationalism, but he said that, if there had to be nationalism, he preferred the form in which it appeared in the English-speaking countries, typified by the phrase: "Right or wrong, my country!" He did not agree with the sentiment, but he thought a German, or a French patriot would never admit that his country could be wrong.

Another illustration of the theme was the story of the young girl in Ecuador, who was cycling down a steep hill when her brakes failed. The cycle went faster and faster, and she almost lost heart. But then she said to herself: "I am an Ecuadorian," and this thought gave her the strength to hold on and control the bicycle until the road flattened out. Bohr commented: "If instead of Ecuadorian you say American, or German, or British, the story is not funny."

He had, of course, his share of absent-mindedness. In the early discussions he always had a cigar (later it became a pipe) which he tried to light while talking, not having it in his mouth. This took a lot of matches, and soon he would pat his pockets and say: "Have you got a match?" Someone would produce a box, which Bohr pocketed after using one match, and in a minute the process would repeat itself—"Have you got a match?"

I treasured, for a long time, as a souvenir a piece of chalk which was blackened at one end. Evidently Bohr had confused the chalk with the cigar, which he held in the same hand.

When he arrived in London during WWII, after his famous flight from Stockholm, he was for a few days on his own—Aage Bohr followed later. When he had to go to meet an important person, the wise secretary of the Atomic Energy Office wrote the address and directions on six pieces of paper, and said: "Professor Bohr, if you put one of these into each of your pockets, you are sure to find one when needed!"

Bohr could understand it when others were not very practical. Pauli told me about the day, after the discovery of hafnium, when the institute had an open day and an exhibition to attract public interest. Before this started Bohr was running around putting finishing touches to the arrangements, when he saw Pauli standing rather forlorn in a corner. Bohr stopped, looked at Pauli and said: "Pauli, you are more suitable to be exhibited than to exhibit!"

I hope that these little stories help his many friends and admirers to recall the beloved personality of Niels Bohr.

Rutherford and Bohr

A s every physicist knows, our present knowledge of the structure of matter, of atoms and nuclei, is built on the foundations laid by Ernest Rutherford and Niels Bohr early this century. Each of them was a great physicist in his own right, but their contributions to physics, particularly in the important period from 1912 to perhaps 1920, owe much to their mutual interaction; we are dealing with a "two-body problem," to use modern jargon. Later their paths separated: Rutherford's primary interest was the atomic nucleus, and once he had identified its role he left the further exploration of the atom to others, whereas Bohr concentrated on the theory of the atom, and for its study developed the quantum theory. This was the "old" quantum theory, spectacularly successful in explaining the hydrogen spectrum but lacking logical consistency in mixing classical and quantum concepts. He maintained an interest in the nucleus but made major contributions to nuclear physics only from 1936 on.

However, even while their fields of work were separate, their relations developed into a close friendship, and in their correspondence and in their frequent meetings they continued discussing the current problems of physics, and of the world. Here, I offer an impression of this friendship between two very different personalities, as it emerges from their correspondence.

They were, indeed, very different in many ways. Rutherford was outspoken, outgoing and direct. He liked his physics and his experiments simple and described his work in simple and concise language. He had a loud and booming voice, and it is said that in the days when counting circuits tended to be sensitive to noise his collaborators' equipment went wrong whenever he entered their research rooms. He made decisions easily and firmly, and once a matter was decided did not give it any further thought. He could be rude, and even unreasonable on occasion, but when he had cooled off would put matters right with a handsome apology.

Bohr spoke in a quiet voice, hardly above a whisper, and listeners had trouble understanding him in any language. In what he said or wrote he was always conscious of the many limitations and conditions that restrict the validity of any statement: as he liked to express it, "Truth and clarity are complementary." He was always concerned about hurting any person's feelings, and he found great difficulty in making his plans, only too often changing them almost as soon as they had been arranged.

Yet with these great differences there were also many important similarities. Both were capable of enormous enthusiasm for a promising idea in physics. Both were not deflected by unimportant details, but could give painstaking attention to detail when it mattered. Both regarded mathematics as an important tool in formulating and applying the laws of physics, but never as an end in itself. Rutherford was fond of making disparaging remarks about theoreticians who were too attached to formal mathematics, so much so that he is sometimes believed to have been opposed to theory altogether. (His respect for Bohr, his friendship with R. H. Fowler, and his enthusiastic reception of Gamow's results, disprove this idea.) Bohr was too polite for such remarks, but restricted himself to the minimum amount of mathematics in his own work. Both were untidy lecturers but would fascinate and stimulate the audience. These similarities no doubt caused their mutual respect and patience with each other's idiosyncrasies.

Niels Bohr arrived in Cambridge in 1911 to work with J. J. Thomson. He was a shy but enthusiastic young man of 26 who had written his Ph.D. thesis on the electron theory of metals, and expected J.J., the discoverer of the electron and the author of some papers on electrons in metals, to be interested in his work. In this hope he was disappointed. Thomson probably never read his thesis, and certainly was not interested in discussing it, but suggested some rather routine research work that Bohr tackled without enthusiasm.

Then, during a weekend visit to friends in Manchester, he met Rutherford, the physics professor in Manchester, who was then 39. They met again when Rutherford visited Cambridge. In these encounters they took an immediate liking to each other and Bohr asked to be allowed to work under Rutherford. He stayed in Manchester from March to July 1912.

The first problem that Rutherford suggested to him was the slowing-down of charged particles in passing through matter. This was a very important problem because the main method for determining particle energies then was by using their range. A study by Darwin had assumed the electrons of the medium to be free, and Bohr soon decided that this was an unwarranted approximation. He therefore had to make assumptions about the forces acting on the electrons. At first he took these to be linear in the displacement, a model then favored as it leads to harmonic motion and thus explains the existence of sharp spectral lines.

However, Rutherford was then developing his ideas about the nuclear atom, and Bohr, who was impressed by Rutherford's results, found that the assumption of Coulomb forces from a central charge gave better results for hydrogen. For his problem he had to know the orbit of the electron before the arrival of the passing particle, and in hydrogen it was natural to assume a circular orbit. What could one assume for the motion of two electrons in helium? Bohr convinced himself that there were no stable solutions to this dynamical problem, because one of the electrons could "fall" into the nucleus, while the liberated energy would be carried away by the other. The difficulty reinforced a suspicion developed earlier in his studies of electrons in metals, where he showed that classical physics could not account for any magnetic properties of materials.

While continuing to work on the slowing-down problem, which retained his interest for the rest of his life, his thoughts now turned to the structure of atoms, and to ways in which the inadequate classical picture might be modified by using Planck's quantum of action. A first manuscript showing the beginnings of these ideas was written before he left Manchester.

Bohr returned to Copenhagen full of enthusiasm. His farewell letter to Rutherford (24 July 1912) said: "Your criticisms and suggestions have made so many questions so real for me." Further progress was slow because of a heavy teaching load. But he kept working both on the slowing-down problem and on his new ideas. He kept in touch with Rutherford. When he worried about slow progress Rutherford comforted (11 November 1912): "Don't hurry with your second paper; it is not likely that anyone else is working on this." He also shares his thoughts on other topical problems in physics with the young Bohr. He and his pupils were discussing the new experiments of von Laue. After saying that the Laue spots can be explained by interference, nothing to do with wavelength (!), he says in the next letter (24 February 1914) that X-rays must be regarded as some kind of wave motion, "but I personally cannot escape from the view that the energy must be in concentrated form." In other words, his intuition favored the quantum theory.

Meanwhile Bohr's work developed into the well-known Bohr model of the atom, eventually published in three famous papers (called "the Trilogy" by Rosenfeld). When the first paper was sent to Rutherford he was impressed by the good agreement with many facts (20 March 1913), but at once put his finger on the main contradiction in the quantum theory: How does the electron decide what frequency to emit? He added, as a minor criticism: "I think in your endeavour to be clear, you have a tendency to make your papers much too long and a tendency to repeat your statements in different parts of the paper. The paper could be cut down without sacrificing anything to clarity. Long papers have a way of frightening readers." He added a postscript: "I suppose I may use my judgment in cutting out any matter I consider unnecessary?"

He did not yet know Bohr well enough to realize his attitude to his writings, in which he would fight over each word as a she-bear over her cubs. Bohr had brought to England a translation of his thesis, but failed to get it published because it was considered too long. He would not shorten it. So the English version of his important work on the electron theory of metals existed only in a few copies until it appeared in his *Collected Works* in 1972.

Rutherford's letter crossed with one in which Bohr sent an amended and expanded version of the paper (21 March 1913). Rutherford tried again (25 March 1913): "The additions are excellent and reasonable but the paper is too long. Some of the discussions should be abbreviated. As you know it is the custom in England to put things very shortly and tersely, in contrast to the German method, where it appears to be a virtue to be as long-winded as possible."

In April 1913 Bohr visited Manchester again to discuss his paper with Rutherford. There is no record of this discussion, but the paper emerged practically uncut: Rutherford communicated it to the *Philosophical Magazine*.

Rutherford evidently realized the importance of Bohr's work and one would have expected him to mention this in letters to the many other physicists with whom he was in correspondence, but the only letter I could find was one sent on 19 April 1913 to K. Fajans, the physical chemist in Munich: "Bohr of Copenhagen has been working at the general theory of atoms built on my model and appears to have made definite progress... . He is a very capable fellow, and there will soon appear a very interesting paper by him in the *Phil. Mag.*"

In further correspondence about the remaining two papers Bohr was evidently aware of Rutherford's critical attitude. On 13 June 1913 he sent the third paper, saying: "I hope I have better succeeded in making it short."

In the summer he visited Manchester again for talks with Rutherford. During that autumn the British Association meeting in Birmingham was to have a discussion on radiation, at which H. A. Lorentz would be present, and Rutherford arranged for Bohr to be invited. At first enthusiastic, Bohr wrote in August (27 August 1913) that he would not be able to leave Copenhagen, and regreted he could not come to Birmingham. But evidently he was persuaded, and in October (16 October 1913) he wrote that he was very glad he did come to the meeting. This episode illustrates Bohr's frequent hesitation in making plans and his inclination to change them at short notice, which will recur. Meanwhile Bohr's position at Copenhagen was only temporary and involved too much teaching and lab demonstration. Rutherford wanted him to come back to Manchester as a reader, and broached the subject in a very gentle way (20 May 1914): "I am glad that...your position in the University of Copenhagen is now being considered...we are now advertising for a successor [to Darwin's

Readership] at £200. Preliminary inquiries show that not many men of stature are available. I should like to get a young fellow with some originality in him."

In due course Bohr accepted the readership, for a year at first. He stayed for a second year, after Rutherford had asked the University of Copenhagen for further leave of absence for him (15 May 1915): "I think the experience that Dr. Bohr is gaining of the work in a larger university will prove of much value to him.... I should much regret the loss of his services to the university here."

From this second, very productive, Manchester period we have an assessment by Rutherford of Bohr, in reply to G. N. Lewis in California who was considering Debye and Bohr for a temporary lectureship (13 December 1915). After some favorable remarks about Debye, Rutherford wrote: "I regard Bohr as one of the coming men in mathematical physics, and I think he has a better grasp of physics than any of the mathematical people I have come across. He is a man of great originality and, as you know, his work has already attracted wide attention, and I am confident will do so even more in the future.... He is thoroughly *au courant* of all the modern physical problems, and has an extraordinarily wide knowledge of experimental, as well as of theoretical, physics. He is a pleasant fellow, speaks English quite well, and is quite a clear and interesting lecturer."

Bohr returned to Copenhagen in 1916 to take up a new professorship for which Rutherford wrote a strong letter of recommendation (16 March 1914), ending: "Finally I would like to state that in my opinion Dr. Bohr is one of the most promising and able of the younger mathematical physicists in Europe today. I think that any university would be fortunate who is able to acquire the services of such an original and fruitful investigator."

World War I made correspondence between Manchester and Copenhagen difficult, but an exchange of letters continued, with Bohr always in the position of the pupil writing to his teacher. Reporting on his work (29 November 1916) he said he had been "remembering your often repeated advice as to the greater use of working out special problems than making philosophy."

In 1917 Rutherford wrote (11 December 1917) that in spite of many wartime duties he occasionally found some time for research. "I have got, I think, results that will ultimately prove of great importance. I wish you were here to talk matters over with." He described the early experiments on nuclear disintegration of alpha-particle bombardment. This was the beginning of his famous experiments that showed that alpha particles could knock protons from many nuclei, thus causing a transmutation of one chemical element into another. "Regard this as private."

As soon as the war was over Rutherford set about getting a chair created for Bohr in Manchester, but wanted to make sure that Bohr would be inclined to take such an offer seriously (17 November 1918). The chair would be independent, but "on account of my grey hair, I would probably continue to be director... . I wish I had you here to discuss the meaning of some of my results." It is hard to get firm proofs. "Counting scintillations is hard on old eyes." (He was then 46.)

To Bohr (15 December 1918) this letter, "besides being the source of the greatest pleasure to me, has at the same time been the object of sorrowful considerations." Bohr felt obliged to stay in Copenhagen and work for the development of physics in Denmark, after the university had done for him what they could. "At the same time I feel very strongly that the result will never be the same as if I could work with you."

Rutherford pleaded with him (11 January 1919) to keep an open mind until they could talk things over, but the problem soon was made academic by Rutherford moving to Cambridge. Their correspondence now settled down to a pattern of exchanges about developments in physics and local news. (For example, on 18 February 1920 Rutherford reported that Aston talked about isotopes and J. J. Thomson said he did not believe the results about chlorine. "You can imagine that I enjoyed myself thoroughly between the two, but I have little doubt that Aston is quite correct.")

After some exchange of letters it was arranged that Rutherford would visit Copenhagen and be present at the formal opening of Bohr's new institute. There were some delays in the correspondence (Rutherford [13 July 1920]: "You are as bad a correspondent as I am, and have not the excuse of age!") Bohr was eager for the visit (27 July 1920): "You can hardly imagine with what expectation and pleasure your visit and lectures are looked forward to by all Danish scientists, and my wife and I only regret that Lady Rutherford will not be coming too." Rutherford did go in September 1920, but the new institute was not yet finished.

The teacher–pupil relation is symbolized in these letters by Rutherford's quite early changing his salutation from "Dear Dr. Bohr" to "Dear Bohr" and often "My dear Bohr," whereas Bohr kept to a respectful "Dear Professor Rutherford." In 1920 he even began using "Dear Sir Ernest" (Rutherford was knighted in 1914). The address "Dear Rutherford" does not appear until 1929.

A visit by Bohr to Cambridge was planned for April 1921, after the Solvay conference. There was much correspondence about dates, arrangements for lectures, etc., spiced with Rutherford's remarks about his work (6 February 1921): "My work on knocking H out of atoms goes on apace. Several atoms succumb each week." Bohr still sent his papers for Rutherford to communicate to the journals, and often left it to Rutherford to deal with the proofs. On one occasion (19 February 1921) Rutherford commented: "I have made a few minor corrections in English... and would have liked to make more, but I know of old how difficult it is to reconstruct the meaning to your satisfaction." Bohr's note disagreed with a theory by Langmuir, and Rutherford added: "I always felt that Langmuir's ideas were much too descriptive for the year of our Lord 1921. They might have served in 1911. See you in Brussels" (at the Solvay conference).

But Bohr could not go to Brussels, or to England, as he was suffering from overwork and was ordered a complete rest. Rutherford expressed his regrets (30 March 1921): "I am afraid you are too unselfish and give too much of your time and energy to help other people." This was in March.

In May, Rutherford enquired about Bohr's health (3 May 1921): "I trust that Mrs. Bohr is in proper control of you and will prevent you doing any serious work for the next three months... . I would feel inclined to go for a holiday in the open air to a place where you never see a scientific man and where you do nothing but eat and read light novels."

In August (22 August 1921) Bohr thought he might be able to go to Cambridge in October or November, but he wanted to wait a fortnight to confirm this, "being sure that you will understand that this form of the answer is dictated by the wish not to disappoint you once more by late alterations."

In September (16 September 1921) he seemed well enough but could not go in October because he was working on a new paper, and also felt obliged to catch up on his university duties neglected while he was unwell. Meanwhile he sent another note for *Nature*, remarking hopefully "...if you should not find the language too bad I do not think I shall need a proof."

But Rutherford did not agree (21 September 1921): "I am afraid your English is deteriorating with absence from this country." He sent the note on, but was anxious to see a full paper to understand the precise assumptions made in assigning electron numbers to the various shells. He added domestic news: his daughter Eileen was to marry R. H. Fowler. "The couple were with us during the vacation, and I derived much interest and amusement from their ways. You and Mrs. Bohr are young enough to sympathize with them." Once again the emphasis on age.

In another episode Bohr reported (24 October 1921) that Jacobsen in Copenhagen had obtained results on the "straggling" of alpha particles that differed from those obtained by Henderson in Manchester. But in the next letter (26 October 1921) he wrote "in a very shameful mood," Jacobsen's work was hurriedly prepared and not reliable.

In December 1921 Bohr was appointed an honorary member of the Royal Institution. Rutherford wrote (9 December 1921) that as a professor in that institution he was delighted "that it has seen fit to honor one of my old friends." Bohr answered belatedly (4 February 1922) and, as usual, with excuses. He acknowledged Rutherford's letters "and all the honor which has been shown me by English societies, which I feel I have only deserved very little, but for which I am very thankful not least on account of this testimony of your friendship for me." A delicate way of saying this was no doubt due to Rutherford's influence.

Bohr's long-delayed visit to Cambridge finally took place in April 1922, following the usual exchange of many letters about dates, with Rutherford urging (11 February 1922): "Please let me know at once the titles and dates for the lectures." Bohr wrote five weeks after the visit (27 May 1922), apologizing for his lateness in sending thanks "for the kindness you and Lady Rutherford showed me during my stay in your house. I feel it very difficult to express myself, but I am thinking very often of Cambridge and of your home and laboratory, and the way you looked after me in small and big things in both places."

On the question of whether Kramers, then Bohr's assistant, could be invited to spend a term in Cambridge, Rutherford said (5 June 1922) he would like to do so, but funds were short. "I must be just before I am generous."

To Rutherford's telegram of congratulations on Bohr's receiving the Nobel Prize in late 1922, Bohr replied (20 November 1922): "I have felt so strongly how much I owe to you, not only for your direct influence on my work and your inspiration, but also for you friendship in these 12 years since I had the great fortune to meet you in Manchester." He was sad at not having written for so long. "I took your rebuke for only writing seldom as not the smallest sign of your friendship." He ended, "Yours ever, Niels Bohr, who will do his best to improve in writing."

This letter crossed with one from Rutherford, following up his telegram (25 November 1922). It was now his turn to apologize for writing late, because he was busy. On the prize: "I knew it was merely a question of time, but there is nothing like the accomplished fact."

Bohr's theory had predicted the existence and properties of a new element, and this was discovered by Hevesy and Coster in Copenhagen in 1922; it was called hafnium. Rutherford wrote (8 January 1923) to congratulate Hevesy on this "admirable example of the cooperation between theory and experiment." He added: "I am sure it is highly gratifying to Bohr, even if he tries to suppress it." He knew, of course, that this letter would be shown to Bohr, with its gentle teasing of Bohr's modesty.

There was controversy about the discovery, with some chemists claiming they had found the new element earlier. Rutherford sympathized with Bohr's Institute (4 February 1923). He called Scott, who made such a claim "a well-known rather elderly chemist" and said the *Times* editorial on the matter "is very wrong and silly. One need pay no attention to such irresponsible utterances." Bohr had sent him a letter by Hevesy and Coster for *Nature*. They now found that Scott's preparation contained no trace of hafnium, so the letter needed to be reworded (10 February 1923). Rutherford made the change, and also made changes in one paragraph, which contained poor English (15 February 1923): "I think it now reads very well and is certainly a document of an explosive character... I quite agree ... that Urbain has not a leg to stand on."

In February 1923 Rutherford's first grandson was born, and this again induced thoughts about age (28 February 1923): "I ought to feel incredibly ancient now that I have turned into a grandfather, but I still have some kick left in me." A few weeks later (26 March 1923) he had evidently forgotten that he had written about this. "Have you heard about Eileen's son? Great amusement over my transformation into a grandfather."

An embarrassing situation arose when Bohr was offered an honorary degree by Manchester, but was unable to come to Manchester on the proposed day. As it happens he was also offered an honorary degree by Cambridge, and thought he could not get there either, and asked for advice from Rutherford (21 March 1923). Rutherford explained (26 March 1923) that honorary degrees were not conferred *in absentia*, and declining the invitation to the ceremony amounted to declining the degree. Bohr, always most punctilious about etiquette, was horrified (2 April 1923) and agreed to come to Cambridge. He asked for Rutherford's help in reconciling Manchester. A kind word from Rutherford resulted in an agreement to hold a ceremony later in the year. As far as the letters show, this seems to have been postponed again to May, when, however, Bohr suffered again from overwork and had to cancel his trip. It seems the Manchester degree was never conferred, after all. But Bohr did go to Cambridge, and afterwards sent his usual eloquent letter of thanks (23 June 1923).

Rutherford was still trying to get Bohr for a full-time position in Cambridge, and in July 1923 the Royal Society, no doubt on his instigation, offered Bohr a research professorship. Bohr was tempted, but felt he could not desert Copenhagen and Danish physics (3 August 1923). A compromise solution whereby he would share his time between Copenhagen and Cambridge was explored but, as Rutherford suspected, this proved unacceptable to the Royal Society. Rutherford did not lose hope (14 August 1923): "If the whole matter should break down, there are possibly other more private arrangements that might be made, but the stumbling block is the lack of funds in Cambridge to finance new ventures." He never overcame this "stumbling block."

When in 1924 Bohr was again ordered to rest, Rutherford wrote (18 July 1924): "You know that it is my opinion that you work far too hard for your health, and you would do just as much good work if you took matters easily. This is the advice of a grandfather, but nevertheless good, as I have found in my own experience."

Things continued on an even keel. Bohr was elected a Foreign Member of the Royal Society, which Rutherford welcomed, particularly as this was during his presidency (1 June 1926): "You will soon be getting a little wearied in having to write letters of thanks for these scientific distinctions."

Several further visits to Cambridge included one following the Faraday lecture in London. Rutherford had heard about Bohr's intentions for that lecture (19 November 1929): "I hear you are on the war path and wanting to upset the Conservation of Energy .... I will wait and see before expressing an opinion, but I always feel 'there are more things in Heaven and Earth than are dreamt of in our philosophy.'" His caution was justified—Bohr's views on energy conservation were not confirmed.

After a visit in 1930, and the usual letter of thanks from Bohr, Rutherford wrote (9 June 1930): "I can assure you that we were all delighted to see you, and that you need not be over-modest about your lectures, which went off excellently."

In 1931 Rutherford was elevated to the peerage about the same time that his daughter, Eileen Fowler, died after the birth of a child. Bohr wrote about the peerage (21 January 1931): "To all scientists, and especially to your old pupils, this great distinction will stand as a symbol of the new epoch in science which you have created, and which has so deeply influenced the life of us all.... Thinking of those twenty years in which you and your family have shown us so much friendship, my wife and I send you and Lady Rutherford many thoughts of gratitude and deep sympathy in your great sorrow."

Rutherford replied separately to the congratulations (24 January 1931): "...good of you to write such a cordial letter about my recent transformation. As you may guess it is not the type of distinction that appeals to me very much, but I am sure it was meant as a recognition of the importance of science to the community and thus was very difficult to refuse."

There was another letter about Eileen's death. Rutherford wrote (27 January 1931): "Just a line to thank you for your kind note of sympathy from your wife and yourself. Fowler gave some lectures at the Poincaré Institute and has now returned to [his] house. Mrs. Cook—as before—will look after the house and children, at any rate for the moment. Fowler has stood the strain well so far. The baby, in the charge of a Truby King nurse, is flourishing." The matter-of-fact tone of this letter must have been hiding great grief.

In 1932, the year of momentous developments in Cambridge, Rutherford wrote (21 April 1932): "It never rains but it pours." He commented on the discovery of the neutron, and on Cockcroft and Walton's experiments on disintegration with accelerated protons. "I am very pleased that the energy and expense in getting high potentials has been rewarded by definite and interesting results.... Actually they ought to have observed the effect a year or so ago, but they did not try it in the right way."

In September 1932 Rutherford visited Copenhagen. He asked Bohr (1 June 1932) if a grant toward expenses could be arranged, and this was done. After the visit he sent a letter of thanks (29 September 1932): "I look back on my stay with you and your good wife with the greatest pleasure." A very different style from Bohr's letters on such occasions, but no less enthusiastic.

A little later (19 October 1932) he ended a letter about physics: "Please give my love to Margaretta and remembrances to your boys." This was the first time he referred to Mrs. Margrethe Bohr by her first name, though he got it slightly wrong, but it was correct on the next occasion (13 November 1932).

In October 1933 the two met at the Solvay conference in Brussels, and both had unfortunate experiences on the way home. Bohr had a bad cold while staying in Paris, Rutherford had some trouble with his knee, and Lady Rutherford was unwell. Rutherford did not enjoy the conference (13 November 1933): "I have come to the conclusion that conferences on the Continent are much too harrassing for anyone over the age of 60, and it is going to take a good deal of persuasion to get me to go to another. There is too much of an idea of getting value for your money by turning a conference into a tread-mill.... I thought the conference a success, but much too big and fatiguing for elderly people like myself who believe in action rather than in talk." A little later (5 December 1933) he reported that his knee was better, "and I hope soon to kick my students not only as a painful duty but with pleasure."

In 1934 a great tragedy befell the Bohr family when their eldest son, Christian, was drowned on a sailing trip. Presumably Rutherford wrote on this occasion, but his letter is not in the scientific archives. He wrote to Hevesy (13 August 1934), "we were all very disturbed about the loss of Bohr's eldest son. I hope it will not have too much of a reaction on them." And later (23 October 1934): "I am very sorry to hear of your impression of the effect of the tragedy on Mrs. Bohr. It may only be temporary and time is a great healer."

In 1934 Kapitza was not allowed to return to Cambridge from his summer holiday in the U.S.S.R., to everybody's concern, and particularly to Rutherford's. He wrote to Bohr (28 January 1935): "Don't believe what you hear. Governments are as a whole pretty bad, but I think the U.S.S.R. can give them all points on mendacity." He kept Bohr informed about developments in this affair.

Meanwhile Bohr again had to cancel a visit because he was overworked. Rutherford wrote (3 May 1935) he hoped Bohr would soon be himself again. "You are much too young to be feeling tired. You must leave that to old buffers like myself—but I am glad to say I have not reached that stage yet."

Bohr acknowledged Rutherford's good wishes on his 50th birthday (22 May 1935): "A 50th birthday is a very small occasion which is mostly apt to remind oneself that one is no longer quite young, and still has not reached any age which in itself calls for a veneration from younger ones. Just in this respect I feel every year greater admiration for the marvelous power with which you set an example to the world of untiring creative activity."

In October 1935 Bohr seems to have had more trouble with his health. Rutherford, slightly teasing, wrote (31 October 1935) he was sorry to hear of jaundice and rheumatism. "Did you examine your jaundiced family by the spectroscope to see whether the "D" lines stand out plainly?" It had been decided to give up hope of Kapitza's return to Cambridge and instead to help him start his work in Moscow. This "has meant a great deal of extra work, as I have to keep the Managing Committee, the University, the Royal Society, the DSIR and the USSR all in grand fettle. It is a strange team to be driving along the road."

At the end of 1935 Bohr made his important contribution to nuclear physics, the concept of the compound nucleus. He visited Cambridge early in 1936. Rutherford immediately saw the point of his idea and gave a very adequate summary in a letter to Max Born (22 February 1936). But he was again impatient for quantitative detail. In a letter to Hevesy (25 May 1936): "I am looking forward to Bohr's paper on nuclear constitution.... I gathered that he was publishing a more detailed account at once, but I suppose we shall have to wait, and he will take a good long time to think matters over before the paper finally appears."

The last letters, in 1937, dealt with the arrangements for a meeting in India, to which Rutherford hoped Bohr would accompany them. Rutherford died before the date of the trip.

In a very moving letter to Bohr after Rutherford's death, Kapitza wrote (7 November 1937): "From his words I always gathered that he liked you the most amongst all his pupils, and to be sincere I was always a little jealous of you. But now it is gone."

It is not necessary to decide if Kapitza was right, or if there were others of whom Rutherford was equally fond, but from the letters it is clear that he and Bohr had become very close friends, and for each it had become very important to exchange views with the other about physics and about the world. Rutherford is reported to have said once that he could do research at the North Pole, as long as he had a laboratory and equipment, but I do not believe he meant that his research would go on efficiently if he was isolated. There are a few scientists who can work on their own without any need for personal contacts, but they are few indeed, and they are always in danger of getting out of touch with the development of their subject. The Rutherford and Bohr types thrive on contacts. They are kept going by their own initiative, but they must share their knowledge and their discoveries with friends and colleagues.

There is no means of knowing whether our physics today would be any poorer if Bohr and Rutherford had not met, but they certainly felt sure that they were helped along by each other. I think, therefore, we should be grateful for this unique relationship.

J. Robert Oppenheimer,
1904–1967

Robert Oppenheimer achieved great distinction in four very different ways: through his personal research; as a teacher; as director of Los Alamos; and as the elder statesman of post–war physics. These various roles belong to different periods, except that of teacher, which overlaps in time with several of these periods. We may therefore review these contributions separately, while following a chronological order.

J. Robert Oppenheimer[1] was the son of Julius Oppenheimer, who had immigrated as a young man from Germany. The father was a successful businessman, and the family was well-to-do. His mother Ella, *née* Freedman, was a painter of near-professional standard, and both parents had a taste for art and music.

As a boy Robert showed a wide curiosity and the ability to learn quickly. He went to the Ethical Culture School in New York, a school with high academic standards and liberal ideas. He went as a student to Harvard in 1922, and, in spite of following a very broad curriculum, which included classical languages as well as chemistry and physics, he completed the four–year undergraduate course in three years and graduated *summa cum laude* in 1925.

In spite of the breadth of his interests, he was quite clear that his own subject was physics. During his undergraduate years he profited greatly from contact with Percy Bridgman, an eminent physicist, who himself had wide–ranging interests, and whose publications dealt with topics far beyond the field of his own experiments; they included, for example, philosophical questions.

---

[1]There has been controversy whether in "J. Robert" the "J" stood for "Julius." P. M. Stern quotes evidence that this was the case. We use the style Oppenheimer used with the explanation that the letter J. "stood for nothing."

J. Robert Oppenheimer in 1960. (Photograph published courtesy of AIP Emilio Segrè Visual Archives, Physics Today Collection.)

After graduating, Oppenheimer went to Europe, and during four years of travel he established himself as a theoretical physicist.

Quantum Mechanics

1925 marked the beginning of an exhilarating period in theoretical physics. During that year Heisenberg's first paper on the new quantum mechanics appeared, and Dirac started to develop his own version of Heisenberg's theory in a paper which appeared in the same year. Schrödinger's first paper on his wave equation was published early in 1926. Up to that time the principles of the quantum theory had been grafted on to the classical equations of mechanics, with which they were not consistent. The resulting rules sometimes gave unique predictions which agreed with observation; sometimes the answers were ambiguous, and sometimes the rules could not be applied at all. The new ideas showed how to obtain a logically consistent and mathematically clear description, and it looked as if all the old paradoxes of atomic theory would resolve themselves.

This inspired a period of intense activity, during which all atomic phenomena had to be reexamined in the light of the new ideas. Oppenheimer's quickness in grasping new ideas enabled him to play a part in this process. His first paper was submitted for publication in May, 1926, less than four years after he entered Harvard, and less than a year after Heisenberg's first paper on quantum mechanics appeared. It shows Oppenheimer in full command of the new methods, with which he showed that the frequencies and intensities of molecular band spectra could be obtained unambiguously from the new mechanics. A second paper, submitted in July, is concerned with the hydrogen atom, and by this time he is making full use of the apparatus of matrix mechanics developed by Born, Heisenberg and Jordan; of the alternative techniques of Dirac; and of Schrödinger's wave mechanics. These two papers were written in Cambridge, and he acknowledges help from Ralph H. Fowler and Paul Dirac.

In the second paper he raises the question of the continuous spectrum, and discusses the question of how to formulate the normalization of the wave functions for that case. This was the beginning of his interest in a range of problems that would occupy him for some time.

In 1926 Max Born invited Oppenheimer to come to Göttingen, where he continued his work on transitions in the continuous spectrum, leading to his first calculations of the emission of X-rays. He also developed, with Born, the method for handling the electronic, vibrational and rotational degrees of freedom of molecules, now one of the classical parts of quantum theory, known as the "Born–Oppenheimer method." Robert Oppenheimer received his Ph.D. degree in the Spring of 1927.

Oppenheimer remained in Europe until 1929, spending some time with Paul Ehrenfest in Leiden, and with Wolfgang Pauli in Zurich; the influence of both these men helped to deepen his understanding of his subject. He continued to work on radiative effects in the continuous spectrum, which he recognized as one of the most important and difficult problems of the time, and found ways of improving the approximations used, which still serve as a pattern for work in this field. Amongst his minor papers, one deals with electron pick–up by ions, a problem which requires the use of nonorthogonal wave functions.

In 1929 he accepted academic positions at both the University of California, Berkeley, and at the California Institute of Technology, and between 1929 and 1942 he divided his time between these two institutions. The list of his papers during this period might almost serve as a guide to what was important in physics at that time. He was now at the top of his form in research work and he knew what was important, so he did not waste his time on pedantic detail. In some of these papers he struggled with key problems which were not yet ripe for solution, such as the difficulties of the electromagnetic self–energy, or the paradox of the "wrong" statistics of the nitrogen nucleus (wrong because before the discovery of the neutron nuclei were believed to consist of protons and electrons). But on others he was able to take important steps forward. He saw the importance of Dirac's idea that one could avoid the difficulty of negative energy states for electrons by assuming them all filled except for a few holes, which were then positively charged particles. He showed, however, that Dirac could not be right in identifying these particles as protons, since they would have to have the same mass as electrons. Thus he essentially predicted the existence of the positron three years before its discovery by Carl Anderson.

When cosmic–ray experiments showed serious contradictions with theory, he studied the possibility that this might indicate a breakdown of the accepted quantum theory of radiation. When the discovery of the meson resolved the paradox he took great interest in the properties of the new particle. He also developed, in a paper with Carlson, an elegant method for investigating electron–photon showers in cosmic rays. In the 1930s the cyclotron and other accelerators opened up the atomic nucleus to serious study, and Oppenheimer played a role in asking important questions, and in answering some of them. His paper with Volkoff shows a very early interest in stars with massive neutron cores.

During his California period Oppenheimer proved to be an outstanding teacher of theoretical physics. He attracted many pupils, both graduate students and more senior collaborators, many of whom, under his inspiration, became first-rate scholars. His virtues as a teacher were those which also distinguished him as a researcher: his flair for the key question, his quick understanding, and his readiness to admit ignorance and to invite others to share his struggle to find the answer. His influence on his pupils was enhanced by his perceptive interest in people, and by his habit of extending informal and charming hospitality. After his marriage in 1940 his wife, the former Katherine Harrison, helped maintain this easy and warm hospitality.

Oppenheimer still maintained a great breadth of interests, adding even Sanskrit to the languages he could and did read. At first his interests were exclusively academic, and he showed little interest in political questions, or in the national and world events of the day. But in the mid–1930s he became acutely aware of the disturbing state of the world—unemployment at home, Hitler, Mussolini, and the Spanish Civil War in Europe. He became interested in politics, and, like many liberal intellectuals of the day, became for a time involved with the ideas of left-wing groups.

The list of publications by Oppenheimer and his group shows a break in 1941, and this marks almost the end of his personal research (the exception being three papers published after the war), but by no means the end of his influence on the development of physics.

Atomic Energy: Los Alamos

The change of direction in Oppenheimer's work was the result of his involvement with atomic energy. After the discovery of fission, Oppenheimer, like many others, had started thinking about the possibility of the practical release of nuclear energy. With his quick perception he understood the importance of fast neutrons for any possible bomb. In 1940 and 1941 the idea of releasing nuclear energy was beginning to be taken seriously. A number of groups in different universities were working on the feasibility of a nuclear reactor, and others were exploring methods for separating uranium isotopes. The latter work would ultimately lead to the production of the light isotope (U 235) in nearly pure form, which is capable of sustaining a chain reaction with fast neutrons. The reactor work led to the production of plutonium, which can be used for the same purpose. While these efforts were well underway by the beginning of 1942, there was no coordinated work on the design of an atomic weapon—its critical size, methods of detonating it, etc. Oppenheimer had attended some meetings at which such matters were discussed, and early in 1942 he was asked to take charge of the work on fast neutrons, and on the problem of the atomic bomb.

On the theoretical side Oppenheimer assembled, at Berkeley, a conference of first-rate theoreticians, including Edward Teller, who on that occasion first suggested the possibility of a thermonuclear explosion. The work continued in a theoretical group led by Oppenheimer at Berkeley. Experimentation on the relevant nuclear data was divided among many small nuclear physics laboratories, and this hampered progress, since it was difficult for these groups to maintain adequate contact, particularly in view of the secrecy with which the whole project had to be treated.

When, therefore, the United States government brought atomic energy research under the auspices of the Army and put Colonel (later General) Leslie Groves in charge of the project, assigned the code name "Manhattan District," Oppenheimer suggested to Groves that the effort be concentrated in a single laboratory. This should include the work on the theory and application of nuclear physics, as well as on the chemical, metallurgical and ordnance aspects of the project. In this way the different groups could work together effectively.

Groves accepted the proposal, and on Oppenheimer's advice chose the site of a boys' boarding school at Los Alamos, New Mexico; a region Oppemheimer knew and loved—he had a ranch there. The remoteness of the site made access and transport difficult, but also had the advantage of reducing contacts with the outside world, and therefore the risk of leaks.

Groves not only followed Oppenheimer's advice in the creation and location of the laboratory, but also selected Oppenheimer as director. This was a bold decision, since Oppenheimer was a theoretician with no experience of administration or of organizing experimental work. Events proved Groves right, and the work of the laboratory was extremely effective. In the view of most of the war–time members of Los Alamos, its success owed much to Oppenheimer's leadership.

He attracted a strong team of first–rate scientists, who came because of their respect for Oppenheimer as a scientist, and because of his evident sense of purpose. Inside the laboratory he was able to maintain the completely free exchange of information between its scientific members; in other words, to compensate for the isolation of the laboratory and the restrictions on travel which its members had to accept there was none of the "compartmentaliza-tion" required in other atomic energy laboratories for the sake of security. Oppenheimer was able to delegate responsibility and to make people feel they were being trusted. At the same time his quick perception enabled him to remain in touch with all phases of the work. When there were major prob-lems, or major decisions to be taken, he guided the discussions of the people concerned in the same spirit of a cooperative search for the answer in which he had guided discussions with his students. In work he did not spare him-self, and in response he received a sustained effort from all his staff.

It seems that the laboratory was set up just in time, because when the design of the plutonium bomb was ready, enough plutonium was available for the first bomb. The plutonium bomb demanded a greater design and de-velopment effort than the uranium bomb, since the more intense neutron background required a much more rapid assembly from subcritical condi-tions to the final highly critical configuration. Failing sufficient rapidity, a stray neutron is likely to set off the chain reaction when the assembly is only just critical, resulting in explosion of very poor efficiency.

When the test of the first bomb at Alamagordo demonstrated the force of the new weapon, all spectators felt a terrified awe of this enormous new power, mixed with pride and satisfaction at the success of their endeavors. Initially some were more conscious of the one emotion, some of the other. Oppenheimer, whose attitude toward his own faults was as unmerciful as it was toward those of others, if not more so, admitted later that he could not resist feeling satisfaction with the key part he had played in the work. Many accounts have quoted the verses from his Sanskrit studies of the Bhagavad–Gita which went through his mind at the time of the test, the first referring to the "radiance of a thousand suns," and the other saying, "I am become Death, the destroyer of worlds." In his awareness of the technical achievement, he clearly did not lose sight of the seriousness of its implications.

None of this was public knowledge until 6 August 1945, when the first uranium bomb was dropped on Hirishima. The repercussions of the decision to use the bomb to destroy a city will continue to occupy historians for a long time. Oppenheimer played some part in this decision: he was one of a panel of four scientists (the others being A. H. Compton, E. Fermi, and E. O. Lawrence) who were asked in May 1945 to discuss the case for the military use of the bomb on Japan. They were told that it would be impossible to cancel or delay the planned invasion of Japan, which was sure to be very costly in lives, unless Japan surrendered beforehand. Their opinion, which Oppenheimer supported, was that a demonstration on an uninhabited island would not be effective, and that the only way in which the atom bomb could be used to end the war was by actual use on a "military" target in a populated area. Today, in retrospect, many people, including many scientists, deplore this advice and the use of the bomb. Oppenheimer commented in 1962:[2] "I believe there was very little deliberation...The actual military plans at that time...were clearly much more terrible in every way and for everyone concerned than the use of the bomb. Nevertheless, my own feeling is that if the bombs were to be used there could have been more effective warning and much less wanton killing..." He remained for the rest of his life acutely conscious of the responsibility he bore for his part in developing the weapon and in the decision to use it.

Aftermath of the Bomb: Princeton

At the end of 1945 Oppenheimer returned to California. This did not mean, however, returning to an ivory tower. He was by now a national figure; his advice was much in demand, and he was also very seriously concerned with the issues raised by the invention of atomic weapons. He took part in the drafting of the "Acheson–Lilienthal report" which proposed international control of atomic energy. Most of the language of this report is undoubtedly Oppenheimer's and so, probably, are many of its ideas. The authors of this report wrote it in a generous spirit: international control of the new weapon would be used to ensure peace and to prevent any nation's threatening another with the formidable new weapons. It probably never had much chance of becoming a political reality. A proposal embodying the outline of the report, but hardly its spirit, was presented to the United Nations by Bernard Baruch as the "Baruch Plan," but nothing came of it.

In 1946 the Atomic Energy Commission was set up under the McMahon Act, which provided for civilian control of atomic energy. The first proposal, the May–Johnson Bill, which would have led to military control, was defeated, largely because of opposition from scientists, although Oppenheimer was prepared to accept it. The commission appointed a General Advisory Committee, with Oppenheimer as Chairman, and he served in that capacity until 1952. The committee did more than give technical advice; it had great

[2]United States Atomic Energy Commission, "In the Matter of J. Robert Oppenheimer." *Transcript of Hearings before the Personnel Security Board*. Washington D.C. 1954.

influence on the policy of the commission. Oppenheimer's role as Chairman was not to dominate opinion, but to clarify the issues, to help people formulate their thoughts. In addition to the General Advisory Committee, he served on numerous other committees concerned with policy questions relating to atomic weapons and defense.

In October 1947 Oppenheimer moved to Princeton, N.J., to become director of the Institute for Advanced Study. Until then the institute had been a kind of retreat for great scientists and scholars who wanted to get on with their studies in peace. Under Oppenheimer's regime the population of the institute grew in numbers, and included many young scientists, mostly as short–term members for a year or two. These in turn included many visitors from other countries. Oppenheimer was an active member of the physics department and usually presided at seminar meetings.

Under Oppenheimer's influence the physics group became one of the centers where the current problems of modern physics were most clearly understood. Many colleagues came to discuss their ideas with Oppenheimer, though to do so meant exposing one's thoughts to penetrating scrutiny, and sometimes to withering criticism. Oppenheimer now had less time for physics than in prewar days, and had to form his judgments more rapidly. He was fallible, and there were occasions when he violently and effectively attacked some unfortunate speaker whose ideas were perhaps not proved, but were worth debating; there were other instances when he hailed as very promising ideas which later proved barren.

The early Princeton years were a time when there was again a buoyant optimism in physics. The theory of electrons and their electromagnetic field had been stagnant for many years because of the infinities predicted by quantum theory for the field energy of a point charge. The discovery of the "Lamb shift" in the hydrogen spectrum showed that there were some questions to which theoretical answers were needed, and attempts to find the answers showed how one could bypass the troublesome infinities. Schwinger, Feynman and Dyson developed consistent formulations for the new form of the theory, and it was hoped that these formulations could be extended to the proton and neutron, and to their interactions with the newly discovered meson field. It was a time of intense debate and discussion, much of which took place at small ad–hoc meetings of theoreticians, at which Oppenheimer was at his best in guiding discussions, and in helping people to understand each other (and sometimes themselves). The phrase he used in an interview to describe the work at the institute—"What we do not know we try to explain to each other"—is very appropriate for these sessions. He had always had a remarkable gift for finding the right phrase, and he had now become an absolute master of the epigram.

While he did not resume personal research on any substantial scale (he was coauthor of three papers on physics after the war, one of them being a criticism of somebody else's theory) his participation in meetings at the institute and elsewhere was still a major factor in the development of ideas in physics.

As director of the institute, Oppenheimer was responsible also for establishing policy in other fields, including pure mathematics and history, among others. Here the breadth of his knowledge was a unique qualification. He did not, of course, take part in the work of the other groups as he did in physics, but he could understand what was being done, and could comment in a manner respected by the experts.

Throughout the postwar period, he wrote and lectured extensively. At first the subject was predominantly atomic energy and its implications, and the plans for its international control. Later he became more concerned with relations between the scientist and society, and from this, with the problem of conveying an adequate understanding of science to the layman. In his Reith Lectures on British Radio on "Science and the Common Understanding" he attempted to set out what science was about. The language of such lectures was probably not easily followed in detail by the nonscientist, but it had a poetic quality which to many listeners brought the subject closer.

The "Oppenheimer Case"

In December 1953 Oppenheimer was informed by the government that his security clearance, i.e., his access to secret information, was being withdrawn because of accusations that his loyalty was in doubt. He exercised his right to ask for hearings, and was exposed to the gruelling experience of more than three weeks' quasijudicial hearings during which all of his past was subjected to detailed scrutiny. The charges were based in part on his 1949 opposition to a crash program for developing the hydrogen bomb, and in part on his contacts or associations in the late 1930s and early 1940s with supposed communists and fellow–travellers. All of these contacts had been known to the Atomic Energy Commission many years before and had then not been considered sufficiently threatening to impede his clearance.

It is impossible to understand how these charges could be raised without first remembering the atmosphere of hysterical fear of communism of the Joseph McCarthy era, and then noting that Oppenheimer had made many enemies, who were delighted at this opportunity to curb his influence. Some of these enemies were people he had bested in public debate, when his devastating logic had not only shown them to be wrong, but had made them appear ridiculous. Others were people interested in military policy who feared his influence, which could act contrary to their interests.

The hearings before the three–man "Personnel Security Board" were originally intended to be confidential, but eventually the transcript was published.[2] It remains an interesting historical document. The board found that Oppenheimer was "a loyal citizen" but also, by a two–to–one majority, that he was to blame for opposing the hydrogen–bomb program, and later was lacking in enthusiasm for it.

The report of the board went to the Atomic Energy Commission. The commissioners did not uphold the board's (majority) decision censuring Oppenheimer for his views on the hydrogen bomb—this would have caused a powerful reaction in the scientific community—but confirmed the withdrawal of his clearance, in a majority verdict, mainly on grounds of "defects of character." This conclusion was opposed by one of the commissioners, the physicist Henry Smyth, who wrote a minority report in favor of Oppenheimer criticizing the arguments of his colleagues.[3]

Oppenheimer continued as director of the institute, and with his writing and lecturing. On many occasions audiences at his lectures gave him ovations clearly intended to express their sympathy for him and their indignation at the treatment he had received.

In 1963, when the McCarthy era was an embarrassing memory, when many of the people who had conducted the Oppenheimer investigation had been succeeded by others, and when tempers had cooled, it was decided to make a gesture of reconciliation. Oppenheimer was given the Enrico Fermi Award for 1963, a prize of high prestige awarded by the Atomic Energy Commission. The award is usually conferred by the President, and John F. Kennedy had the intention of doing so when he was assassinated. It was then conferred by Lyndon Johnson, and Oppenheimer acknowledged it with the words he had intended to say to President Kennedy: "I think it is just possible ... that it has taken some charity and some courage for you to make this award today."

He knew for almost a year that he had throat cancer, and he could contemplate this fact and talk about it as lucidly as he could discuss a conclusion in physics. He died in 1967.

---

[3]United States Atomic Energy Commission, "In the Matter of J. Robert Oppenheimer." *Text of Principal Documents*. Washington D.C. 1954.

The Growing Pains of Robert Oppenheimer

Robert Oppenheimer was one of the most famous scientists of his time, not so much for his contributions to theoretical physics, although these were substantial enough, but as mentor of a large number of successful theoreticians, and above all for his direction of the Los Alamos Laboratory at which the atom bomb was developed and produced during WWII.

The impressions he made on superficial acquaintance were very different at different times and on different occasions, and sometimes appeared hard to reconcile. The shy and introverted young boy and student became the successful director of a vast enterprise, successful very largely by his skill in dealing with people. The young researcher, whose professors had doubts about his scientific future because he had difficulty in expressing himself in writing, became famous for always finding the *mot juste* to sum up a situation. His lightning speed of understanding, which often produced the answer or a comment before his questioner had completed half his explanation, would on occasions make him miss the point of an important new idea, and he could meet presentations of such ideas by young colleagues with withering criticism. The ivory-tower scholar who did not know or care much about public affairs found his social conscience in the late 1930s by taking an interest in liberal and left-wing groups (an interest which formed part of the accusations in the security hearings against him), and later became a national figure in the debates about international relations and defense policy.

This is a review of *Robert Oppenheimer: Letters and Recollections*, edited by Alice Kimball Smith and Charles Weiner. Cambridge: Harvard University Press, 1980. For the complete reference, please see the Acknowledgments section.

The present volume traces the development of this complex personality. The editors, Alice Kimball Smith and Charles Weiner, are singularly well qualified for their task. Mrs. Smith was at Los Alamos from the start until the end of the War, and therefore well-acquainted with the atmosphere of the place and with many of its members. She later made a study of the part played by scientists in the post–war period in influencing legislation toward civilian control of atomic energy. Dr. Weiner is a physicist concerned with recent scientific history, who has recorded interviews with many distinguished physicists.

In describing themselves as editors of Oppenheimer's letters and recollections, they do not really do justice to their work. The letters are supplemented, where there are gaps, not only by passages of Oppenheimer's recollections taken from later interviews, but also by comments from friends, colleagues and pupils, painstakingly collected and sensitively evaluated. In many places the editors' comments, based on this material, are more informative than the letters. They not only identify many of the people and events referred to in the letters, but also summarize events and developments not mentioned in the letters or which took place in periods from which no letters have survived. The book is therefore really an intermediate between edited correspondence and a serious biography.

The young Oppenheimer was not the type to keep copies of his letters, and not many of the recipients preserved them. The fact that there remain enough letters from the pre-War period to provide a clear picture of Oppenheimer's attitudes and development during that time is largely due to the few friends who kept the letters and made them available. One of these is Herbert W. Smith, his English teacher in the Ethical Culture School; letters to him are most numerous of any of the correspondents. In the earlier part Oppenheimer sends him essays and verse; his evident desire for approval makes Smith a kind of father figure.

Almost as numerous are the letters to Oppenheimer's brother, Frank, in which the elder brother gives much advice and exhortation, and shows great warmth and affection, though sounding a little patronizing at times. Amongst other sizeable collections of letters are those to Francis Ferguson, whom he first met in his final year at the Ethical Culture School and with whom a close and sometimes tempestuous friendship developed. Of a later period, the physicist George Uhlenbeck, who was a close friend, preserved many interesting letters.

The first three chapters are based on these letters and a few others, supplemented by other information collected by the editors. From his home background we follow Robert to the Ethical Culture School, where he was an excellent scholar in all subjects, and where his interest in science was first aroused through mineralogy. After leaving school and after a year's convalescence from dysentery contracted during a trip to Europe, he visited New Mexico in the summer with his teacher, Herbert Smith. This was the beginning of his deep attachment to the southwest. In 1922, aged 17, he entered Harvard. He spread his studies over a wide range of subjects, mostly on the literary side but including chemistry, in which he graduated in 1925, and some physics. The number and diversity of courses, and the amount of reading he managed, must have meant hard work; at the same time he learned to make contact with people and form friendships.

In his final year at Harvard an important influence was Percy Bridgman, in whose laboratory Oppenheimer did some experimental work, and his decision to go for postgraduate study in physics probably owed much to Bridgman. He decided to go to Cambridge. His studies there did not work out too well. He was not accepted to work under Rutherford, as he had hoped, but did some experimental research under J. J. Thomson, which did not make much progress. However, he had the chance at Cambridge of meeting many distinguished people, including Niels Bohr and other theoreticians, and it became clear to him then that he wanted to do theoretical work. During the Cambridge period he had times of very deep depression and unhappiness, but he was able to deal with his own disturbed state more effectively than any doctors he consulted.

After a year at Cambridge he moved to Göttingen, a congenial environment, to work with Max Born. This was the exciting time when quantum mechanics emerged, and the beginning of Oppenheimer's passionate devotion to theoretical physics, and also of his constructive contributions to the subject. He qualified in one year for a Ph.D, and returned to the United States in 1927. After a semester at Harvard and one at the California Institute of Technology, he spent another year in Europe, visiting Ehrenfest, Kramers and Pauli. Then he moved to California, holding appointments at Caltech and Berkeley, dividing his time between the two places. He liked this arrangement because at Caltech there were many theoreticians from whom he could learn, while at Berkeley he was practically the only modern theoretical physicist, and thus could contribute more to teaching and research. This unusual arrangement continued until he became involved with atomic energy in 1942. It was here that he started to gather a group of graduate students, who usually followed his annual migration—and that he developed his power as a teacher and leader in research. It was in the California years that he first became concerned with the world around him and with politics. It was in that period that he got married.

He was now an established and respected member of the physics community, as expressed in his election to the National Academy in 1941.

In 1942, when work on the atomic bomb was becoming a serious project, he was asked to take charge of work on fast-neutron fission, which was related to the problems of design and performance of a bomb. The period from this time until the end of the War is dealt with in chapter 4. Here the task of the editors changes. It is no longer a question of searching out the few surviving letters and filling gaps, but rather to select, from amongst the voluminous files of official correspondence, the letters which are most informative of Oppenheimer's attitude and ways.

In coordinating the experimental and theoretical work carried out in different universities, he immediately showed his characteristic ability for summarizing, for stating the main issues clearly and for formulating questions. It soon became clear that efficiency demanded that this work be concentrated in a single center, and he became involved in the planning of the new laboratory and the selection of a site. At some stage General Groves, head of the "Manhattan District" (the code name for the atomic weapons project), decided to appoint Oppenheimer the director of the new laboratory. This was a daring decision, considering that Oppenheimer was a theoretician, not experienced in administration and had the reputation of being rather disorganized in practical matters. Yet the decision was amply justified by its success. Chapter 4 helps to understand why. In addition to his lucidity in recognizing and expressing the main issues, Oppenheimer managed to deal with people in a manner which made them feel that they were respected, and gave them the confidence that their views and their needs were taken into account. The letters give examples of the many large and small problems that the director had to handle, and the pressure under which he was working. One of the surprising facts is how much mutual respect, and indeed affection, developed between Oppenheimer and Groves, in spite of the enormous disparity between their personalities.

Chapter 5 covers the three months from the end of the War to Oppenheimer's departure from Los Alamos. The letters are now concerned with post–War plans for Oppenheimer, positions for the staff of Los Alamos and the organization of atomic–energy research, and reflect the burden of responsibility that scientists carried because of their work on the bomb. The last document by Oppenheimer is his speech at Los Alamos, in which he discusses the future of international relations and the idea of international control of atomic energy.

The editors decided, wisely, to terminate the collection of letters at this point, since from then on the glare of publicity which fell on him, the great volume of correspondence and his membership of numerous official committees, changed the style of his letters, which tended to become more official and more guarded. They do not give the kind of insight which makes the letters reproduced in the book such fascinating reading. The book therefore does not, except in a brief résumé, deal with the post–Los Alamos Oppenheimer, with his directorship of the Institute for Advanced Study in Princeton or with the security hearings usually called the "Oppenheimer Trials," but there are other sources to cover these matters. However, the period it does cover, the growth of an unusually gifted and sensitive person to maturity, and to his success in a gruelling job, is of absorbing interest.

Heisenberg's Recollections

Physics went through a major revolution in the first quarter of this century. It was started by Max Planck and Niels Bohr, whose use of the idea of the quantum of action brought order into a wide range of phenomena concerning the structure of matter and the nature of atoms, but was as yet incomplete, in some parts arbitrary, and even self–contradictory. The final stage of this development, which completed our understanding of atomic physics, owes as much to Werner Heisenberg as to any other single person, although another development, started independently by Schrödinger, turned out to be a different expression of the same basic laws. Many others, of course, made important progress along the path opened up by the pioneers.

Heisenberg's greatest contribution was to recognize that the contradictions of the old quantum theory were the result of asking questions to which there was no physically meaningful answer. Physics deals with observable phenomena, and in the atomic domain the indivisibility of particles and the existence of quanta put a limit to the accuracy of possible observations, if they are not to disturb unduly the very phenomenon under observation. The lesson was, therefore, to avoid asking questions about atoms to which no experiment could give an answer, and to express the laws of physics by reference to observable quantities. Heisenberg succeeded in carrying out his program, and the set of laws he arrived at form one of the starting points of modern theory.

Physics and Beyond is an account of his thoughts, including the ideas that led him to quantum mechanics; this part will be of absorbing interest to anyone interested in the history of physics. The book is not an autobiography: details of the author's life are mentioned only where they are needed to pro-

This is a review of *Physics and Beyond: Encounters and Conversations,* by Werner Heisenberg, translated by Arnold J. Pomerans. New York: Harper & Row, 1971. For the complete reference, please see the Acknowledgments section.

vide the background to the thoughts and discussions, which are the real sub-ject of the narrative.

The thoughts described are not, however, limited to physics and to the epistemological problems that were bound up with the beginnings of quantum mechanics, but range widely over philosophy and religion, reactions to life under the Nazis before and during the War, work on atomic energy, and the reorganization of German physics after the War. It seems to be a thesis of the book that attitudes toward the more abstract and the more concrete of these problems are interconnected. This is stressed in the title of the German original "The Part and the Whole" (*Der Teil und das Ganze*), which in trans-lation has mysteriously, and to me rather disappointingly, become "Physics and Beyond."

However, the nature of this connection is not spelled out, and the reader is left to find it for himself. Is there a recognizable connection between Heisenberg's thoughts in very different fields? I find the answer to this ques-tion very elusive. Was he better able to see what accepted ideas of physics had to be discarded because he grew up in a post–war world in which young people were questioning accepted social values? There is some plausibility in the thought, yet others, such as Einstein, Planck, or Bohr, came to equally revolutionary thoughts from very different personal backgrounds.

Were his views on politics or public affairs conditioned by his reasoning powers as a scientist? Scientists probably have something in common in their attitudes toward such problems in that there are some fallacies and some kinds of confusion to which they are not prone. But beyond that there is room for vast differences in approach. One should probably not look for such sim-plified connections. It remains that the thoughts on physics, on music, on world and national affairs are the thoughts of one person, and all go to make up the "whole" of the mind behind the book.

To escape the tedium of continuous abstract argument, Heisenberg pre-sents conversations with friends and colleagues, describing, for greater live-liness, the setting in which each conversation took place, whether in the study of one of the participants, on a bicycle ride through the mountains, or when they were watching the boats in the port of Copenhagen at night.

Even so the conversations are necessarily simplified and telescoped. The author makes it clear in the preface that they are not meant to be accurate recordings. In fact, it is clear from the preface, and from reading the conver-sations, that the other people appear only as a backdrop to the author's thoughts. It could not be otherwise because it would be a hard task indeed to present adequately the thoughts expressed years ago by another person whose views were different from one's own. The influence of other people's arguments on Heisenberg comes out clearly enough, but one should not look for a faithful picture of Bohr, Pauli, or other actors in the narrative—in spite of their participation in sailing trips and mountain walks, they do not really come alive.

What does come out very clearly is the extent to which in our time scientific ideas do not materialize suddenly in the mind of one person, but grow from interactions with others. This applies not only to the early stages, when teachers help one to understand what is already known, but also later when many colleagues, younger and older, struggle in discussions for an understanding of the facts and of the meaning of new concepts. The influence of environment on a scientist does not, of course, diminish our respect for one who, like Heisenberg, has taken a major step forward.

The presentation in the chapters dealing with physics is nontechnical. For the reader who is a physicist this is not a disadvantage, since he will recognize the precise meaning of what is being discussed. I find it hard to judge whether the explanations given are adequate to make the arguments meaningful to a lay reader, but I believe that it is possible, without a knowledge of the technical details, to sense the essence of the story: the difficulty, and the excitement, of finding and formulating the new laws of nature, which retain the successes of the older forms, but incorporate the new and apparently contradictory findings; the struggle to find new concepts appropriate to the newly gained knowledge, but contrary to our intuition, which has grown from experience on the scale of everyday things and needs correcting when we are concerned with atoms or still smaller objects.

Another topic of great interest is the author's attitude toward the Nazis and toward the War. Heisenberg reports the horror, and the sense of frustration, with which he watched the growth of the Nazi movement. Some of the reported conversations showed how he, who as a boy had been active in the German youth movement, which stood for a rethinking of conventional values, against materialism and against hypocrisy, was pained by appearing now as a representative of the older generation, in opposition to the "adventurous" spirit of the young National Socialists.

But he clearly saw the disastrous nature of the new movement. He saw its anti-intellectual nature, he was horrified by its anti-Semitism, and he realized that its aggressiveness would sooner or later lead to war and disaster. Yet in arguing with a Nazi youth leader among his students, he presents the young man's views with a certain amount of sympathy. His answers stress the faults of the new regime, but the student's criticism of the old state of affairs which in his mind justified any drastic change is left uncontradicted. One guesses that Heisenberg, who evidently abhorred the methods and the inhumanity of the Nazi "revolution," may have felt a measure of sympathy for its aims.

As the pressure of the Nazi regime mounted, with its interference in university affairs and with irksome rules, and as the ranks of university staff became more and more depleted by dismissals, Heisenberg's thoughts turned inevitably to the possibility of resigning and emigrating. He asked Max Planck, the respected old physicist, for advice. The chapter started by the dialogue with the Nazi student is then completed by a dialogue with Planck, and Heisenberg's own struggle with the problem on the train back. In both only the arguments for staying are marshaled; evidently the case for leaving is so obvious it does not have to be explained. In the end Heisenberg decides to stay, his strongest argument the need for keeping a group of people of good will together, who will not be able to influence what happens under the Nazis, but who will be available to help rebuild "after the holocaust." Here, in a conversation during a trip to the United States in 1939, it appears that he was quite clear that there would be war, and equally that Germany was bound to be defeated.

War did come and soon Heisenberg was drafted into work on atomic energy. This chapter is of special interest because ever since the end of WWII there has been controversy about the German atomic energy project. Some writers claimed that the German scientists deliberately avoided making atomic weapons, for moral reasons, while others say that they did not take the possibility as real since they had not realized how small, and how powerful, atomic bombs were going to be. Heisenberg describes his attitude toward the moral problem in a conversation—this time specifically labeled as imaginary— with C. F. von Weizsäcker, in which they agree that it would be morally wrong to use atom bombs in any circumstances, and therefore feel relieved that there does not seem to be any realistic possibility of making them. He comments later, "...we overestimated the technical effort involved."

In view of this statement, it is very puzzling to read further on that in 1954 negotiations about permission for West Germany to start working on nuclear power were influenced favorably by the fact that "Germany had made no attempts to build atom bombs during the War, although she had not lacked the necessary skills and knowledge." This contradicts the earlier quotation. As a result we do not know where Heisenberg stands on the question whether the German scientists could not, or could and would not, work on the making of atom bombs.

Heisenberg describes how, to the group of German atomic scientists then interned in England, the news of Hiroshima seemed at first unbelievable: their disbelief again confirms the impression that they thought the project impractical. When finally convinced of the reality of the development, whose difficulty they had overestimated, they were—like everybody else—shaken by the thought of the death and destruction caused by this manifestation of the progress of science. There follows another conversation about the moral issue and the position of the scientist.

One of the last chapters deals with the beginnings of the work on the "unified field theory" which Heisenberg initiated in 1957 and which he and his collaborators still pursue with enthusiasm. This is still a controversial subject in that a great many physicists doubt the value of the results already obtained in this work and the prospects of its developing into a valuable discipline.

There was also controversy over this between Heisenberg and Wolfgang Pauli, with whom he had often collaborated in the past, and whose judgment and critical powers Heisenberg, like all other physicists, respected highly. This is recorded in another set of conversations, Pauli being at first highly skeptical and critical, but then convinced of the promise of the new theory and enthusiastically contributing ideas to its development. Pauli left for a visit to the United States full of this new enthusiasm but, to Heisenberg's intense disappointment, returned in his old critical spirit and fiercely attacked Heisenberg's position at an international conference. Pauli died a few months later. The whole episode was so short that many physicists were not aware of Pauli's period of optimism, and did not know what to make of a paper bearing both his and Heisenberg's names which had been circulated by Heisenberg and which was later repudiated by Pauli.

Niels Bohr, Werner Heisenberg, and Wolfgang Pauli. (Photograph published by permission from the Niels Bohr Institute and courtesy of AIP Emilio Segrè Visual Archives.)

An explanation of the nature of the argument would be too technical for the book, so the reader is not told what this whole episode is about, but the chapter will give him a feeling of the controversies that can arise even in an exact science, and the emotions that can be engendered even by the "dry" subject of quantum field theory.

Werner Heisenberg, 1901–1976

I n the great revolution in fundamental physics which started with the ideas of Planck and culminated in the impressive breakthrough of the mid–1920s, and in the completed framework of quantum mechanics, Heisenberg made many important contributions. Two of these, the basic treatment of quantum transitions and the formulation of the uncertainty principle, were so original and impressive that they came close to the popular image, so often unrealistic, of great new concepts growing out of the thoughts of a single individual of genius. In the two years from 1925 to 1927, which saw the emergence of a new set of principles in physics, since then refined and widely applied but not substantially changed, the steps taken by Heisenberg were large and decisive.

Early Life

Werner Heisenberg was born on December 5, 1901 in Würzburg. In 1910, the family moved in 1910 to Munich, which he came to regard as his hometown. He retained throughout his life an intense affection for this city. His father, August Heisenberg, was the son of a master locksmith, an outstanding craftsman and designer. The grandmother, though described as "a simple person," is remembered for her wisdom and human qualities, and as a person often in demand as an adviser to neighbors with problems. August Heisenberg was professor of medieval and modern Greek philology, a man of great energy and with a happy disposition. The mother, Annie, *née* Wecklein, was the daughter of a headmaster of the Maximilian Gymnasium in Munich.

Werner distinguished himself early at school; his ability was particularly evident in mathematics and in other formal subjects such as grammar, and

This article was written with Sir Nevill Mott, F.R.S. and published in *Biographical Memoirs of the Fellows of the Royal Society*, **23** (1977). For the complete reference, please see the Acknowledgment section.

none of the work seemed to require of him any great effort, although he was ambitious and enjoyed doing well. He showed an early interest in physics, and was particularly fascinated by the possibility of applying mathematics to practical situations. His school education, however, was interrupted in the spring of 1918, when boys of 16 were called up for auxiliary service, and he was sent to work on a farm. This involved long hours of hard physical labor, and this left him too tired to read at night the books he had brought with him which included some of the works of Kant.

When World War I ended in November 1918, conditions were chaotic, with political authority passing back and forth between different groups. At one period Heisenberg and other young boys worked as messengers for a force which was to bring law and order back to the city. The duties were not strenuous, and there was plenty of opportunity to make friends and to make progress with reading, in which philosophy played an important part. He remembered later the impression made by Plato's *Timaeus*; in spite of his respect for Plato, the statements there about the smallest parts of matter and their shapes seemed to him wild speculations.

The time on the farm and the later messenger service provided many contacts and later friendships with young men of the same age, and many long discussions. This was the time of the tremendous growth in Germany of the Youth Movement, consisting of a great variety of groups of young people who were dissatisfied with what their elders had made of the world and who were impatient with old customs and old prejudices. They were anxious to create a better world. Heisenberg describes in *Der Teil und das Ganze* his first meeting of such a group with considerable affection, although he found the talks confused and contradictory, lacking in orientation. He was worried by the lack of synthesis, and relieved when the gathering turned from debating to the music of Bach, restoring his confidence that clarity was possible. He quotes this night as an experience which had a considerable influence on his later thinking.

But many of the conversations with his friends touched on the idea of atoms, and he was fascinated by their philosophical and practical implications. After leaving school he came across Weyl's *Space, Time and Matter* and became attracted and worried by the complexity of the mathematical arguments and the underlying physical concepts. This, he says in *Der Teil und das Ganze*, confirmed him in the decision to study mathematics at the university. However, in an interview with the mathematics professor Lindemann, in which Heisenberg requested admission to the seminar—perhaps an apparently immodest request from a new student—he was turned down, and his mention of Weyl's book prompted the comment from the professor that, having read this, he was spoilt for mathematics anyway.

After this disappointing start he applied, on his father's advice, to Sommerfeld and was immediately accepted. He comments on this first interview that the forbidding and martial impression made initially by Sommerfeld, with his handlebar moustache, very soon gave way to a feeling for his kindness and helpful authority. Sommerfeld was at that time not only a great theoretical physicist with an extensive experience in all parts of his subject and an intimate knowledge of its frontiers, but also probably its greatest teacher. So Heisenberg became a student in Sommerfeld's Institute. He soon made friends with Wolfgang Pauli, who was somewhat senior to him and whose depth of understanding and power of criticism he learned to appreciate. This was the beginning of a lasting friendship.

Music remained a strong interest. He was a very gifted pianist, and the parents of a friend, in whose house he used to play trios with his friends, tried to persuade him that he ought to become a musician rather than a scientist.

Early Physics in Munich, Göttingen and Copenhagen

In 1920, when Heisenberg started his studies under Sommerfeld, the most burning questions in fundamental physics related to the quantum theory of atoms. Niels Bohr had used Planck's concept of the quantum of action to explain the spectral lines of atoms in terms of the energy differences of "quantum state," and had formulated rules for determining these states. Sommerfeld had refined and elaborated these rules. Bohr's spectacular success in accounting for the spectrum of the hydrogen atom had been followed by many other demonstrations that the new theory contained a good deal of the truth. One could understand the details of X-ray spectra, of the spectra of alkali atoms and of ions with one external electron. Also Bohr's "Aufbauprinzip" came close to explaining the nature of the periodic table. Even the fine structure of the hydrogen spectrum seemed to confirm the use of his quantum rules to obtain allowed solutions of the relativistic equation of motion (though this agreement was later seen to be somewhat fortuitous). The Franck–Hertz experiment had confirmed the existence of discrete energy levels, and Stern and Gerlach were soon to verify the quantization of angular momentum. These new insights and the techniques for dealing with the theory were expounded in the first (1919) edition of Sommerfeld's text *Atomic Structure and Spectral Lines*, to be followed by three further editions as the subject expanded.

Yet there were serious weaknesses in the theory. Conceptually, it was very unsatisfactory that the Newtonian equations of motion could be used only for the stationary orbits, which meant selecting from the possible solutions the ones satisfying the quantum rules, and there was no way of describing the transition from one such orbit to another during the emission or absorption of light. There were also conceptual difficulties with the effect of "space quantization," i.e., the possible orientations of the angular momentum. If an external field was reduced to zero and then a field in a different direction applied, the orbits would suddenly have to select amongst the new discrete directions. For problems involving several interacting degrees of freedom, or several interacting particles, the rules did not give simple answers, though it was not clear whether this was a basic failure of the theory or whether it could be remedied by a more refined formulation, or a more skillful evaluation, of the rules. Later Pauli (1922) found the electronic ground state of the hydrogen–molecule ion H_2^+, perhaps the most ambitious application of the old quantum theory, and did not get the right energy. A glance at even the last edition of Sommerfeld's text reveals little of these inadequacies and contradictions; Sommerfeld was an optimist, more concerned with what had been, or could be, done than with difficulties.

However, he was well aware of the unsolved problems, and Heisenberg soon was put to work on some of these. Three papers completed in Munich between December 1921 and August 1922 report attempts to deal with such problems.

Sommerfeld went for the winter semester 1922 to 23 as a visiting professor to the University of Wisconsin and arranged for Heisenberg to spend this period with Max Born in Göttingen. Born was enthusiastic about his new student, and they soon were engaged in joint work. The first joint paper, was on the general subject of the quantum theory of the atom. It attempted to describe an atom with many electrons by assuming the motion to be strictly periodic, so that the revolution periods of all the electrons would be commensurable. It was a gallant attempt to fit quantum theory to a problem that, in its state at that time, it was not capable of handling. The next paper showed that one did not get correct answers for the excited states of the helium atom if one treated the interaction between the two electrons by perturbation theory, adapting to this problem the perturbation methods familiar in astronomy.

One might have thought that the new ideas and the attempts to extend them would have kept a student of two years standing fully occupied, but during the same period Heisenberg was thinking about fundamental problems in hydrodynamics. One paper written in Munich attempted to calculate the parameters of the vortex trail behind a moving plate described by von Kármán, from simple physical considerations. In a note accompanying this paper, Prandtl, a great authority in fluid mechanics, found this paper by a newcomer very instructive, although he pointed out that the arguments were only qualitative, and could be regarded as convincing only in determining orders of magnitude.

A more substantial investigation of the problem of turbulence had been started in Munich, evidently at Sommerfeld's suggestion, and Heisenberg continued with this in Göttingen. When he returned to Munich for the summer semester in 1923, this was completed and submitted as a dissertation. The substance was published later. The paper discusses both the stability of laminar flow and the nature of turbulent flow. The stability problem is investigated by using the Orr–Sommerfeld equation, a fourth–order linear differential equation for small disturbances from laminar flow. This becomes singular when the local speed of propagation of disturbances equals the local flow velocity. In the case considered, the two–dimensional Poiseuille flow between parallel plates, it is found that laminar flow is stable below a critical Reynolds number, which is only estimated, because in that region the approximations used in the evaluation are not adequate. For higher Reynolds numbers there is instability, but when the Reynolds number becomes very large the laminar flow becomes again less unstable, in the sense that only disturbances of very long wavelength tend to grow. This result was new and unexpected. It was not until many years later that Heisenberg's ideas on this point were accepted and confirmed by further theoretical work. The discussion of the turbulent state itself is by now very dated, and probably was, at its time, no great advance on current ideas.

Heisenberg returned to the subject of turbulence much later, in 1945. This was prompted by circumstances which will be described later. By this time the statistical approach to turbulence initiated by G. I. Taylor, Kolmogoroff and others had led to a new understanding. Heisenberg and von Weizsäcker arrived at very similar conclusions independently, before they knew about the earlier work. Heisenberg described these ideas in the language of a physicist on the basis of statistical mechanics. He showed how energy is exchanged between the different degrees of freedom (vortices of different sizes). For the larger–scale irregularities the coupling is mainly by the non–linear terms in the Euler equations, so that the energy is conserved, but degraded to smaller–scale vortices. For very small vortices viscosity plays an essential part, and ultimately the energy is converted into heat. Heisenberg proposed a simple way of describing the region in which viscosity is dominant. His ideas attracted a good deal of attention, but were not confirmed by experiment. So far no better solution to this problem has been found.

It was Sommerfeld's wish that he should use the paper about turbulence as his dissertation, and it certainly was amply sufficient for this both in substance and in quality, though no doubt he could have chosen some of the work on the quantum theory of atoms equally well. Nevertheless, the degree examination was not plain sailing. The requirements of German universities included, besides the dissertation, a thorough oral examination in three subjects, a custom no doubt related to the fact that the D.Phil. was the first examination normally taken by a physics student, there being no equivalent of the Bachelor's degree. Heisenberg had never taken the laboratory course very seriously, and this had irritated the professor of experimental physics, Wilhelm Wien. When it came to the oral examination in experimental physics, he asked the candidate in succession about the resolving power of the Fabry–Pérot interferometer, that of the microscope, of the telescope, and about the principle of the lead accumulator. When Heisenberg was unable to answer any of these questions, Wien wanted to fail him in physics. However, the rules required a joint mark in physics, including both the experimental and the theoretical part, and there ensued a difficult negotiation between Wien and Sommerfeld, with the result that the candidate was awarded a third class ("rite") in the oral. An interesting footnote on this episode is provided by the details of Heisenberg's paper on the uncertainty principle, in which his discussion of the resolving power of a microscope was still inadequate, as we shall see.

In the autumn of 1923 he went again to Göttingen, at the invitation of Max Born, who had secured an additional grant for the purpose, as the one regular assistantship was already held by F. Hund. There followed a period of further work on the open questions of quantum theory, much of it in collaboration with Born, and in fields in which Born had particular interest and experience.

While some of the applications of the theory gave satisfactory answers, the conviction was growing that change was needed. This was brought out by an attempt, with A. Landé, to deal with the higher multiplets in the Ne spectrum. Heisenberg decided to try a modification of the quantum rules. A proposed modification of the rules involved attributing to each quantum state of the atom two values of the angular momentum quantum number, namely $\mathcal{F}+1/2$ and $\mathcal{F}-1/2$ if the angular momentum is $\mathcal{F}\hbar$. The need for such an assumption arose from the difficulty of accounting for the number of levels in an atom consisting of a core of known angular momentum and an outer electron; it was later superseded by the inclusion of spin. However, the boldness of the step of talking about pairs of quantum numbers foreshadows some of the thinking that led later to the introduction of matrices. The paper showed a deep insight into the problems of atomic theory, and it was, at Born's suggestion, submitted, and accepted, as a "Habilitationsschrift" to gain the "venia legendi," the authority to lecture, in the University of Göttingen. To acquire this status at the age of 22, and barely one year after the doctorate, was very unusual.

After reaching this distinction, and a summer vacation spent with his friends of the Youth Movement in a walking tour of the old towns and the countryside of South Germany, he went to Copenhagen as a research associate. He had met Niels Bohr first at the Göttingen seminar in the Spring of 1922. The affection between the two men and their mutual respect had deepened. Heisenberg knew how deeply Bohr understood the problems of physics, and how much he could learn from him. Their conversations, often conducted on walks in the country, also ranged over many other fields of human life and human affairs, and here Bohr's wisdom and warmth made a deep impression on Heisenberg. This is very clear from the somewhat schematic account of such conversations in *Der Teil und das Ganze*.

Not surprisingly, the emphasis of the Copenhagen period is again on fundamentals. He attempted to make the correspondence principle (which was really a way of thinking rather than a quantitative rule) more definite by adding the condition of spectroscopic stability. This says that the sum of the intensities of the Zeeman components of a spectral line is insensitive to the value of the magnetic field. He made another attempt to sort out the multiplet structure. A very important step forward resulted from the collaboration with H. A. Kramers. Kramers had been thinking about the problems of the emission, absorption, and scattering of light by atoms in dense matter and its relation to dispersion. He aroused Heisenberg's interest in these problems and the result was a joint paper which was of great importance for two reasons: the first was that it clearly recognized the analytical structure of the dispersion curve (the variation of dielectric constant with frequency), by which the existence of an absorption line at one frequency makes a known contribution to the refraction at another. The idea of these "Heisenberg–Kramers relations" later led to the concept of "dispersion relations," which have become a major tool in the description of the scattering of radiation and of elementary particles. The second function of this paper was to help Heisenberg to understand which of the quantities characteristic of atomic transitions were relevant to the interaction of the atoms with radiation, and therefore had to appear in a consistent theory of quantum phenomena.

Quantum Mechanics

In the Spring of 1925 Heisenberg was back in Göttingen. He was by now convinced that one needed a theory which abandoned the classical description of the electron orbits inside the atom, and instead concerned itself with quantities that were accessible to observation. In this he was no doubt influenced by the success of Einstein's theory of relativity, in which the paradox of the constancy of the velocity of light is resolved by realizing that the simultaneity of distant events is not a meaningful concept, because it does not correspond to any imaginable observation. However, if the concepts and laws of Newtonian mechanics were to be thrown overboard, new rules would have to be put in their place which would lead to firm predictions for the values of observable quantities. Heisenberg tried hard to obtain the relationships between the amplitudes (the strengths and phases) of the oscillations corresponding to the various transitions in the hydrogen atom which one knew, or could guess, from the correspondence principle. However, this attempt led to mathematical problems of such complexity that he was unable to find the required relations. In order to escape from this frustrating complication, he decided to look at a simpler system, for which he chose the anharmonic oscillator. In retrospect one is inclined to ask whether the harmonic oscillator would not have been still simpler and adequate for the purpose. Evidently he feared that it would be too simple; in particular, the equality of the frequencies of all allowed transitions might have caused difficulty if one was to label transitions only by their frequency, as was suggested by the starting point of a generalized Fourier series.

He had reached about this stage when he suffered a violent attack of hay fever and had to ask for leave of absence to travel to Heligoland, a haven for hay fever sufferers because of the restricted vegetation. The irritation then subsided and because of this, and perhaps also because of the absence of other distractions, progress was rapid. He was able to see the relations that had to be imposed to ensure that the transitions would be in line with the correspondence principle. Having written down that condition, he realized that there was no further freedom left. In particular, there was no way of ensuring the conservation of energy, unless this were already a consequence of the new equations. At this stage, he saw no valid principle which would settle this question, so he had to calculate all the contributions to the energy. We would say today that one needed to show that all the nondiagonal matrix elements of the energy were zero, and the diagonal elements equal to the known energy levels. When the first few terms came out as required, he became so excited that he had difficulty concentrating on the remaining calculation without algebraic errors. When the final result confirmed energy conservation fully, he knew that this completed a proof of the consistency of the whole scheme. The fact that this result had come out of the scheme without having been put in explicitly indicated that the whole approach had something in it which was very reasonable physically.

The paper containing these results was sent for publication after Heisenberg's return to Göttingen. It reached the *Zeitschrift für Physik* on 29 July and was published in September 1925. It is fair to regard this paper as the start of a new era in atomic physics, since any look at the physics literature of the next two or three years clearly shows the intensity and success of the work stimulated by this paper. His ideas were widely accepted. On his return from Heligoland, Heisenberg stopped in Hamburg to tell his friend Pauli about the new results, and Pauli, inclined to be critical, recognized the promise of the new scheme immediately. Max Born in Göttingen was soon won over, though perhaps a little sad that the vital step had been taken while Heisenberg was away from Göttingen. News of the work spread even before the paper was published and Heisenberg was invited to visit Holland and Cambridge during the summer, and lecture about the new ideas.

Meanwhile, Born and Jordan examined the mathematical structure of Heisenberg's work. They recognized that the rules which Heisenberg had invented for combining the different transition amplitudes were none other than the rules of matrix multiplication, well known to mathematicians. (Heisenberg must have felt as if he had been told that he was speaking prose.) This made it possible to use many theorems and techniques from the mathematical literature. More important, Born and Jordan formulated the main quantum rule in Heisenberg's scheme in terms of the "commutation law" $pq-qp=h/2\pi i$, and thereby greatly simplified the presentation of the scheme.

On Heisenberg's return to Göttingen, after the summer the discussions were resumed, and led to a joint paper by Born, Heisenberg, and Jordan, which completed the structure of matrix mechanics, as the new system came to be called. The rules of the new mechanics were now established, and in principle almost all atomic phenomena were covered by it, but of course much work was needed to extract quantitative answers from the new equations which were not always easy to handle. The one main exception which limited the range of the new mechanics was the case of continuous spectra, which, in principle, were included, but which it would have been most difficult to evaluate by matrix methods.

The Born–Heisenberg–Jordan paper, which was submitted for publication in November 1925, extended the formalism to several degrees of freedom, discussed degeneracy, developed perturbation theory, mentioned the replacement of summations in the matrix multiplication by integrations in the case of continuous spectra. It derived the matrices for the angular momentum components and thereby corrected a guess in the basic Heisenberg paper, where incorrect statements on angular momentum quantization had been obtained from the assumption that there was a matrix for the angle variable. The paper also discussed the Zeeman effect and looked at the quantization of the electromagnetic field.

The impact of these papers on the scientific community was profound and immediate. Few theoretical physicists had any doubt that here at last was the key to a consistent description of atomic phenomena, and that all the paradoxes and inconsistencies resulting from the old mixture of classical mechanics with nonclassical quantization rules could now be eliminated. A measure of this reaction is given by the speed with which others started to elaborate, extend and particularly to apply the new methods. One of the first con-

verts to the new view was Dirac, who had heard Heisenberg's presentation in
Cambridge, or heard about it from R. H. Fowler, even before the paper was
published, and who immediately developed a full formalism of his own, which
largely duplicated the contents of the Born–Jordan and the Born–Heisenberg–
Jordan papers, while using his own original and powerful methods and his
own distinctive notation. Pauli solved the matrix equations for the hydrogen
atom and showed that it reproduced the Balmer formula exactly, without any
ad hoc restrictions on the possible states. It would not be possible here to
enumerate all the papers published in early 1926 which helped to consolidate
the new mechanics and to accumulate evidence that it was capable of accounting for all atomic phenomena. We may take, as a characteristic symptom of the rapid acceptance of the new system, the opening words of a paper
by F. London dealing with a minor didactic question, and received for publication in March 1926, which starts simply "In quantum mechanics...", thus
assuming, evidently with justification, that the reader was aware of the existence of the new kind of mechanics that had grown in the past six months or
so.

Quantum Mechanics or Wave Mechanics? The Uncertainty Principle

In May 1926 Heisenberg was back in Copenhagen, now settled for a time
as Docent (lecturer). As always, there were prolonged and intense discussions with Niels Bohr, particularly about the physical interpretation of quantum mechanics. Heisenberg was satisfied that the mathematical formalism of
the new mechanics was in good shape, but he was anxious to clarify the
connection with the physical concepts and the interpretation of the new ideas.
Bohr was always more interested in these questions than in the details of the
formalism, so Heisenberg found the arguments with Bohr extremely welcome and stimulating.

About this time Schrödinger's papers on wave mechanics had begun to
appear. It was soon clear that the equations of wave mechanics led to the
same results as the matrix form of quantum mechanics, and soon Schrödinger
gave a proof of the mathematical equivalence of the two systems. Moreover,
wave mechanics had the great attraction of expressing the problems in terms
of differential equations of a familiar type, so that the determination of the
hydrogen spectrum became a matter of a few pages of fairly simple calculation, as compared with Pauli's *tour de force* in getting the same results from
the matrix equations. However, Schrödinger's own point of view was to interpret the wave equation as dealing with quantities that had physical reality,
and thus to regard the dynamics of an electron inside an atom as continuum

physics, like the vibrations of the air in an organ pipe or the resonant modes of the electromagnetic field in a cavity. Bohr and Heisenberg could not accept this view, because it was incompatible with the indivisibility of the electron and with the many other phenomena in which the discrete nature of particles manifested itself, and also with the Planck radiation law, the starting–point of quantum physics.

Heisenberg turned to a serious study of problems involving several particles. This was a natural generalization of quantum mechanics, where the simple soluble problems which had been used as the first test cases all related to one-electron situations. In addition, many–body problems were likely to constitute a severe test of the continuum interpretation of wave mechanics. In the first of three papers on this subject, written in Copenhagen in June 1926, he starts with a criticism of wave mechanics, which reads oddly in a paper in which he later uses wave functions in one of his arguments. The explanation is that his objections relate to Schrödinger's physical interpretation, and not to the formalism of wave mechanics, the advantages of which he acknowledges. The paper emphasizes the importance of the symmetry of states in the case of identical particles, and the fact that transitions between states of different symmetry are forbidden as long as the particles react identically to all forces. It also shows that for a correct enumeration of the states of a given symmetry one should not count states differing by a permutation of the particles as distinct. This leads to the Einstein–Bose prescription for counting states. It is also made clear that the choice of totally antisymmetric states leads to Pauli's exclusion principle. At this stage Heisenberg believed that this choice could account for both the Pauli principle and the Bose–Einstein statistics. One important step in this paper is the recognition that the para and ortho states in the helium spectrum can be understood as states of different orbital symmetry. Six weeks later a second paper was complete. It applies the results of the previous paper to He and Li$^+$, using perturbation theory for the interaction between the two electrons, with the warning that this is not likely to be a satisfactory approximation for the ground state. The results give satisfactory agreement on the sign and approximate magnitude of the displacement of ortho relative to para states. The work of this paper is carried out entirely by the wave mechanical formalism. The third paper of this series completes the treatment of symmetry and also applies it to the band spectra of diatomic molecules, thus explaining the mysterious alternation in the intensities of the rotational bands. This argument later proved to be of great importance in giving information about the properties of nuclei.

Meanwhile, the discussions about interpretation continued. During a visit to Munich in the summer of 1926 Heisenberg had heard Schrödinger talk to Sommerfeld's seminar about wave mechanics and defend the continuum point of view. Heisenberg's insistence that there were too many facts which were inconsistent with this viewpoint did not convince Schrödinger, and apparently did not even make much of an impression on Sommerfeld. Later Schrödinger was invited to Copenhagen and there were intense debates with Bohr which led to no agreement. Heisenberg was in no doubt that the continuum viewpoint was untenable, but he was very dissatisfied with the degree of understanding that had been achieved. His most important step forward had been the result of the decision to base the description only on observable quantifies. In this he had been influenced strongly by the example of Einstein's theory of relativity. In this spirit, he had eliminated the description of the electron orbits within an atom from physics. Yet he had found Einstein, in a conversation in April 1926, strongly opposed to this radical step. Indeed, one could not dispense completely with the concept of electron orbits since, for example, the cloud chamber tracks of an electron established a trajectory.

The resolution of the difficulty was finally found in February 1927, and Heisenberg recalls that he was greatly helped by a remark which Einstein made during their debate: "Only the theory can decide what is observable." This meant one had to look within quantum mechanics itself for the limitations. The difference between the trajectory in the cloud chamber and that within an atom was quantitative, the first requiring only a rough specification of position and velocity. These thoughts led to the important paper on what is now called the uncertainty principle. (The term used in the paper is "Ungenauigkeit"—inaccuracy.) This paper, completed in March, starts by quoting Born's statistical interpretation of wave mechanics. He says he was worried by this at first, because it seemed to require a separate hypothesis relating the square of the wave function to the probability density, whereas he was sure that the laws of quantum mechanics were complete enough to predict the outcome of experiments and therefore left no freedom of choice in their interpretation. In the paper he shows that he has accepted Born's view, and he adds a list of references showing that the statistical interpretation had been suggested on many earlier occasions.

The main thesis of the paper and the arguments for it are too well known to require restating. It uses, as one of its illustrations, the idea of a gamma–ray microscope. A late addendum to the paper acknowledges that, as pointed out by Niels Bohr, the analysis of the gamma–ray microscope is incomplete without mentioning the dependence of the resolving power on the aperture, since a small aperture would provide knowledge of the direction in which the photon had been scattered, and hence would allow one to correct approximately for the change of momentum of the electron. This was an amusing oversight, in view of Heisenberg's trouble with the resolving power in his D.Phil. oral!

This paper formed the keystone which completed the structure of basic quantum mechanics, and it was immediately accepted as such by most physicists. Einstein never became reconciled to it, and in many debates with Bohr proposed counterexamples which, when Bohr had found the resolution to each such paradox, greatly helped to deepen the understanding of the uncertainty principle. Schrödinger also retained the hope that some new development might avoid the need for this limitation of our concepts. On the other hand, Bohr devoted much effort to further clarification and specification of the principle. The majority of physicists accepted the arguments in Heisenberg's paper as convincing and as an essential part of the contents of quantum mechanics which needed no further discussion. Only recently a tendency has developed in some quarters of separating the interpretation from the rest of quantum mechanics and questioning the validity of what is then called the "Copenhagen interpretation."

Leipzig

Heisenberg's distinction had not gone unnoticed, and in 1927 he was offered chairs in the Universities of Leipzig and Zurich. He decided to go to Leipzig, because he felt attracted by the opportunity of collaborating with Peter Debye. He became the Ordinarius (full professor) of Theoretical Physics in October 1927, at age 26. Besides Debye, his colleagues included G. Wentzel, who had taken up a chair of Mathematical Physics a little earlier. He was an active contributor to atomic theory. When he left later to take up an appointment in Zurich, he was succeeded by F. Hund, whose deep understanding of atomic spectra Heisenberg had learned to respect in Göttingen. In Leipzig Heisenberg started to build up a department. At his first seminar on atomic theory he had an audience of two, but soon students and other collaborators arrived and many senior visitors. The first assistant was Guido Beck, and the first research student Felix Bloch, who solved the problem of electrons moving in a periodic potential, suggested by Heisenberg as a thesis topic, and thereby started the modern theory of metals.

The duties of the chair were quite heavy. It was then normal for the professor to give the main undergraduate lectures in theoretical physics, usually in a four-or six–semester cycle, and to set problems, which he had to mark. There were also the usual faculty and committee meetings, but in spite of these demands on his time he was always accessible to his students. One could not have blamed him if he had started to take himself seriously and had become a little pompous, after having taken at least two decisive steps that changed the face of physics, and after reaching, at so young an age, the status of professor, which made many older and lesser men feel important, but he remained as he had been—informal and cheerful in manner, almost boyish, and with a modesty that verged on shyness.

The weekly seminar was preceded by tea, and for this the professor went out to a nearby baker's shop for a collection of pastries. After a strenuous discussion, and at other free periods, the whole group would descend to the basement where there was a ping–pong table. Heisenberg played extremely well, and his ambition to beat everyone was then more obvious than any ambition to be a great physicist. The arrival of a Chinese physicist, who could hold his own against Heisenberg, was quite a sensation. But physics was the serious matter, and problems, difficulties and new ideas would be debated intensely. Heisenberg was able to help his pupils particularly through his powerful intuition. He usually did not pay too much attention to the mathematical details of their work, as long as he could see where they were going, but he had to understand the physics of their problem and in the process to help them understand it. Once he had seen the problem clearly, he usually was able to guess the answer, and usually he was right.

F. Hund lists the following as members of the Leipzig group for various periods: K. Ariyama, E. Bagge, G. Beck, F. Bloch, H. Dolch, M. Draganu, H. Euler, U. Fano, S. Flügge, Th. Förster, Y. Fujioka, G. Gentile, B. O. Grönblom, V. Guillemin, E. Hellmig, Grete Hermann, E. Hückel, F. Hund, D. R. Inglis, M. Iskraut, H. A. Jahn, G. Kellner, S. Kikuchi, B. Kockel, W. Kroll, H. Ludloff, D. Lyons, R. C. Majumdar, B. Milianczuk, B. Mrowka, R. S. Mulliken, R. Peierls, G. Placzek, A. Recknagel, S. Rozental, F. Sauter, Th. Schmidt, A. Siegert, J. C. Slater, H. Stier, I. Supek, K. Umeda, E. Teller, L. Tisza, S. Titeica, J. H. van Vleck, R. Volz, Wang Fo San, S. Watanabe, V. F. Weisskopf, C. F. von Weizsäcker, H. Wergeland, G. Wentzel, G. C. Wick, A. Wintner, A. Wolf, C. Zener. Even this long list has gaps, and one should add J. H. Bartlett, P. Y. Chou, O. Halpern, W. V. Houston, I. I. Rabi, J. Steward, G. Wataghin, A. H. Wilson, and probably others.

It is natural that Heisenberg's own output of papers during this period did not maintain the pace of the previous years; a further distraction was his lecture tour of the United States in 1929. He had made leave of absence for this a condition for accepting the Leipzig chair. But of course he continued working. Quantum mechanics was now essentially complete, and the next task was to work out its consequences and to see how it would explain the many mysteries, paradoxes and contradictions in atomic physics. This was what his collaborators were mostly engaged in, and it was an exhilarating experience for them to see how easily the solutions to the old puzzles fell into place. Heisenberg himself followed the same approach.

His first Leipzig paper solves the mystery of ferromagnetism. The behavior of iron and similar metals suggested that the elementary magnets, now known to be electron spins, interacted through a strong force, P. Weiss's "molecular field", which tended to align them. But this force had to be much stronger than the magnetic interaction between the spins, the only known force affecting the spins directly. Heisenberg saw here an analogy with the helium spectrum which he had studied earlier. The Pauli principle requires the total wave functions for two electrons to be antisymmetric. For parallel spins the spin wave function is symmetric, hence the space part must be antisymmetric in the electron coordinates, and similarly the wave function for electrons of opposite spin is symmetric in the positions. The relative spin direction therefore influences the distribution of the electrons in space and this affects their electrostatic interaction, which depends on their relative position. Heisenberg had shown that this connection explained the energy difference between para and ortho helium, and it could again be invoked to explain ferromagnetism. In the helium atom the para state, in which the electron spins are opposite, lies lower than the ortho state of the same configuration. In the ferromagnets the effect must be in the opposite direction, and he showed that this could happen for sufficiently large atoms. It was not possible to determine all the energy levels of a many–electron system, but he could find the mean and the mean square of the energies belonging to a given total spin. If one assumed that these levels were distributed according to a Gaussian law, the parameters were tied down. This led, quantitatively, to two conclusions: (1) ferromagnetism was possible only if each atom had at least eight neighbors; (2) the substance at high temperatures would be paramagnetic, as in Weiss's model, become ferromagnetic on cooling, but again paramagnetic at extremely low temperatures. Heisenberg saw immediately that the second result was due to the hypothesis of a Gaussian distribution and should not be believed, but he regarded the first as significant.

The main thesis of Heisenberg's paper, that ferromagnetism is due, via the Pauli principle, to the electrostatic interaction between the electrons, was a major insight, which has stood the test of time. Whether he was right in using, for metals, a model of electrons tied to atoms, rather than itinerant electrons, is still a matter of controversy, though the model is certainly correct for some magnetic insulators. The Gaussian approximation has been forgotten, and much more powerful mathematical techniques are now used in its place.

Electrodynamics

Another major study of the consequences of quantum mechanics was the joint work of Heisenberg and Pauli on quantum electrodynamics. Schrödinger had already given, in one of his fundamental papers on wave mechanics, an expression for the absorption rate of radiation by an atom, and from this one could deduce the rate of spontaneous emission by using Einstein's relations. Dirac had shown that the quantization of the radiation field gave the right answers for the energies of light quanta and that the transition probabilities obeyed the Einstein relations.

But it was still necessary to treat the whole electromagnetic field, including electrostatic terms, consistently as a quantum system. Heisenberg and Pauli encountered a number of difficulties in doing so. Some of these were formal in character. These formal difficulties can be overcome by adjustments of the formalism and Heisenberg and Pauli developed one particular way of doing so. This, like most other such devices, makes prominent use of the gauge invariance of electrodynamics, the importance of which was brought out clearly in the papers. However, in addition to these formal difficulties, there is the more substantial problem of infinities in the self–energy of an electron and of other charged particles. This was a well-known trouble of classical electron theory, if the electron was assumed to be a point charge. H. A. Lorentz had tried the hypothesis of a finite electron radius, but this was hard to reconcile with relativity, which does not admit the existence of rigid objects. At a time when so many difficulties of classical physics had disappeared, one could reasonably hope that also this one would go, since the new concepts of position and motion involved in quantum mechanics might give the point electron a different character. However, the trouble was still there. It remained a serious obstacle to progress until the work of Tomonaga, Schwinger and Feynman in the 1940s showed that the theory gave finite answers to any order in perturbation theory if one asked questions only about observable quantities, and avoided dividing the electron mass into a mechanical part and a field contribution. It is still an open question whether this "renormalization" completely disposes of the difficulty, or merely evades in certain calculations a fundamental inconsistency of the theory that will show up ultimately.

In a later paper, Heisenberg discusses the self–energy problem in a very optimistic spirit, and expresses the hope that it is just due to asking the wrong kind of questions. This was, of course, the attitude that had helped him get the quantum theory out of its deadlock, and in a sense it expresses a plan which was implemented twenty years later by the idea of renormalization. However, it took a great deal more quantitative understanding and insight to get there.

Heisenberg's interest extended to many other parts of physics in which quantum mechanics could make a contribution. The cosmic radiation and its effects was one such field, other work related to more practical questions, no doubt raised by discussions with Debye about his experiments. We refer in particular to his early work on nuclei. Three papers, written shortly after the discovery of the neutron by Chadwick, were based on the hypothesis that nuclei consisted of neutrons and protons. This idea occurred to several people independently at about the same time, and as it was a fairly natural idea, it is not worth arguing who should have the credit for suggesting it first. However, he did not stop there, but attempted, at that early stage, a dynamical description. Although he put aside the question whether or not the neutron should be thought of as made of a proton and an electron, he clearly visualized it as composite, and hence postulated an exchange interaction between neutron and proton analogous to that between a proton and a hydrogen atom. He took it for granted that any "direct" interaction (which would not interchange the roles of neutron and proton) would have to be repulsive, and must therefore be weak if nuclei are to hold together. He included, however, an interaction between two neutrons which, by analogy to the force in a hydrogen molecule, could be attractive. Most of these ideas have been overtaken by later developments, but the idea of an exchange force as part of the internucleon force has remained. The main reason it was accepted was that, together with its generalization by Majorana, it gives a saturating force, making the binding energy of a large nucleus proportional to its mass number, not to its square. Heisenberg does not explicitly refer to the fact that the exchange force is favorable for saturation, but it is possible that he was aware of the connection.

In dealing with exchange forces he introduced operators analogous to the spin operators, and these are in fact the isospin operators now in constant use. He introduced these only as formal devices—the idea of isospin space and the related symmetry came later.

Life under Hitler

Early in 1933, Adolf Hitler came to power in Germany and the ideology of his party soon started to pervade life. The universities, as other state–controlled organizations, were purged of Jews and of political undesirables. The rules made exceptions for those of Jewish origin who had fought in World War I, but in practice the only effect was to delay their dismissals.

Heisenberg was, like almost all other academics, deeply shocked by so many people being unjustly deprived of their livelihood in Germany and the universities being deprived of many able members. He was worried by the destruction of many other values and standards and by the anti–intellectual attitude of the régime. However, he did his best, at first, to see the positive side of the changes, while being fully aware of the adverse features. In *Der Teil und das Ganze* he sketches a conversation with a student who is a leader of the Hitler Youth. This, like the other conversations in the book, is not meant to be a verbatim account, and may contain the essence of a number of different conversations. One feels that, while firmly maintaining his refusal to have anything to do with Nazi gatherings or other activities, he can see something to admire in the ideas of the young student. The student himself deplores the antisemitism and other destructive features of his movement, but insists that its aim is to create a better world, to fight corruption and dishonesty, and to restore respect for Germany. He emerges as quite a sympathetic character, even if his views are mistaken. One wonders to what extent this conversation symbolizes a debate within Heisenberg's mind. A. Hermann in his biography reports that in March 1933, on holiday with Niels Bohr and other friends, Heisenberg "...emphasized with his characteristic optimism the (apparently) positive aspects of the new political developments" in his country.

However, the disastrous aspects soon started to dominate to such an extent that he and a few colleagues began to talk of resignation. Heisenberg went to seek the advice of Max Planck on this. Planck's advice, as recalled in *Der Teil und das Ganze*, was to stay. Resignations of any number of distinguished professors would not affect Nazi policy. Heisenberg would have to emigrate and, while he undoubtedly would find a position abroad, he would take this away from someone else who was forced to emigrate. The present régime was bound to end in disaster, and people like Heisenberg would be needed as leaders after the disaster.

Heisenberg decided to stay, partly because of Planck's arguments. No doubt he was also influenced by his emotional attachment to Germany, evident from his account of the romantic wanderings of his youth with his friends. In his later account in 1969 he says of the decision: "It involved unavoidably making compromises and accepting later just punishment—perhaps even worse." He had agreed to sup with the devil, and perhaps he found that there was not a long enough spoon.

Many others faced the same problem and there were hardly any who decided to leave, other than those dismissed, or expecting to be dismissed. Of those who remained, Max von Laue will above all be remembered for his uncompromising stand, his proud aloofness and his refusal to cooperate with the régime. To take that line was not in Heisenberg's character. He was too involved in directing a group of young collaborators, in teaching, in explaining. For him, von Laue's kind of withdrawal might have been harder to bear than emigration.

The year 1933 was not one of unmitigated distress. He was awarded the Max Planck medal by the German Physical Society and shortly afterwards the Nobel Prize. The Prize gave him great pleasure, but he noted that Schrödinger and Dirac received only a joint prize, whereas he had one to himself, and that Born had no share in this honor. In a letter to Bohr he says that this gives him a bad conscience. Max Born also mentions a letter from Heisenberg saying so in very generous terms. He tried to maintain in his

Institute the old atmosphere, in spite of the obstacles which consisted not only in the loss of senior collaborators like Felix Bloch, but also in press attacks, mainly instigated by Stark and Lenard, the two obscurantist physicists who had always opposed relativity and quantum theory as "Jewish physics," and who, as old sympathizers with the party, now enjoyed power and influence. Occasional visits to Copenhagen were a welcome return to normality. Besides the usual heated debates about the latest ideas in physics, the Copenhagen meetings were an occasion for seeking help, jobs or invitations for dismissed physicists.

In 1935 Heisenberg was proposed for the chair in Munich as successor to Sommerfeld, who was retiring. This was an attractive appointment, both because it meant succeeding his respected teacher, and particularly because of his fondness for the city. However, the proposal resulted in an intensification of the attacks by Stark and Lenard and by the *Völkische Beobachter*, the party newspaper. A counterattack in the form of a memorandum signed by 75 professors, including Heisenberg, about the parlous state of theoretical physics in Germany failed to change the official attitude; in the end the authorities ruled against Heisenberg's appointment in favor of a nonentity.

After this, the attacks on Heisenberg continued and became more virulent. Heisenberg's mother approached the mother of Himmler, with whom she was slightly acquainted, and on the advice resulting from this Heisenberg approached Himmler directly to request a decision whether Stark, in his views expressed in the journal of the S.S., had the approval of its head, in which case Heisenberg would have to resign from his chair. If that was not the case, he asked for protection from these attacks. It took a year before there was a reply that the attacks were not approved and that instructions had been issued not to repeat them.

This was in the summer of 1938. By this time Heisenberg had found added strength and support: in January 1937 he had made the acquaintance of a young girl, Elisabeth Schumacher. Their mutual understanding, which began with a common love for music, soon broadened and they were married in April. Twins, a son, Wolfgang, and a daughter, Anna Maria, were born in 1938.

In the Institute research work continued. Heisenberg followed up subjects of previous studies, such as cosmic rays nuclei and others. Jointly with H. Euler, a student of outstanding ability, he worked out the nonlinear corrections to electrodynamics due to the virtual creation of electron–positron pairs. Apart from these specific applications, he kept thinking about the fundamental difficulties of the theory, connected with the infinite self–energy and similar divergences, and the failure to account for the properties of nucleons. He attempted to discuss the limits of applicability of the theory, but did not make much progress in that direction. One, more specific attempt was to assume that the description in terms of a Hamiltonian and a state function had to be abandoned, and that the only quantity one could define would be the scattering matrix, or "S matrix." This idea was, in a way, following the precedent of his creation of quantum mechanics, avoiding the difficulties by eliminating from physics redundant concepts which might in principle be unobservable. The S matrix would describe only the states of the incident particles in a collision process and the states of those emerging after the collision, and also stable bound states, but would not refer to the intervening states. This way of looking at physics was taken up by many others in later years and had its enthusiastic followers. To date nobody has succeeded in deriving rules which would be adequate to determine the S matrix, without reverting to the full Hamiltonian description. The general rules obeyed by the S matrix, which result from this approach, are now in common use, but the attempt to express the complete laws of physics in terms of the S matrix is not being pursued widely today.

The papers on the S matrix were written in 1943 and 1944. By that time many other things had happened.

War Time

By the spring of 1939 it was clear that war was unavoidable. Heisenberg bought a country house at Urfeld, in the Bavarian foothills of the Alps, as refuge for his family (there were now three children) in case of need.

In the summer he visited the United States again and lectured in Michigan and Chicago. There many of his old friends and colleagues tried to persuade him to leave Germany because of the impending disaster, in which his presence could not achieve anything. To this Fermi added the warning that the recently discovered uranium fission might contain the possibility of military applications, and that in war time scientists might find themselves working for their government on such schemes. In *Der Teil und das Ganze* Heisenberg remembers these arguments and his reply: he felt committed to staying; above all, leaving would be disloyal to the young people of his group, who relied on him for guidance in keeping science going and whose responsibility it would be to rebuild science after the war. These young people could not find places abroad as easily as he could, and he would feel like taking advantage of his privileges. As regarded atomic energy, this was a long–term problem and the war was likely to be over before such a project could be completed. According to his account of these conversations, he expressed the conviction that Germany would lose the war, because Hitler's policies had not only isolated Germany politically, but weakened its scientific and technical resources. In the recollection of his colleagues, he appeared to foresee a German victory.

Was this a failure in communication, or did the views appear to him, or to the others, in a different color in retrospect? One knows the fallibility of human memory.

When war did come, Heisenberg found himself assigned to a study of the possibility of atomic energy. For the next five–and–a–half years this problem took most of his time and energy. He developed the theory of a nuclear reactor, adopting Harteck's suggestion to separate the uranium and the moderator physically, in order to reduce the loss of neutrons by resonance absorption. By the end of the war, the whole project had not achieved much in practical terms. Laboratory experiments had indicated that a system of uranium metal and heavy water of a suitable size could sustain a chain reaction, and there were data on the optimum shape, size and spacing of the uranium elements and on the minimum size of the reactor. Unquestionably, a little more time would have been enough to produce a successful experimental reactor, but this would still have been far short of producing any practical result such as usable power, or the quantities of plutonium that might have led to a weapon.

Since then two related questions have been asked repeatedly: why did Heisenberg work for the Hitler government on atomic energy, and why was the total achievement of the German uranium project so slight?

On the first question it is reasonable to assume that he wanted Germany to win the war. He disapproved of many facets of the Nazi régime, but he was a patriot. To desire the defeat of his country would have required far more rebellious views than he held. Even if he had wanted to refuse cooperation, however, this would not have been easy under a régime which did not tolerate conscientious objectors as readily as did Britain and the United States. Most citizens of most countries at war participate in the war effort when called upon, and the few who do not require exceptional courage and exceptional strength of conviction. It might have been different if the work had come close to the making of atomic bombs, but it never did.

Some writers, particularly R. Jungk and W. Kaempfert in the *New York Times*, asserted that Heisenberg and the other German scientists deliberately refrained from developing atomic weapons for ethical reasons. This view is not supported by Heisenberg's accounts. In *Der Teil und das Ganze* he says: "We knew then that one could, in principle, make atomic bombs, and knew a realizable process, but we regarded the necessary technical effort as rather greater than, in fact, it was." This left them in the fortunate position of being able to report honestly to the authorities about the state of knowledge, without expecting to be instructed to make a serious effort to develop a bomb. In his postwar summary of atomic energy work in Germany he says, similarly, that they were fortunate in not having to make a decision whether to cooperate in making a bomb.

It is then a little confusing, though, to see his remark in *Der Teil und das*

Ganze, in the context of the negotiations in 1954 in which West Germany was permitted to start a nuclear reactor program: "The fact that in war time no attempt was made in Germany to construct atom bombs although the knowledge of the principles existed, probably had a favorable effect on these negotiations."

As regards to the other question, the basic reason for the slow progress was that the work was not pushed along with urgency, and once it was accepted that in the circumstances of the war it would not lead to a weapon, this was undoubtedly a sensible decision, both for Heisenberg and for the German authorities. It was known that nuclear reactors could certainly produce power, but to realize this was also a very major technological project, and its military advantages not crucial enough to justify a crash program. Hence the authorities never instructed their scientists to make an all–out effort and did not give them the support and the resources which this would have required.

Although Heisenberg admits that they were inclined to overestimate the technical difficulties, he rightly points out that even the "Manhattan District" in the United States, working with greater industrial resources, without serious shortages and without interruption by air raids, were not able to complete the first bomb until after the end of the war in Europe. The view that Germany could not consider making a bomb for use in this war was therefore sound.

One might argue that the magnitude of the effort needed could not be known at the beginning; in the West, too, it was not known just how big an effort was involved until a great deal of research had been done. In that case, the initiative came from the scientists, who, fearing that Germany might be on the way to produce atom bombs, persuaded the authorities to give support to an all–out program. This could not have happened in Nazi Germany so easily. The relations of the scientists to their government were not such that they could easily have made their views count, and although most of them probably wanted to see a German victory, they did not feel that the war was their own personal contest to the same extent as did most of their counterparts in the West. In this sense there is some truth in the statement that the lack of desire of the scientists to be involved in making a weapon prevented Germany from trying.

According to David Irving, more progress could have been made with the available effort if it had not been for a division of the work among different groups, acting under the orders of different branches of the government. This gave rise to demarcation disputes over the limited amounts of uranium metal and heavy water then available. In his very thorough study of the German work, Irving points to a number of errors, misjudgments and omissions, both in the scientific work and in the practical organization and planning, and attributes to Heisenberg a share for some of these. No doubt such mistakes also slowed down the pace, but it has to be remembered that there is no large technical and scientific project in which everything is done the right way, and in which one could not in retrospect make up a similar list of mistakes.

Heisenberg was at first a consultant to the Kaiser–Wilhelm Institute in Dahlem, near Berlin, where much of the research work was being done and during that time continued to live in Leipzig, where also some neutron–scattering experiments were done in the Physics Institute. The Kaiser–Wilhelm Institute was then under military control, but after a reorganization in 1942 it was returned to the Kaiser–Wilhelm Gesellschaft and Heisenberg was made director (his title was "Director *at* the Institute," since Debye was officially still the Director *of* the Institute, on leave of absence). He was also made a full professor at the University of Berlin. When he moved to Berlin in the Spring of 1943 his family went to live in their country house in Urfeld. But later that year the air raids on Berlin became more intense and the laboratory was evacuated to Hechingen. The reactor experiment was placed in a cave under the small town of Haigerloch.

There the arrangements were made for another test, ready for the uranium metal and heavy water, which arrived in February 1945. There was hope that the test assembly might be large enough to start a chain reaction with thermal neutrons, but while it came close to being critical, it did not make it. It was clear that a small increase in size, or a better geometry, would have achieved criticality, but time had run out.

It must have been clear for some time that the end of the war was near and that there could not be any expectation of practical results before the end. The great effort to complete the test was due to the hope that the achievement of the first nuclear chain reaction would have been an impressive step, to the credit of German science, and helpful for the chances of reconstructing German science after the war. It was not known, of course, that Fermi had succeeded in setting up a chain reaction in Chicago in December 1942.

Soon afterwards Allied troops occupied the area. Heisenberg and a number of his senior colleagues were, for a time, interned in a country house near Cambridge. During their enforced idleness and in the absence of any library, Heisenberg and von Weizsäcker looked for problems one could think about from first principles, and this led to Heisenberg's renewed interest in turbulence, which we have mentioned already.

It was here that they heard in August 1945 the news of the atom bombs dropped on Japan, and thus the first news of the scale of the American atomic energy work. The house was "bugged," and it is reported that the group of German scientists were incredulous at the first brief and vague radio reports; they were incredulous, not of the possibility of making atom bombs, but of the idea that the United States might have made the tremendous industrial effort to complete this enterprise in a few years.

One wartime episode deserves special mention. Although the German atomic energy work was not, in the short term, aimed at producing weapons, Heisenberg and his colleagues were worried by the fact that they were working on a program of which the ultimate outcome was likely to be a weapon. Was it right for scientists to become involved with this kind of work? They conceived the idea of asking the advice of Niels Bohr and in October 1941 Heisenberg, who had arranged to be invited to lecture in occupied Copenhagen, visited his old friend Bohr. There is no clear record of what was said, but they parted with Bohr very angry and Heisenberg disappointed. Heisenberg's impression of the conversation was that he told Bohr of the possibility of a weapon, but stressed the enormous technological effort which it required, so that scientists would be right in advising their governments that this was not a practical proposition and could therefore keep away from atomic weapons development work. According to his account in *Der Teil und das Ganze*, and in an interview with David Irving, Bohr was so shocked by the statement that it was possible to make atom bombs, that he did not take in the rest of the explanation. He quotes Bohr as saying that it was unavoidable that in wartime scientists would work on the development of weapons and that this was probably justified. In Bohr's recollection, Heisenberg was trying to find out what Bohr might know about fission. He also had the impression that the Germans were working very hard on the uranium problem, and that Heisenberg thought it might decide the outcome of the war if the war was prolonged. Nothing said later changed Bohr's mind in the least on this score.

Either account could be inaccurate. Niels Bohr was known to his friends and pupils as being better at talking than at listening, and he could well misunderstand what people tried to tell him. Heisenberg had no very clear recollection of the details—in the interview with Irving he says "[the conversation] *probably* began with my question whether it was right that physicists devoted themselves to the uranium problem in wartime," (author's italics).

Whatever was said, there was never any chance of mutual understanding, as Heisenberg realized later, according to the interview with Irving. Bohr was a citizen of a peaceful country, occupied without provocation by the armies of a hateful régime, and Heisenberg, although an old friend, was also a member of the occupying power. There was no way in which the two could have carried on a dispassionate discourse about the ethics of war work.

Perhaps Heisenberg found it difficult to see this kind of situation from the other person's point of view, to understand what Blackett called "the view from the other side of the hill." This caused a great deal of difficulty for him in the immediate post–war years. There were many instances of remarks which caused resentment, though nobody has collected them or checked them in great detail. One typical example is told by the people to whom the remark was addressed: visiting the house of a German refugee physicist during his stay in England in late 1947, Heisenberg commented "the Nazis should have been left in power for another fifty years, then they would have become quite decent." If this is correctly remembered, it was a strange remark to make to a man who had been dismissed in spite of his having served in World War I, and who had lost relatives and friends in extermination camps, in a conversation in which Heisenberg was anxious to re–establish cordial relations. Whatever he may have meant, it is sad but true that the difficult task of rebuilding cordial relations was not always helped by remarks of this kind.

Post-War Science

At the end of the war in Europe, German scientists were dispersed, all laboratories were closed, and many of the cities in which they were located were very heavily damaged. The Allied armies which had occupied the country were trying to get life back to normal. In the British zone a member of the scientific section of the military government was Colonel B. K. Blount, who had obtained a Ph.D. in chemistry in Germany, hence was himself a scientist by training and knew German and many German scientists. He regarded it as his task to encourage the best German scientists to resume their research and teaching. He concluded that Göttingen was, of all places in the British zone, the most suitable to serve as a center for the rebuilding of research. When Heisenberg and the other "atomic" scientists returned to Germany from internment in January 1946, Blount took Hahn and Heisenberg to Göttingen and let them spend a weekend meeting old friends and colleagues, including Planck. He was probably bending the rules a little, because official policy was still to keep these experts under close supervision, in case any of them departed for the East. They agreed with him that a start should be made in rebuilding German science in Göttingen. By February they were all there. At first life was primitive and everything was in short supply. There was some accommodation for Heisenberg, but not for his family, whom he was able to visit only once during that time. However, before long he found a home where the family could be reunited, and gradually life and work returned to normal.

Blount had set up an Advisory Council for Science to help the administration and represent the views of the scientists. After their arrival Heisenberg and Hahn joined this council. Previously the Kaiser–Wilhelm Gesellschaft had been extremely effective in running research laboratories, and it seemed desirable to continue this form of administration. However, there were moves to abolish this organization, partly because the policy of the occupying powers was to oppose any organization operating throughout Germany, and very largely because of the name. Colonel Blount anticipated these difficulties and sponsored the formation, in the British zone, of a new Max–Planck Gesellschaft, to which the assets of the old society were transferred. This proved popular, and after a time the new society was able to extend its activities over the whole of West Germany.

During this Göttingen period Heisenberg devoted much of his time and energy to questions of the best organization of science. He participated in 1949 in the foundation of a German Research Council, which was to advise the new Federal Government of Germany in the way in which the Advisory Council had advised the military government, and he became its first chairman. There were, however, differing views among the scientists about the merits of this. Some felt that after the bad experience with the Nazi government, which had interfered with science in a disastrous way, scientists should be represented by an independent organization not too closely involved with the government. The old Notgemeinschaft der Deutschen Wissenschaft (Emergency Association of German Science), which had recently been reconstituted, was regarded as appropriate for this. For a time both organizations functioned side by side, pursuing similar, but not identical, aims. Whether the division and the strong animosities created by it were predominantly due to differing views on policy, or to individuals seeking influence, is not clear. After a long struggle the two bodies agreed to merge, and in practice this meant that the Notgemeinschaft (later called Forschungsgemeinschaft, or Research Association) had won. Heisenberg had lost this battle in the field of science policy but, nevertheless, he collaborated with the new organization, and one of his major activities in it was to take charge of its Committee for Atomic Physics, whose remit included nuclear physics and atomic energy. He made a major contribution to the decision to start a nuclear reactor program in the Federal Republic, but this also did not happen without friction, which culminated in Heisenberg's withdrawing from the "Atoms for Peace" Conference in Geneva in 1955, where he was to go as leader of the German delegation. This was in protest against decisions by the Adenauer government, who, in Heisenberg's view, had not done enough for the reactor development and had not listened to scientific opinion.

Whereas Heisenberg pleaded strongly for nuclear power, he opposed, equally strongly, any suggestion that Germany should make or acquire atomic bombs. This also involved him in political controversy, but here he and his colleagues had their way when the Federal Republic signed the Nuclear Nonproliferation Treaty.

One noncontroversial project on which he had considerable influence, and to which he was firmly attached, was the foundation of the "Alexander von Humboldt-Stiftung," which provides fellowships for foreign scientists to work in Germany. He became the president of this foundation, in whose aims he believed very strongly. He was an enthusiastic supporter of C.E.R.N., the European high–energy laboratory at Geneva, and served on many committees helping to bring C.E.R.N. into being. Later in the 1960s, he opposed the idea of adding a much bigger "300 GeV" machine, and was skeptical of the proposal to build the intersecting storage rings. This was because he was optimistic that the unified field theory, on which he was working, would produce all the answers, and there was enough evidence available from the existing accelerators to check the theory. He acquiesced in the building of the storage rings, because it was a relatively cheap project and involved new technical principles. Later when the decision to proceed with the 300 GeV machine had to be taken, he did not maintain his strong opposition.

In the autumn of 1958 the Max–Planck–Institute for Physics and Astrophysics, which Heisenberg had directed in Göttingen, moved to Munich, and he was able at long last to return to the city he had always loved. He remained the director of the institute until his retirement in 1970.

The growing demands on his time were not allowed to interfere with his close relations with his family, all of whom shared his love for music, and many his scientific curiosity. Today, of the twins, Anna Maria Hirsch is a physiologist, while Wolfgang, a lawyer by training, works for a foundation concerned with science and politics. Jochen is an experimental physicist at M.I.T. Martin is a professor of biogenetics. Barbara Blum is the wife of a physicist. Christine Mann is a teacher and her husband a psychologist. The youngest, Verena, is a technician in a physiological laboratory.

Post-War Physics

Heisenberg's involvement with administration and science policy could not stop him from thinking about the problems of physics. Once again a group of collaborators gathered around him. Members and visitors of the group in Göttingen included: R. A. Ferrell, H. Lehmann, G. Lüders, W. Macke, K. Nishijima, R. Oehme, K. Symancik, W. Thirring, C. F. von Weizsäcker, W. Zimmermann, B. Zumino and others. Among the many members of the Munich Institute a particularly close colleague was H.–P. Dürr, who took a very active part in the work on unified field theory. He remained active himself, and the years 1940, 1942 and 1945 were the only ones in which he did not have a paper on pure physics published. Some of his papers after the war extended or continued work done earlier. He was still interested in cosmic-ray showers and began to consider the multiple production of mesons, whose

Werner Heisenberg. (Photograph published courtesy of AIP Emilio Segrè Visual Archives, Landé Collection.)

properties had by then been identified. His renewed interest in turbulence has already been mentioned.

In the main, however, he was aiming for the major unsolved problems of physics. Among these there was the challenge posed by the phenomenon of superconductivity. This happens in metals in circumstances to which the known laws of quantum mechanics should certainly be applicable. Yet all attempts at deriving it from the admittedly very complicated equations made since the late 1920s had failed. There had been so many unsuccessful attempts that Felix Bloch remarked "theories of superconductivity can be disproved," and this was jokingly referred to as Bloch's theorem. Heisenberg made another attempt at this problem, taking into account only the electric interaction between the electrons. This took courage, because the low temperatures at which superconductivity is found, and the small difference in energy content between the superconducting and the normal phase, did not appear consistent with a mechanism involving the rather strong electrostatic interaction. In Heisenberg's view these quantities had to depend on some small number appearing in the solution of the equations, or some kind of cancellation, and this had to happen in all superconductors.

A complete, even approximate solution of the quantum mechanical equations for the motion of many electrons in the field of the lattice of ions is quite impossible, and one must therefore proceed by the use of approximations chosen intuitively, or by analogies and models. This Heisenberg did. His papers aroused controversy, since many physicists were not convinced by his explanation. It was not easy to test, since there were no specific conclusions which could have led to predictions about experiments. Eventually the theory had to be abandoned, not because, following Bloch's "theorem," it had been disproved, but because the discovery of the isotope effect proved that the motion of the ions was involved, and simultaneously Fröhlich suggested a way in which the lattice vibrations might give a force of a suitable kind and order of magnitude. Later Bardeen, Cooper and Schrieffer put forward a detailed theory on these lines, which carried conviction and which proved a productive way of suggesting new experiments.

Apart from this unsuccessful attack on one of the 'leftover' problems of nonrelativistic, low–energy physics, he took up the challenge of the front line of fundamental physics, the problem of elementary particles and their interactions. Since the end of the war, experiments with cosmic rays and with accelerators had added (and are still adding) a bewildering variety of new "elementary" particles. If these were all truly elementary, one would expect to deal with field equations specifying a field for each species of particle and a separate coupling term for each type of interaction between them. Systematic studies of the behavior of these particles have shown a number of regularities, some exact, some only approximate, so that the number of independent parameters required to specify these fields and their interplay is somewhat reduced, but it remains so complex that it becomes quite unreasonable to believe that there are not some simpler underlying laws from which all the complexities of elementary particle physics could, in principle, be derived. It is indeed very natural to believe that several particles may be manifestations of the same field, just as a basic description of nuclei need not introduce all species of nuclei as separate entities, but would refer only to neutrons and protons and their interaction. In that case, one pictures the nucleus as being built of sub–units, but the relationship need not be of this nature. In any case, one would then be dealing with fields which interact with themselves or with each other, and this is expressed by nonlinear terms in the field equations. The development in which Heisenberg was interested is often simply referred to as "nonlinear field equations." This term is somewhat misleading, since all the conventional descriptions of fields and their interactions also contain nonlinear coupling terms. The difference is that, in talking about "nonlinear equations" one pays attention to those consequences of the equations which could not be obtained by a small perturbation of the linear equations for free fields.

One may now wonder how far the number of fields can be reduced, if indeed the present type of field theory is capable of dealing with the situation. Heisenberg boldly decided to try the most extreme possibility of assuming only a single field with four components, specifying spin and isospin variables. The equation for such a field would, in its simplest form, contain one nonlinear term, whose coefficient is the only parameter in the theory. The discussion of this equation was his main concern from about 1950, and the progress of the work is reported in about twenty papers (not counting reviews and conference reports), some written jointly with collaborators, including particularly Hans–Peter Dürr.

Anyone with less courage than Heisenberg might have seen, from the start, enough objections to abandon this project. For example, the theory contains no provision for the electromagnetic field, which must therefore be derived somehow from the basic field. This means that there is no provision in the basic equation for the law of charge conservation, which would have to result, somehow, as a consequence of the structure of the equation. Again, there is no free parameter to specify the magnitude of the weak interactions, which are known to be very many orders of magnitude weaker than the "strong" couplings. Some of the conditions that have to be imposed on the weight functions occurring in the solution of the equations cannot be met, unless the weight functions can take negative as well as positive values, and one is therefore led to an "indefinite metric," which allows negative probabilities. This in turn would give physically meaningless results unless one added a rule that states with negative weight cannot occur either as initial or as final states of any process.

The basic equation also contains no provision for "strangeness," an important quantum number which was found to play an essential part in the classification of particles and their interactions. The theory tries to incorporate this in terms of a "broken symmetry." This means that the ground state of the vacuum for the whole universe is one of a set of possible and equivalent states, each with the same energy, and each with lower symmetry than in the fundamental laws. The analogy often quoted is that of an ideal ferromagnet, which, in its ground state, is magnetized in some direction, thus breaking the isotropy of the underlying laws of physics. This isotropy is still preserved in the sense that the magnetization can have any direction, and the actual direction is accidental or determined by previous history.

In this ferromagnetic analogue model a new feature appears when the coupling between the spins responsible for the magnetization is strong. Then an atomic excitation, which can travel through the medium, has different properties according to the way it aligns the atomic spins over which it passes, and thus this excitation may have more states than it would have in the absence of coupling. The theory assumes that an exact solution of the nonlinear field equations would lead to a similar additional freedom, which could account for the different values of the strangeness. No simple explanation seems to exist in this framework as to why the strangeness is conserved except for the "weak" interactions, whose strength is many orders of magnitude less than that of the "strong" interaction between elementary particles. There is no room, as in the conventional description, for a separate "weak" coupling constant, since there are no disposable parameters in the theory, and the "weakness" of certain interactions would have to be the result of calculational complications. Again the mathematical difficulties of such a nonlinear quantum field theory are extremely severe, and once again one has to proceed by intuitive reasoning, analogies and models.

The extent to which such reasoning is convincing is very subjective. On the whole, the majority of theoreticians have remained rather skeptical, and the work on Heisenberg's unified theory has been kept going mostly by Heisenberg, Dürr and other members of his group, though some of the methods and ideas developed in the course of this work have played an important part in current developments.

Once again, Heisenberg sought encouragement from Pauli, who at first was very skeptical. At one time Pauli thought he could disprove the whole idea by proving that one mathematical statement, which was essential for the scheme, was incorrect. In trying to prove this, he discovered that that particular statement was in fact in order, and he then reconsidered his attitude to the whole scheme and for a time was optimistic and even enthusiastic. Heisenberg's impression of their joint understanding was embodied in a draft of a joint paper, but by the time this reached Pauli he had reverted to his former negative point of view, much to Heisenberg's distress. This change of attitude occurred while Pauli was visiting the United States, and in *Der Teil und das Ganze* Heisenberg suggests that it was the "sober pragmatism" of the American physicists which resulted in this change of mind. He had in fact feared this result and had tried to persuade Pauli to cancel the trip to America for that reason. After Pauli's return there was a debate between the two following Heisenberg's report of the 1958 C.E.R.N. Conference on High Energy Physics and Pauli attacked the theory with all the strength of his well-known critical ability.

It is too early for any definitive assessment of this work. The mainstream of theoretical development has passed it by, but since the more commonly accepted ideas have not achieved any full formulation or basic confirmation, this does not rule out the possibility that Heisenberg's work might have been along the right lines. The latest ideas to have attracted general attention, the "gauge theories" of weak interactions, also involve treating several seemingly disparate fields as part of the same physical entity, which is in the spirit of Heisenberg's scheme, though one cannot discern any direct relationship between the procedures.

Perhaps the opposition to the scheme was increased by Heisenberg's optimism in his claims for what had already been established. He was always inclined to let his intuition lead, rather than follow, the mathematical formalism, and we have all benefited from this approach when it governed his basic work on quantum mechanics. Who is to say that in his last project he may not still turn out to have been a prophet in the wilderness?

After his retirement in 1970 most of his writings were reviews or essays on more general topics. In 1975 his health began to give way, and he died on February 1, 1976. His wife and seven children survive him.

He was a great physicist, whom science will remember with gratitude and respect. He was also an attractive personality, whom many who knew him will remember with affection. In this memoir we have tried to present him as a real person, including in our account also those parts of his life which have aroused controversy. Respect or affection would not have much value if they required an expurgated view of the real person.

We have taken much information from the 1976 biography by A. Hermann, (Hermann, A. 1976. Heisenberg. Hamburg: Rowohlt; English translation, Werner Heisenberg, Leck/Schleswig: Clausen & Bosse) as well as from Heisenberg's recollections. We have used the German edition of the latter, since the English translation is not always reliable. G. K. Batchelor advised us on the significance of Heisenberg's papers on turbulence. We are indebted to many others who provided useful information or constructive criticism, in particular to H. A. Bethe, F. Bloch, B. K. Blount, A. Bohr, R. H. Dalitz, S. Goudsmit, A. Hermann, Mrs. Anna Maria Hirsch, F. Hund, W. Jost, and V. F. Weisskopf.

A Heisenberg Biography

W erner Heisenberg was undoubtedly one of the greatest physicists of this century. In 1925, at the age of twenty-three, he wrote the paper that laid the foundations of quantum mechanics on which all subsequent generations have built. This was not just an extension or elaboration of the work of others, but an unexpected, radical new departure, which abandoned the basic notions of the old "classical" physics, such as that of electrons moving in orbits, replacing them by a much more abstract description.

In the public mind many advances in science are attributed to famous scientists, but in most cases the famous discoverer has completed a structure that was already developing, and without him someone else would sooner or later have done the same. Heisenberg's paper was so original that, if he had not been around, it might have taken a very long time for the idea to occur to some other physicist; so this is one of the cases in which the personal attribution is justified. It is true that less than a year later Erwin Schrödinger published his theory of wave mechanics, which turned out to be identical in content to Heisenberg's quantum mechanics. But we needed both points of view to develop a real understanding of the physical world.

Heisenberg was born in 1901 into an academic family. He was in his teens in Munich after the First World War, when a strong youth movement was emerging, and he became an enthusiastic member. Long hikes and campfire discussions of poetry, philosophy, religion, and music appealed to him. He learned to appreciate the beauty of nature: love for the countryside was part of the patriotism that was a strong emotion throughout his life. He became a

This is a review of *Uncertainty: The Life and Science of Werner Heisenberg,* by David C. Cassidy. New York: W. H. Freeman, 1991. The book review is reprinted with permission from *The New York Review of Books.* Copyright © 1992 Nyrev, Inc. For the complete reference, please see the Acknowledgments section.

youth leader, and formed a close friendship with a group of younger boys with whom he continued his long walks well after he became a famous and established scientist.

Another early interest was music; he was a highly gifted pianist, and it was even suggested that he might choose music as a career. But he was more strongly attracted to mathematics and physics; as a schoolboy he was fascinated by the theory of relativity. At the University of Munich he first tried to enroll in the study of mathematics, but he was put off by the mathematics professor, Ferdinand Lindemann, and he joined instead the theoretical physics group under Arnold Sommerfeld, then the greatest teacher of the subject. Sommerfeld immediately recognized Heisenberg's unusual ability. Another student in the group, Wolfgang Pauli, a year his senior, became a lifelong friend.

Heisenberg was introduced to the intricacies of the quantum theory of Niels Bohr and Arnold Sommerfeld, which had successfully explained many facts about atoms, but had also had many failures and contained internal inconsistencies. He wrote his first paper in 1922, barely a year after entering the university. During Sommerfeld's absence on leave he went to Göttingen to work with Max Born, and later returned there as Born's assistant. He met Niels Bohr, and visited him in Copenhagen. This became another close and lasting relationship, in spite of the difference in their ages.

The next few years were a time of great activity and great confusion in atomic physics. New discoveries of phenomena and regularities showed up the inadequacies of the existing quantum theory and the need for new ideas. Bohr, Born, Pauli, and Heisenberg were among the people struggling to find a way forward, and, by intense correspondence and personal discussion, assisted and criticized each other in these attempts, but to no avail. This continued until the situation was revolutionized by Heisenberg's paper of 1925.

All those who were able to understand the novel and difficult ideas in his paper recognized that here was the long–awaited breakthrough. The Bohr–Sommerfeld theory, accepted before Heisenberg's paper, described electrons in the atom as revolving around the nucleus in orbits, like planets around the sun, as in classical mechanics; but only certain selected orbits were allowed. Radiation was emitted when an electron jumped from one orbit to another, and the energy loss of the electron determined the frequency (color) of the radiation. The theory predicted correctly the existence of sharp lines in atomic spectra, and their positions in the case of hydrogen, the simplest atom. But it encountered difficulty in dealing with more complex atoms, as well as many other problems. Heisenberg discarded the concept of orbits, which could not in principle be observed—this was made more precise later through his uncertainty principle—and he proposed that the physicist should deal only with observable things. This meant concentrating not on single orbits, but on the emitted radiation, which comes from a jump between two orbits, so that he was talking of two states of the atom at a time.

Many physicists started to elaborate and apply the new theory. In Göttingen Born and his collaborator Pascual Jordan perfected the mathematical formulation of Heisenberg's scheme; they recognized that Heisenberg had reinvented a system known to mathematicians as "matrix calculus," but unfamiliar to most physicists. Pauli in Hamburg managed to apply the new theory to the hydrogen atom, and showed that it gave the correct answers, while Paul Dirac in England reproduced and extended the theory in his own original presentation.

Schrödinger's wave mechanics started from a very different approach, but it also gave correct results and appeared at first to be an alternative theory. It soon proved to be the same as Heisenberg's, although expressed in a different language. Heisenberg was at first reluctant to accept it, but soon saw that Schrödinger's equations were easier to handle for most purposes, and he started to use them himself. Schrödinger, however, believed that the waves in his theory were tangible waves, like waves on water, or sound waves. To Bohr and Heisenberg this view was untenable. The waves spread all over space, and Schrödinger's interpretation could not account for the fact that when we look for an electron, we always find it concentrated at one point. Also, if there are several electrons, the wave function must give us the probability of their all having various positions, and this cannot be done by a real wave in space. But after many heated discussions Bohr and Heisenberg failed to convince Schrödinger. The correct view, as finally expressed by Max Born, was that the intensity of the waves determines the probability with which the electron will be found at a given point in space.

Thus physics cannot specify the exact position of a particle; its position is a matter of chance, with only probabilities being the subject of the physicist's description. This conclusion led Heisenberg to his "uncertainty principle," which has to do with the accuracy with which different attributes of a physical object can be known; the more precisely we want to know the position of a particle, the more uncertain must be its velocity, because the act of observation causes an unknown change in the velocity. Bohr had reached similar conclusions, and was a little annoyed with Heisenberg, because he would have preferred to state the argument in his own way.

In October 1927, when he was not yet twenty-six, Heisenberg was elected professor of theoretical physics at Leipzig, and became the successful leader of a group of students and more senior collaborators. His infectious enthusiasm and his simple and unpretentious manner no doubt reflected the qualities that had early engaged his sympathies in the youth movement. (As one of his students in 1928 and 1929 I had occasion to appreciate these qualities.) He continued to make original and important contributions to physics, although after the early 1930s his projects were less successful.

After Hitler came to power in 1933 Heisenberg found the effects of the régime on academic life, on science, and on life generally deplorable, but he decided to remain in Germany because of his patriotism, and because he felt a duty to do what he could to mitigate the régime's evil effects. During the war he worked on the German atomic–energy project, which failed to get any results. His actions during that period have caused intense controversy, which continues, and to which I shall have to return.

After the war he conceived the ambitious idea of finding a single equation which would describe all the particles of physics and their interplay, but though he claimed at times partial success, the scheme never got off the ground. He was, however, instrumental in rebuilding German physics, and in arguing for Germany's having nuclear power, but not nuclear weapons. He also participated in the creation of CERN, the European nuclear research center in Geneva.

David C. Cassidy has made a monumental effort to give us in 669 pages, including eighty-eight pages of notes, a full, personal, and scientific biography. The early chapters cover family background, childhood, and adolescence. They present a lively and, as far as I can judge, very fair picture of the young Heisenberg and his environment. With Heisenberg's start in physics, the story includes many accounts of the physics to which he was exposed, and of his own work. Unfortunately, Cassidy's account of the physics is much too difficult for the nonphysicist reader. He uses technical terms and jargon without restraint, and includes equations that most lay readers could not follow even if they knew what the symbols meant. For the physicist, on the other hand, there is not much to learn. Most physicists will be familiar with the concepts developed in 1925; they might not know all the details of the arguments in quantum theory before Heisenberg's paper, but what is said is much too brief to satisfy any curiosity. Besides, the physicist will notice many errors.[1]

---

[1] A few examples: the author misunderstands what was wrong with Heisenberg's description of a gamma-ray microscope in the paper on the uncertainty principle—the argument did not concern the reason for the uncertainty, but a detail about the resolving power of a microscope. The account of Bohr's refutation of Einstein's criticism is garbled. The argument of the famous Einstein–Podolsky–Rosen paper, which raised what can be regarded as a difficulty for the interpretation of quantum mechanics, is completely misunderstood though this error is common among popularizers. "Fermi...developed a theory of the neutron-proton force..." Fermi did nothing of the kind. He proposed a successful theory of beta-decay, which was then applied by others in a misguided attempt to explain nuclear forces. We are told that the new elementary particle (the muon), discovered by Anderson and Neddermeyer, "possessed the charge of an electron but carried nearly double its mass"; but its mass about fifty times the mass of the electron.

The atmosphere of the years of confusion before 1925, with their frustration, their cooperation and disagreements between the theorists, comes across very well, even though it is not possible to understand what they were arguing about. But Cassidy's account of the debates with Schrödinger does not distinguish clearly (1) the debate over Schrödinger's wave mechanics as opposed to Heisenberg's matrix mechanics (which was soon settled when both came to be used interchangeably) from (2) Schrödinger's claim of the tangible reality of the waves (which it took him a long time to abandon).

On the controversies about Heisenberg's behavior during the Hitler period, Cassidy is clear and reasonable. The first question asked by many is: Why did Heisenberg stay in Germany, when he was opposed to much of what the system was doing, and why did he make so many compromises with the system? Cassidy's answer, with which I agree, is that Heisenberg was strongly patriotic. He also had some sympathy for the idea of a national renewal put forward by Hitler, since the social system did not seem to be working under the Weimar republic. But he is on record as strongly disapproving of the sacking of "non-Aryans" and liberals, and the attacks on "Jewish," as opposed to "German," physics.

He often sought the advice of Max Planck, the elder statesman of physics, who felt it his duty to remain in his post as head of the Kaiser Wilhelm Society, the main body supporting research, to use his influence to prevent the Nazis' worst excesses. In fact there was little he could do.

Cassidy criticizes Heisenberg for trying to help his own students and others in his immediate circle and not anybody outside, but there were limits to what one could do under the system without great personal risk. Once he had decided to remain in Nazi Germany, he could hardly avoid making concessions to the regime. Not everybody could be like Max von Laue, who kept out of sight and made as few concessions as possible (it is said that he never left the house without carrying two packages or a briefcase and a package, so he had no hand free to give the Hitler salute). This was not a possible way of life for Heisenberg, who wanted to continue being active as a leader in science.

He never openly opposed the Nazi system, but he was outspoken when it came to the attacks on modern physics by Johannes Stark and other fanatical supporters of "German" physics, even though this exposed him to virulent personal denunciation. This culminated in an attack on him in a Nazi newspaper, and he was, as a result, turned down for the appointment to the chair in Munich, for which he was the favored candidate. He complained to Himmler, whose organization sponsored the paper, asking if this attack represented the official line, in which case he would feel obliged to resign his chair in Leipzig. It took a long time before the reply came that such an attack did not have

In the front row are shown Sir Rudolf E. Peierls (left) and Werner Heisenberg (right), and in the rear row, left to right are G. Gentile, George Placzek, Giancarlo Wick, Felix Bloch, Viktor Weisskopf and F. Sauter. The photograph was taken in Leipzig in 1931. It is published here courtesy of AIP Emilio Segrè Visual Archives, Rudolf Peierls Collection.

official approval, and would not be allowed to happen again. Indeed the attacks on him ceased, to his gratification, though this result was not achieved without cost. While he was allowed to teach relativity, he had to promise not to mention the name of Einstein. Cassidy thinks that in this struggle, Heisenberg was concerned solely with his personal honor and standing, but it is hard to tell whether or how far he was also thinking of the standing and reputation of his subject and his profession.

Another important controversy centers on the question why the German scientists did not produce an atom bomb. Robert Jungk[2] and Walter Kaempfert of *The New York Times* have claimed that the German scientists were inhibited by moral scruples. Others, including Sam Goudsmit in his book *Alsos*[3] (though he later withdrew some of his statements) and Jeremy Bernstein in his April 1991 review in these pages of Victor Weisskopf's autobiography,[4] claim that the failure was owing to incompetence. Bernstein quotes Paul Lawrence Rose's assertion that Heisenberg did not know the difference be-

[2]Robert Jungk, *Brighter Than a Thousand Suns* (Harcourt Brace Jovanovich, 1958).
[3]S. Goudsmit, *Alsos* (Tomash Publishers, 1983).
[4]"The Charms of a Physicist," *The New York Review*, April 16, 1991, pp. 47–50.

tween a bomb and a reactor, and he mentions, though without endorsing it, a rumor that the Germans planned to drop a reactor as a weapon.

Cassidy rightly dismisses most of these stories. Heisenberg and the other German scientists, after beginning research on an atom bomb, genuinely believed that there was no chance of completing the manufacture of a weapon during the war, and indeed even the American project did not produce the first bomb until after the end of the war in Europe. As a result German scientists did not press the government for a crash program, and that explains the limited resources that were allocated to the project and its slow pace.

It is true that they overestimated the difficulties, because they never made a careful estimate of the critical size needed for a fast chain reaction, and therefore tended to overestimate the size of a weapon. It is also true that there were many errors of judgment in their work. For example, they excluded graphite as a moderator to slow down the neutrons because of a mistaken finding that it absorbed too many neutrons (not, as Cassidy says, that it was ineffective in slowing them down). This failure is often blamed on Walther Bothe, who is alleged to have made "wrong" measurements. In fact his measurements were correct; but the "pure" graphite he was using contained impurities in amounts that were too small to be detected chemically, but were fatal for the absorption of neutrons. In the same situation, Fermi and his collaborators in the United States guessed that further purification would improve the results. Another bad judgment was Heisenberg's insistence on using uranium plates, rather than rods or cubes of uranium, in the face of theoretical and experimental demonstration that rods or cubes were more effective.

While the Allied scientists were driven by the fear that Nazi Germany might get there first, the Germans arrogantly felt sure nobody could do better than they. But their view that Germany could not make a weapon before the end of the war was no doubt right. They were relieved that they did not have to make a moral judgment.

Would they have gone ahead if they had believed it possible to make a weapon? It is not easy to answer this hypothetical question, and Cassidy does not give a definite answer, but he implies that they probably would have done so, and I believe this is right.[5] Several people report that Heisenberg made the strange statement that on finding out about the death camps and other atrocities, he decided that he did not want Germany to be defeated, because the hate generated by these events would result in the complete destruction of

---

[5]An interesting comment was made by C. F. von Weizsäcker, Heisenberg's pupil and friend, in an interview on a BBC Horizon program on the "German Bomb." He said that Heisenberg would not have wanted to work on a bomb, but would not have been prepared to risk his life by refusing to do so.

Germany. He therefore felt he should donate his efforts to his country. It is not clear what he meant by this, since at that stage there was no way in which his work on atomic energy could have assisted the war effort.

Heisenberg said after the war, "It was from September 1941 that we saw an open road ahead of us, leading to the atom bomb." This was no doubt based on the knowledge that a reactor would produce plutonium, which was suitable material for making weapons. But he must have been aware that this open road was still extremely long. The Germans had not yet succeeded in making an experimental chain reaction (indeed they never did so), and if they succeeded in that, it was still a long way to a production reactor, which would have required more heavy water and uranium metal, both in short supply, as well as plutonium extraction facilities and bomb design. When in 1942 Heisenberg indicated to the authorities that there was a possibility of a bomb, he was more concerned with maintaining support for the work of the German scientists than with a real expectation of such an outcome.

The opinion of Heisenberg and other German scientists about the feasibility of making a weapon during the war is made clear in the "Farm Hall transcripts," the record of the conversations between German scientists, including Heisenberg, during their internment in England after V.E. Day, in a house that was bugged by British intelligence. When the first news came of the atom bomb dropped on Hiroshima, most of them refused to believe this, claiming that the Allies were bluffing. Even when more detailed statements convinced them that this was an atom bomb, Heisenberg kept saying that he could not understand how the Americans had been able to make a weapon in the time available to them. The transcripts also dispose of the idea that the Germans did not make a bomb for moral reasons: they discussed at great length the reasons for their failure, but they did not claim that moral scruples stopped them from working on the bomb. When Weizsäcker tentatively suggested that such scruples had prevented success, Otto Hahn firmly contradicted him.

The question of Heisenberg's visit to Bohr in 1941 is another matter for controversy, and here we are not likely ever to get a clear answer. Bohr's and Heisenberg's accounts of their conversation differ completely, and we cannot believe either. Heisenberg's recollections were very much influenced by his later thoughts, and Bohr, who was much better at talking than at listening, would often misunderstand what people said. Cassidy reviews various hypotheses based on Heisenberg's accounts. Perhaps he wanted Bohr to know that an atom bomb was possible, but that the Germans were not working on it. Perhaps he wanted Bohr's advice on whether one should work on it. In a postwar letter to B. L. van der Waerden, Heisenberg said he wanted to avert a crash program by the Allies. The last is most implausible, because he did not then know that Bohr was in touch with the British, and also because he believed that a bomb could not be made in wartime. Bohr said later that Heisenberg came to threaten him with the bomb, and Bohr's wife, Margrethe, said that Heisenberg wanted to sound out Bohr on what the Allies were doing. If, as seems the case, he did not know Bohr was in touch with them, this is not very plausible either.

There was never any chance of a dispassionate conversation between Bohr, a prominent citizen of an occupied country, and Heisenberg, who, though an old friend, was now a representative of the hated occupying power. The inability of Heisenberg to see this was a typical example of his insensitivity, his lack of understanding of other people's points of view. Cassidy quotes a number of cases in which, during his lecture tours in occupied or neutral countries during the war, his remarks were deeply offensive to his hosts.

Some sections of Cassidy's book are difficult to read partly because the author tends to jump around in time. Some deviation from strict chronology is of course necessary. But here it is done much more than seems necessary, so that the reader has difficulty seeing what period is referred to at a given point. This can lead to substantial errors if the author himself gets confused about timing.[6]

Still, Cassidy supplies much interesting information, and he has collected some of it with meticulous care. For example, the chapters about Heisenberg's youth contain not only summaries of many personal letters, but quotations from his school reports and the comments of his teachers. The only material missing was destroyed during the war.

---

[6]Bohr's arrival in England in 1943, and his encounter with Chadwick are mentioned. The text continues, "The... report of a... committee meeting shortly *thereafter* describes Chadwick's impressions" (my emphasis). But according to a note the meeting was on September 10, 1943, *before* Bohr's escape from Denmark!

The Bomb That Never Was

During World War II the scientists and technicians in Britain and the United States who were building an atom bomb feared that Nazi Germany might produce such a weapon first. Yet at the end of the war it became clear that there had been no such threat. Not only was no German bomb under construction, but no extensive effort was being made to develop one. Why was that so?

The first answer was given by the physicist Sam Goudsmit, who was sent to Europe in 1947 on what was called the "Alsos" mission to discover the facts about the German project. In his book *Alsos*[1] he attributed the failure to gross incompetence on the part of the scientists. In particular he claimed that Heisenberg did not know the difference between a bomb and a reactor. This last statement he later withdrew, but he remained convinced that Heisenberg, the most famous German physicist, did his best to develop a bomb, but his best was not good enough.

Another explanation was put forward in 1956 by Robert Jungk in his book *Brighter Than a Thousand Suns*,[2] and supported by Walter Kaempfert of *The New York Times*. According to them, the German scientists refrained from making a bomb for moral reasons. I know of no evidence that Heisenberg made this claim himself in public, and it is not clear whether he really took that position in his interview with Jungk. But he also never repudiated it publicly.

This is a review of *Heisenberg's War: The Secret History of the German Bomb,* by Thomas Powers. New York: Knopf, 1993. The book review is reprinted with permission from *The New York Review of Books*. Copyright © 1993 Nyrev, Inc. For the complete reference, please see the Acknowledgments section.

[1]Henry Schuman, 1947.
[2]Harcourt Brace, 1958.

Heisenberg later gave his own explanation in articles in scientific journals and in his autobiographical book *Physics and Beyond*.[3] While he and his colleagues knew a bomb to be possible in principle, he wrote, they saw that the effort to develop it was far too great to be undertaken by Germany in wartime. (In retrospect this seems reasonable, because even the Americans, who had greater technological resources and were not subject to air raids, did not complete the bomb before V.E. Day.) As a result the German scientists did not ask for a crash program, and took as their goal the construction of an experimental reactor, in order to prove the feasibility of a chain reaction and to prepare the way for a future nuclear power program. They felt relieved that they did not have to make a decision on whether to work on a weapon. It is anybody's guess whether they would have worked on a bomb if they had regarded this as a practical possibility.

I find this explanation convincing; it seems consistent with the evidence I have seen. It is true that Heisenberg was known to allow his memory to be colored by what he would like to be true, but I have learned from experience that one should never discount the possibility that a person might actually mean what he is saying.

The "Farm Hall Transcripts," of recorded conversations of German atomic scientists who were interned in England shortly after the war, firmly ruled out the idea of their moral superiority. The transcripts record the reaction of the scientists to the announcement that a bomb was dropped on Hiroshima. Apart from a tentative remark by von Weizsäcker, none of them mentioned moral objections as reasons for not having made a bomb, although they argued about the reasons for their own failure and the consequences of the bomb for their own future in Germany. Yet the controversy has continued, and in *Heisenberg's War* Thomas Powers produces a new theory.

Before I consider his view, I should recall the history of the development of the bomb both in Germany and in the U.S. and Britain. Both sides started with rather tentative basic research to ascertain the possibilities for larger projects. These studies got underway in 1939 in Germany, and in the Spring of 1940 in Britain and the U.S. In the Summer of 1941 the British "Maud" committee reported that a bomb making use of the separated uranium isotope 235 was feasible and worth pursuing, and the report persuaded the British government to step up the pace of research; a little later the U.S. project was reorganized and the "Manhattan Project" started as a crash program. By this time it was known that bombardment by neutrons would produce a reaction in the abundant uranium isotope U-238, which would eventually lead to the new element plutonium, expected to be an alternative fuel for a weapon. In

---

[3]Harper and Row, 1971.

1942 Enrico Fermi succeeded in building an experimental reactor, thus demonstrating the first man–made neutron chain reaction.

To accomplish this, and go on to produce a bomb, many obstacles had to be overcome. To a physicist, the idea of separating substantial quantities of the isotopes of such a heavy element as uranium first appeared utopian. This had been done only for light elements, for which the percentage differences between the masses of isotopes were much bigger than for the heavy uranium. For the separation, moreover, it would be necessary to use a gaseous compound of uranium; but the only such compound is the very corrosive hexafluoride, an awkward substance to work with. The work on light elements had produced only milligrams of isotopes, so to produce the kilograms needed for a bomb the process previously used would have to be scaled up a million times.

For these reasons the idea of a bomb using uranium 235 was at first not taken seriously, and the question of how much U-235 would be required remained academic, although most scientists who might have addressed the problem would intuitively have guessed that tons were needed. Once one took the possibility seriously and thought about the "critical mass" of U-235 needed to produce a chain reaction one could see it was a matter of kilograms rather than of tons; to separate the isotopes would still need a tremendous effort, but not a prohibitively difficult or expensive one.

The reactor project also faced problems; it required a "moderator," a light material to slow down the neutrons, since slow neutrons are more effective in producing nuclear fission. Of the obvious light elements, carbon, in the form of graphite, was the favorite for use as a moderator, but experiments showed that even pure graphite absorbed too many neutrons to make it suitable. The vital step was the intuition by Leo Szilard and Fermi that commercially "pure" graphite contained impurities in amounts that were too small to show up in chemical analysis but still contributed to a very high absorption of neutrons. By insisting on further purification of graphite, they solved this problem. The primary reason for building the experimental reactor was to establish the reality of the chain reaction. The plan to produce plutonium for weapons came later.

In Germany the pessimism about separating isotopes on a large scale was never overcome. There was some small–scale research on such separation, but the whole matter was regarded as a very long–term prospect. Since it was so academic, nobody, not even Heisenberg, made a serious effort to estimate the critical mass needed to produce a chain reaction. Heisenberg used different arguments at different times, giving answers ranging from "as big as a pineapple"—about 15 kilograms of metallic uranium—to many tons. This vagueness seems surprising, but Heisenberg, though a brilliant theoretician, was always very casual about numbers. When I was his student in the late 1920s the first assignment he gave me was to check whether a recent observation in a spectroscopic experiment could be explained as an example of his uncertainty principle. A simple back–of–an–envelope estimate would have shown that the effect was 100 or even 1,000 times greater than could be explained by his hypothesis. When he first heard, at Farm Hall, of the reality of the Hiroshima bomb, he at last thought through the question of critical mass seriously and came up with a reasonable answer.

In attempting to make reactors, the Germans, too, found that graphite absorbed excessive amounts of neutrons, but did not insist on further purification, and turned instead to heavy water—i.e., water containing the isotope of hydrogen called deuterium—as a moderator that would slow down neutrons. This is more efficient and makes it possible to use a much smaller reactor, but it is hard to obtain. The world's only facility for producing heavy water at that time, in Norway, was destroyed by sabotage and in Allied bombings, so that there was a continuing shortage of it in Germany.

The large-scale production of plutonium seemed to the Germans also a very distant goal. It is true that Heisenberg said after the war: "We saw an open road to the atomic bomb," implying that, since they knew a reactor moderated by heavy water would work, they could see how to produce plutonium (actually neptunium, the intermediate element that decays into plutonium, and was not yet known to them) and hence weapons. But this open road was still a rocky one: it would have required the design of large–scale reactors and methods for cooling them as well as chemical technology for extracting the plutonium from the highly radioactive reactor fuel; finally, the method of detonating a plutonium bomb was the hardest of the problems solved at Los Alamos. Heisenberg's statement was a little naive.

The evidence supports the contention that since he and the other scientists regarded the bomb as a long-term problem with no relevance to the war, they did not ask the Nazi government to support a crash program. Their aim, like the initial aim of Fermi, was to prove the existence of a chain reaction. They pursued this with a very small team, and without great urgency. They still hoped they might be the first to achieve a chain reaction and thought this would be good for the prestige of German science. It seems never to have entered their thoughts that others might be ahead of them. After they first heard the news of Hiroshima, some of the Farm Hall group comforted themselves by saying that, while the Americans had made the bomb, history would recognize that the Germans had made the reactor!

Meanwhile the Allies were afraid of a German bomb. Their intelligence services made an intensive effort to discover what was going on. While working on the British Atomic Energy Project early in the war I was able to convince myself there was no German crash program by checking the list of physics lecturers in German universities published each semester by the *Physikalische Zeitschrift*. This showed that nearly all the physicists were in their usual positions, teaching their usual subjects, a very different picture from that which would have been shown by a similar list for the UK or the U.S. In addition, many messages arrived from German opponents to the regime, a few warning of a German bomb program, but most saying that only a small reactor project was actually underway. Some, notably the publisher Paul Rosbaud and the physicist Fritz Houtermans, claimed that Heisenberg and other German scientists were trying to delay work on a bomb. Yet the Allied authorities continued to worry. When Strasbourg was captured by the Allied forces, documents were found showing the small scale and slow progress of the work, but American concern remained intense. At the end of the war the leading German atomic scientists were rounded up and interned in Farm Hall until early 1946, presumably to prevent their being hired or kidnapped by the Russians.

Thomas Powers's book covers these events in enormous detail. He has collected an impressive amount of information, though much of it is from secondary sources, including such controversial writers as Robert Jungk and David Irving. He describes many episodes that are not widely known. Many of these concern the relations, and the squabbles, among the different U.S. intelligence services, including their travels to and within Europe, with precise details of their departure and arrival dates. Readers who are not fascinated by intelligence operations may not find these accounts very interesting; but many of the stories are surprising and intriguing. Unfortunately it is not easy to decide which of these stories to believe, in spite of their extensive documentation, partly because of the book's unfortunate errors, of which I shall give examples later.

One of the more exciting episodes was the plan by the U.S. to kidnap or assassinate Heisenberg. This is reported with a dramatic effect not unlike that common in TV thrillers, where at dramatic moments the presentation is interrupted by a commercial. Powers, too, builds up tension, only to change the subject for many pages. We read how Colonel Carl Eifler was selected and briefed for the job and then, after long deliberation, was eased out because it was feared he might fall into German hands, and might under pressure reveal vital information about the American project. Instead the intelligence officials chose another man, Morris "Mo" Berg, a former baseball player and an accomplished linguist, who, at least officially, was ignorant of the Manhattan Project. Plans were made to get him, along with a support team, into Germany, or into Switzerland during a visit by Heisenberg to Zurich where he was to give a lecture. But Berg had a mind of his own, and went alone to Zurich while Heisenberg was there.

The author quotes the physicist Philip Morrison, who worked at Los Alamos, as saying that he was doubtful whether at this time, December 1944, there was any sense in putting Heisenberg out of action. If there was a German project that had any chance of success before the war ended, it would have progressed by this time to the stage where the technological problems would have been central and individual scientists would not have been vital.

Berg made contact with Swiss physicists sympathetic to the Allied cause and, with a gun in his pocket, he found a seat among the audience at Heisenberg's very abstract theoretical seminar. So far, there is no reason to question Powers's account. But we are told that Berg was authorized to make his own decision whether to shoot Heisenberg. He did not shoot, it is said, because Heisenberg did not mention anything about an atom bomb in his talk.

I find it wholly incredible that it should have entered anyone's head that Heisenberg could, or would, talk about an atom bomb in an open academic seminar in Switzerland, or that a junior agent of the OSS would have been authorized to decide whether to kill him or not. Surprising things happen in war, especially within secret intelligence agencies, but I would need very strong additional evidence to make me accept this story.

Another episode that I find surprising concerns Niels Bohr, who arrived in America with a crude sketch of a reactor, which Heisenberg had given him during his visit to Copenhagen in 1941. American physicists considered whether the reactor might have been intended as a weapon. The author claims that Bohr himself regarded it as possible that a weapon could be made using slow neutrons. This seems quite unlikely, since Bohr had been the first to point out the impossibility of doing so. Hans Bethe is cited as the author's principal source for this story, but Bethe tells me he does not think Bohr ever held this view.

In the main thesis of his book, stated explicitly in the last two chapters but present in the background throughout, Powers advances his own, new explanation of the failure of the German project. Heisenberg, he argues, deliberately withheld information from the authorities and from his colleagues in order to stop the construction of a German atom bomb. The most important aspect of this deception was that Heisenberg kept stressing that the development of a bomb was too difficult and involved too great an effort to be completed in wartime Germany. No doubt Heisenberg said this on every possible occasion; the questions are, did he believe it, and was it true?

It was indeed true, as I have pointed out earlier. Powers, however, disagrees. He argues that the German work started in 1939, whereas the Manhattan Project began only in the summer of 1942, so German scientists could have been testing their first bomb two years before the Americans did, in 1943. This is a completely mistaken comparison. One cannot launch an enormous technological program based on a new discovery without preliminary laboratory studies of its feasibility and its basic principles. Such studies went on in Britain and the U.S. from about early 1940, and Germany could not do without similar studies either, so there was, in fact, at most a six month difference in the respective starting dates. In this early phase Western scientists, unlike the Germans, had the use of a number of cyclotrons and other accelerators to determine the behavior of nuclei and they also had mass spectrographs to produce small samples of the relevant isotopes. During the later, technological phase of manufacturing a bomb the hugely expensive technology, assembled, for example, at Oak Ridge, Tennessee, and Hanford, Washington, could not have been matched by the Germans.

From the evidence I have seen, Heisenberg certainly believed that the atom bomb was something for the distant future. In fact he was sure that even the Americans could not make one during the war. This is quite clearly demonstrated by his reaction at Farm Hall to the news that a bomb had been dropped on Hiroshima, when he at first firmly believed that the announcement was a bluff, and that the U.S. did not really have an atom bomb. Of all the scientists at Farm Hall he seems to have been the last to accept the reality of the event.

A contrary impression is given by the report, quoted by Powers, that in 1942 Heisenberg, in answer to a question from the Luftwaffe general Erhard Milch, estimated that the Americans could not complete their first reactor before the end of 1942, and the first bomb would take another two years. What is the source for this surprising report? It turns out to be a 1947 interview by Heisenberg with *Der Spiegel*. News magazines are not the most dependable secondary sources, and we know of many instances when Heisenberg's memory deceived him.

Why did Heisenberg overestimate the difficulties? The explanation is connected with his failure to make an accurate estimate of the critical mass, on which I have already commented. Most of the time he relied on an argument that was conceptually wrong, though he quoted different estimates at different times. After Hiroshima, as I have said, he took time to think through the problem rationally, and a week later had a perfectly reasonable figure. This was not, as Powers says, because he had initially used a wrong figure for the "mean free path" of neutrons in uranium 235, for which he could not have discovered a better number at Farm Hall, but because his original argument was logically flawed. Knowing that many "generations" of neutrons were needed to make a chain reaction, he assumed, in estimating the critical size, that each neutron must, on the average, make a certain number of collisions with uranium nuclei, instead of enough collisions to generate one further neutron before escaping.

In his revised estimate he also took into account that the bomb core could be surrounded by a reflector, which would scatter back some of the escaping neutrons and thereby reduce the size of the critical mass needed. According to Powers he must have devoted much thought to bomb design for such a subtle idea to have occurred to him. Yet Powers ignores the fact that reflectors are also used for the design of reactors, on which Heisenberg had been working for a long time; so it would have been quite obvious to him to apply the idea of a reflector to a bomb.

Powers also stresses, as part of his argument that Heisenberg was deceiving his colleagues, that he never explained such points to them; but even if he had any interesting thoughts about bomb design, why should he have discussed them with others when nobody was planning to develop an atom bomb?

Powers also relies on the message from Fritz Houtermans carried to the U.S. by Fritz Reiche, which said that many German physicists were working on an atomic–bomb project, but that Heisenberg was trying to delay the work. We must remember, however, that Houtermans was not a member of the project, and not necessarily familiar with what was going on. The first part of the message was certainly misleading because, as we know, there was no work on a bomb, so it is difficult to accept the second part. I therefore conclude that the new and ingenious explanation of the German failure is without substance.

I earlier mentioned wrong or misleading statements in Powers's book. Many of the facts of which I have direct knowledge are in some way garbled, although some do not affect the substance of the argument. For example, Powers claims that I originally estimated the critical mass of U-235 as many tons; I never did so. The memorandum I wrote with Otto Frisch in 1940 for the British authorities estimating the critical mass is dated in one place in the book as 1940, in another as 1941. The German town of Duisburg (to which some of the uranium ores was traced) is identified in the book as "Duisburg in Belgium." The famous physicist Max Born is said to be a mathematician on one page and a physicist on another, and the well-known nuclear physicist Walther Bothe is called a chemist.

Other errors are more serious. Powers has a persistent tendency to interpret any work on uranium fission as bomb research; he expresses surprise that Bohr allowed his "bomb lecture" to be published in 1941. In fact this is an academic exposition of fission physics which contains the remarks that no explosion is possible in natural uranium and that, with present technical means, it is impossible to purify enough of the rare uranium isotope to cause an explosion. A paper by Frédéric Joliot-Curie, written in 1939, showed that there were secondary neutrons emitted in the fission process, which pointed to the possibility of a chain reaction. This, the author says, "was just another way of saying bomb," but a chain reaction also drives a reactor.

In 1939 the German physicist Siegfried Flügge wrote about the possibility of a nuclear reactor generating power and referred to the need to avoid a runaway chain reaction. This, the author again comments, was just another way of saying "bomb." In fact Flügge evidently referred to the kind of nuclear disaster that occurred at Chernobyl.

In 1977 Goudsmit wrote to me as follows:

> Armin Hermann, whose opinion I asked, claims that I am now "softer" on Heisenberg than I was in my book, written in 1947. I have had time to think. I believe, that we resent that this great physicist, our idol, wasn't any better than we are, that he didn't present the humane leadership we hoped for. Did anyone among our prominent colleagues? Perhaps Laue or Blackett or Franck?

Powers quotes only a phrase of this letter and gives the impression that Goudsmit's attitude had not changed:

> Goudsmit felt nothing but abiding cold fury for the possibility that Heisenberg and his friends might claim any iota of moral responsibility for the failure of Germany to build a bomb. Even after thirty years Goudsmit told Rudolf Peierls he resented the fact that "this great physicist, our idol, wasn't any better than we are."

Powers's prose is lively and readable, in spite of occasional clumsy passages: "Heisenberg was the single greatest physicist remaining in Germany" or "explain...what had to be done to the commander of the Eighth Air Force" (meaning "explain to the commander...what had to be done"). The narrative frequently jumps back and forth in time, so it is not always easy to see what period the author is talking about.

A Physicist Who Enjoys It

M ost of nuclear physics has today become "big science," requiring large and expensive accelerators, with complex detection equipment and computers, used by large teams of research workers. Its senior practitioners have to spend much of their time and their energy on administration and organization. This image of the new scientist does not fit Otto Robert Frisch at all. His very readable reminiscences are those of an individualist, at his best when working and thinking on his own, or with perhaps one close collaborator.

This characteristic comes out very clearly, together with the basic happiness of a man who enjoys what he is doing. Scientists tend to be driven by a variety of motives including, to varying degrees, ambition, a sense of duty to the community, curiosity about nature and its laws, and enjoyment in the design and execution of experiments. All these no doubt have some reality for Frisch, but the pleasure of doing what he does is clearly dominant. Ambition is the least of his concerns.

All this goes for his many other interests. Besides being an outstanding experimenter, with a flair for the simple and decisive experiment, he has many other interests and talents. This breadth probably owes much to his background. He grew up in Vienna, in a family belonging to the cultural tradition of early twentieth-century Vienna. His aunt, the famous nuclear physicist Lise Meitner, was his mentor in much besides physics. Next to physics his strongest line is music. He is a pianist of near–professional standard, and his hearing has absolute pitch. But here again he is less concerned with demonstrating his brilliance at the piano than in the pleasure of making music, and he can enjoy playing with others who may be much below his

This is a review of *What Little I Remember*, by Otto R. Frisch. Cambridge: Cambridge University Press, 1979. For the complete reference, please see the Acknowledgments section.

Otto Robert Frisch in May 1967. (Photograph published courtesy of AIP Emilio Segrè Visual Archives.)

level in skill, as long as they share his love for music and his pleasure in making it.

He also likes drawing, and after every meeting the papers in front of him are covered with portraits or caricatures. Some of these are reproduced in the book. They are all recognizable likeness, and some bring out the personality of the subject excellently. His drawing is not of the standard of his music or his physics, but his enjoyment in the activity is again evident.

Presumably the drawings reproduced in the book are the most successful ones of many. The written characterizations of people, which, in his preface, he compares with the pencil sketches, also vary very much in depth. Niels Bohr and Lise Meitner come very much alive, as does George Placzek, with whom Frisch worked much in Copenhagen. But very little emerges of Chadwick, with whom he was closely associated in Liverpool and later in Los Alamos, beyond an account of their first encounter; and of some other colorful personalities there are only pale shadows.

As a physicist, Frisch studied in the University of Vienna, where he received his Ph.D. (he does not mention the subject of his thesis). After a year in a small firm making scientific instruments he was offered a position in the Physikalisch–Technische Reichsanstalt in Berlin, and after three years moved to Hamburg to work under Otto Stern, the authority on atomic beam research. That much migration was normal at a time when modern physicists formed a close–knit but widely spread family.

But in 1933, when Hitler came to power, he had to be dismissed from his post as a "non–Aryan," and there followed the typical odyssey of a displaced academic of the time. A year at Birkbeck College in London under Blackett gave him the first acquaintance with life in England. This was the time when artificial radioactivity was discovered and Frisch made his first venture into nuclear physics: he invented a clever device to observe very short–lived activities, and was able to discover new species of radioactive nuclei. He stayed in London only a year, and was then invited to Bohr's institute in Copenhagen, where he remained until 1939.

This was a period of great excitement in nuclear physics. Fermi had started using the newly discovered neutron as a probe for studying nuclei, and neutron physics became one of the major subjects in many laboratories, including Copenhagen. Frisch became involved in this work, originally because he had a well–running counter suitable for detecting the activities produced by neutron bombardment. He describes the way the work was done, including the preparation of neutron sources by mixing radium and beryllium. A particularly exciting moment was Niels Bohr's first pronouncement of the idea of the compound nucleus, as a comment after a seminar talk which highlighted the difficulties of the then conventional view in accounting for neutron resonances.

Then, in 1938, came the discovery by Hahn and Strassmann that radioactive barium was one of the products of the bombardment of uranium with neutrons. Frisch was visiting Lise Meitner in Sweden when she received the news of Hahn's result, and there is a fascinating account of the discussion which finally led them to realize that the uranium nucleus was undergoing "fission" (a term chosen by Frisch to describe the phenomenon). Before long he was back in Copenhagen and was doing an experiment which verified the suggested explanation.

By this time war was imminent, and Denmark was not a very safe place for a refugee from Hitler, so Frisch moved to Birmingham. He started some experiments relating to the fission problem which was still one of his main interests, but in a department not well equipped for nuclear physics, and with most of its members working on radar, progress was slow. He continued thinking about the implications and possibilities, including nuclear energy. Bohr had shown that a violent explosion was not possible with natural uranium, but Frisch started to question whether the possibility of separating the uranium isotopes in substantial quantities was really ruled out. He and Peierls concluded that a nuclear weapon did not seem to be impossible, and wrote a memorandum which led to an atomic weapons project being started in Britain. From then on most of his effort was devoted to this project. It was an odd situation that he and several others working on one of the most secret military projects were technically enemy aliens, and he tells of a number of ludicrous consequences of this. It soon emerged that he could work more usefully in Liverpool, where Chadwick's laboratory had a small cyclotron and was not preoccupied with other war research. He worked there, through the period of heavy air raids, until 1943, when it was decided to discontinue the British project during war time and send to the United States all those who could be useful to the American atom bomb work. So Frisch went to Los Alamos, and he gives a vivid description of life and work in this odd secret town. Very sensibly he was not attached to one of the large research groups there, but had a roving commission, helping with experimental troubles wherever appropriate.

One of his other accomplishments at Los Alamos was to learn driving, and he tells the story of his first accident, which brought him a broken rib and other minor injuries. He does not seem to remember another accident which became legendary at Los Alamos. Pulling aside to clear another car on a narrow lane without footpath, he knocked over an Army girl. The girl spent a few days in hospital and complained that Frisch had never visited her or shown interest. This was reported to him, and he promptly appeared in hospital, apologizing and explaining that he had been concerned. "I inquired of the doctor about about your health and was glad to hear that there was nothing seriously wrong with you, mainly the effect of the fright."

After the war he became head of the Nuclear Physics Division of AERE, Harwell, the new atomic energy laboratory. According to his account of this period, he left the running of the division to his deputy, who enjoyed administration, and continued his individual research. Perhaps his diffident description exaggerates his lack of application to the work of running the division, but it is in character that he was not at his best in this kind of responsibility.

He had been in Harwell less than two years when he was offered the Jacksonian Professorship in Cambridge. The book is exceedingly brief about the 25 years he spent in Cambridge before retiring. He explains in the preface that his memory, like that of many others, is more complete about earlier events than about the more recent ones, and that it is embarrassing to write about people with whom one is still in touch. So the whole period, which includes his marriage and the growth of his two children, gets only a few pages. The story ends with a description of another ingenious device, a machine for measuring and evaluating bubble–chamber tracks, which he designed; he is now a partner in the firm manufacturing this commercially.

Three chapters of the book, Atoms, Nuclei and Research Resumed, contain expositions of parts of modern physics by way of background to the accounts of his own involvement with it. Being a physicist myself, I am not qualified to judge the appropriateness of these pieces. They are written with the clarity and simplicity to be expected from the author of *Meet the Atoms*, yet I wonder whether the reader who does not know the background can really assimilate enough information to help him with the biographical chapters which are the real meat of the book, and which seem quite intelligible on their own.

This is a happy book, from which the author's personality and his enjoyment of physics, of music, of life, emerges clearly. It is also a portrait of the pre–war world of physics, of the days of small numbers and small apparatus, of times when a physicist could think of an ingenious experiment today and set it up tomorrow.

Otto Robert Frisch, 1904–1979

O tto Robert Frisch brought to physics the approach of a craftsman. He enjoyed, above all, doing experiments with his own hands, preferably on apparatus of his own design, aimed at simple basic questions. This approach required the ability to think about the important problems of physics simply, but deeply, an ability underlying the two contributions for which he was best known: his share in the explanation of the fission process and that in the recognition of the feasibility of an atomic weapon. The craftsman was also something of an artist, not only in his love for music and his skill and taste as a pianist and a violinist, but in the use of language, which made him an outstanding expositor.

There were many problems, in physics and elsewhere, which he regarded as interesting, and these he pursued persistently, in depth, until he saw convincing answers in simple terms, and he evidently enjoyed himself in doing so. He equally firmly refused to become involved with matters other than those which he had decided were of interest to him.

Otto Robert Frisch was born in Vienna. His paternal grandfather, Moriz Frisch, immigrated from Galicia, and did well as a printer, until generosity and a casual attitude to accounting led to bankruptcy. Otto Robert's father, Justinian Frisch, brought the firm back to solvency, and later worked for other printing and publishing firms. He was also a gifted painter, specializing in water colors. His son's recollections stress his wit, his firm sense of right and wrong, and the many interests to which he introduced the son, including mathematics. It is likely that he also passed on to him his keen sense of language.

The mother, Auguste, *née* Meitner, was the daughter of a lawyer. She was a gifted musician, who became a concert pianist at a very young age, and also studied composition and conducting, but gave up her musical career when she married. Her sister was Lise Meitner, who became a nuclear physicist of great distinction and was to have a strong influence on her nephew.

In the family the boy was always known as "Otto Robert," as if the two names were hyphenated. Later he tended to call himself Robert, except at Los Alamos, where there were too many Roberts, and he was called Otto. This usage seems to have persisted; his book of recollections and some articles are headed "Otto Frisch."

He showed quite early a gift for mathematics, and the ability to see through mathematical or logical arguments at great speed. This later extended to a similar speed in seeing mechanical points. I noticed this when, about 1940, I started teaching him to drive, and at the first lesson, when he had been introduced to the gears, he started downhill, and shifted from first directly into third gear. When I pointed out that this was not the usual way, he replied, "Downhill this should be all right," as, of course, it was.

His ability and interest in mathematics suggested this as his main subject of study at the university, but when he did enter the University of Vienna in 1922, he felt that a career in mathematics would be too dry and abstract for his taste, and he chose physics. Austrian physics courses then had no division between undergraduate and postgraduate study, and no equivalent of the bachelor's degree. The first examination was the Dr. Phil., for which he qualified *cum laude* in four years, which was fairly normal. His thesis work was done under Karl Przibram in the Institut für Radiumforschung. on 'Verfärbung von Steinsalz durch Kathodenstrahlen' (Discoloration of rock salt by cathode rays).

His first published paper, datelined from that institute and read to the meeting of the Vienna Academy on January 27, 1927, was related to the subject of his thesis. It is concerned with the effect of slow electrons on rock salt, of interest in connection with the formation of color centers, on which Przibram was then working. The experiment was difficult because of the effect of space charges on the slow electrons, and required much care and ingenuity. The findings were largely qualitative.

After graduating he spent about a year in a small private laboratory of a firm manufacturing X-ray dosimeters. The head of the firm was an inventor, Siegmund Strauss, who liked to try out his ideas on his young assistant. "He was full of ideas, and it was part of my job to listen and throw out the dud ones." This was evidently an opportunity for practicing his ability to understand and judge such points quickly, and his employer must have been aware of this ability.

In 1927 he was offered an appointment in the Physikalisch–Technische Reichsanstalt in Berlin–Charlottenburg, the German government institution roughly equivalent to the National Physical Laboratory. The offer came as a surprise to him; it was probably due to a recommendation from Professor Przibram. His duties were hardly exciting. They were concerned with an attempt to set up a better way of measuring light intensity. A short paper, published jointly with his boss, related to the measurement of very weak currents, in connection with the instrument they were developing.

However, the period in Berlin was much more interesting than the job he had to do. He used much of his spare time to stay in the laboratory and to try out various ideas of his own. He also was able to attend seminar meetings at the University of Berlin, and to listen to the physicists there, including Max Planck, Max von Laue, Einstein, Gustav Hertz and many visitors. Another important contact was Lise Meitner, who lived and worked in the Berlin suburb of Dahlem. Frisch had lodgings in the same neighborhood, although this meant a long bus ride to work, so as to keep in touch with his aunt, to the benefit both of his physics and his music.

During his third year in Berlin he was given facilities to do experiments in the laboratory of Peter Pringsheim, originally to allow him to try an experiment which was to show that a light quantum can have an intrinsic angular momentum in a transverse direction. The experiment gave a negative result, and eventually James Franck explained to him that he had overlooked the existence of orbital angular momentum. Nevertheless he published a paper on the experiment. He explains, "My paper was already written, and the best I could do was to add Franck's explanation as a footnote." Perhaps the paper was already submitted, and the explanation was added in proof, though it is not marked as an addition in proof (and is not a footnote). However, the contact with Pringsheim led to a temporary grant when the appointment at the Reichsanstalt ran out, and joint work with Pringsheim resulted in a paper which settled a point of principle which had been in doubt in the early application of quantum theory to spectroscopy.

He left Berlin in 1930 to accept a position in Hamburg. The professor there was Otto Stern, a great personality and an outstanding experimenter. This was not through manual dexterity, which he seemed to be lacking, but through a brilliantly clear insight into the nature of the experiment, which allowed him to minimize, and to control, all possible sources of error. His field was the use of molecular beams; he won the Nobel Prize for his work with Gerlach, which proved the directional quantization of atomic angular momenta. Since then he had, with his collaborators, notably I. Estermann, perfected the technique and applied it to many important physical problems.

Frisch was invited to join Stern as an Assistant, a regular appointment in a German university, no doubt on Pringsheim's recommendation. He took rapidly to the molecular–beam technique, with all its technical intricacies. He took part in an experiment on the diffraction of atoms by a crystal surface. This had previously been done qualitatively by Estermann and Stern, but for a quantitative check on the wavelength one required beams in which all atoms had very nearly the same velocity, and this was achieved by means of a "beam chopper," a set of two rotating slotted discs, through which only the atoms of a certain speed can pass. He reports, that a small discrepancy between the wavelength found and the theoretical de Broglie value was traced to an error in the number of slots in the discs.

He describes his role in such experiments as Stern's 'pair of hands,' and this is no doubt correct in the sense that Stern did not like touching the apparatus himself, but one would guess that Frisch played also an active part in the design of the experiments and in considering the factors involved.

Other work of the Hamburg period includes a study, jointly with E. Segrè, of the response of atomic angular momenta to changes in direction of the magnetic field, a problem which had been an intriguing mystery for the old quantum theory, but could be described by quantum mechanics. Joint work with Stern reports anomalies in the specular reflection of atomic beams from crystal surfaces. However, the most important finding was the discovery of the magnetic moment of the proton with the surprising result that this was some two to three times larger than the nuclear magneton predicted by the Dirac equation.

Frisch's last paper from Hamburg, and probably the last paper from the Stern group there, was rather characteristic of his approach. It set out to verify the prediction of quantum theory that an atom, on emitting a light quantum, recoils. At the end of the paper he comments that he had verified the magnitude of the predicted recoil only with limited accuracy and adds: "It would undoubtedly have been possible to obtain, perhaps with even narrower beams, substantially cleaner and less objectionable results, but the experiments had to be discontinued for external reasons."

The date of receipt of this paper was August 22, 1933, and the "external reasons" were the racial laws of Nazi Germany, which compelled Stern, Estermann and Frisch to leave. He had planned to go to Rome with a Rockefeller Fellowship which he had been awarded, but this became impossible because the Fellowship was contingent on having a regular appointment to return to. Eventually Stern, who worked hard to find posts for his displaced collaborators, arranged for Frisch to join Blackett at Birkbeck College in London.

During his time at Birkbeck he was maintained by a grant from the Academic Assistance Council, set up to provide help and temporary support for displaced scholars. This was done in a spirit of impressive generosity, coming at a time of economic depression and a shortage of academic posts, particularly as much of the initiative came from the academics, who might well have been expected to fear the competition of so many newcomers.

Blackett's laboratory did not have the equipment for molecular–beam research, and Frisch joined in the work of the laboratory, in which the cloud chamber was the dominant tool. He set himself the task of developing a cloud chamber with a long acceptance time. He succeeded in constructing this, but it was never used as a research tool.

At that time artificial radioactivity had been discovered by the Joliot–Curies, and several members of the laboratory took up this new line of study. Frisch joined them. He realized that there might exist some very short–lived activities, which could be detected only by moving the sample very rapidly from the α–ray source to which it had been exposed, to the vicinity of a counter. With his love for gadgets he soon had made a mechanical device for achieving this, and with it he discovered two new radioactive isotopes, though not particularly short–lived.

He did not really settle down at Birkbeck College, where his stay clearly would be only temporary. But it was an opportunity to broaden his outlook, to make new friends, and to make his first acquaintance with life in England, where he would later spend so much of his life.

In the summer of 1934 Niels Bohr invited him to join his institute in Copenhagen. Bohr's Institute of Theoretical Physics had, in spite of its name, a long tradition of experimental work, and there was interest in nuclear physics, where Frisch's recent experience and skill could fit in well. This was the first appointment which he was able to keep for a long time, and he soon felt at home there. The atmosphere in the institute, with its unique position in modern theoretical physics, and a first–rate group of experimentalists, was stimulating, and he made many friends. He became quite proficient in Danish, at least in speaking; he complained that understanding the rapidly spoken word remained difficult. He liked to speculate that the peculiar sound of Danish, dominated by vowels, with the consonants mostly swallowed, developed from the shouts of seamen from boat to boat to communicate during a gale. Like most people who had been in Bohr's institute, he always retained a special affection for the institute and for Copenhagen.

At first he continued the type of work he had started in London, discovering two more radioactive isotopes. Then there was a new excitement in nuclear physics when Fermi showed how much could be learnt by studying the collisions of neutrons with nuclei. The new approach was taken up vigorously in Copenhagen, and Frisch took part in it. The next ten papers on the list contain the results of this activity.

At that time there was no way of measuring the energy of a neutron, or producing monoenergetic beams, so the exploration of this new and complicated field of physics had to proceed as if one studied the optical properties of materials without a spectroscope, just using filters whose characteristics had to be deduced from the results. This exploration was carried out in many laboratories, whose results sometimes duplicated, but more often inspired and supplemented each other. It would take a specialized study to ascertain the importance and impact of the contributions of any one person or group, but it is clear that the Copenhagen work, and Frisch's experiments in particular, had an important place in this exploration.

These papers are characterized by the imaginative use of experimental techniques and by a choice of problems relating directly to fundamental issues. The relation to other parts of physics was not overlooked, as, for example, in a paper which asks which of the Maxwell quantities, B or H, is probed by a neutron passing through magnetized iron.

Much of this work was done in collaboration with H. von Halban, who pursued neutron physics with his tremendous energy and sense of urgency, and G. Placzek, a man of great breadth of interest and erudition, at home both in experimental and theoretical physics, a remarkable character described in Frisch's recollections with humor and affection.

In all this joint work Frisch seems to have been landed with the more onerous tasks, such as grinding beryllium into a fine powder for the α–Be neutron sources. Presumably the reason was that he was good at doing such rough mechanical jobs, and was willing to do what was necessary.

It was a great experience for him to watch Niels Bohr resolve the paradox of the large absorption cross sections of nuclei for neutrons and the sharpness and closeness of resonances by his picture of the compound nucleus, which after its publication in 1936 dominated most of the thinking about nuclei.

Another great discovery which shook nuclear physics was the discovery of fission, in which Frisch became closely involved. He was spending Christmas 1938 with Lise Meitner, who, after being dismissed from her job at the Kaiser–Wilhelm Institute in Berlin–Dahlem, had settled in Stockholm. She received a letter from Otto Hahn, with whom she had collaborated during most of her career, reporting that one of the products of the collision of neutrons with uranium was barium, an element of about half the atomic weight of uranium, rather than the transuranic elements Fermi had thought he had formed by neutron capture in uranium. Hahn concluded that this must mean that the uranium nucleus had split into two large pieces.

Frisch and Lise Meitner discussed this apparently incredible information. They began to realize that this must be due to the mutual electrostatic repulsion of the charges in the nucleus, which tries to drive the nucleus apart, whereas the surface tension tends to hold it together. If by an external influence, such as the impact of a neutron, the shape of the nucleus becomes distorted, becoming elongated instead of spherical, the electric forces can become dominant, and the nucleus will split. Frisch suggested the term "fission" for this process. They also estimated the energy acquired by the fragments as a result of the repulsion, and found the staggeringly large figure of

200 MeV for the total energy. This agreed, in order of magnitude, with the energy difference, computed from the mass defect, between a uranium nucleus and two nuclei of about half the mass.

These conclusions were written up in a letter to *Nature*, which has become a classic. Its final text was drafted over the telephone after Frisch had returned to Copenhagen. It was shown to Bohr just as he was leaving for a visit to the Institute for Advanced Study in Princeton.

Frisch decided[1] to try proving the interpretation of the process by detecting the fission fragments, and it took him only two days to do so, since the great ionizing power of the heavy fragments makes them very easy to detect and to identify. He recalls that just as he completed the experiment, he received a telegram announcing that his father had been released from a German concentration camp and would be able to join him. The experiment was also reported in a letter to *Nature*, sent off jointly with the other one. It was five weeks before they were published, about the normal delay at the time. If Bohr had been in Copenhagen he could no doubt have persuaded the editor to give these letters some priority.

Meanwhile the news of the fission process was causing a sensation also in America. Bohr reported to a meeting what he knew, and also the paper by Hahn and Strassmann about their experiments arrived in the U.S. Several physicists, unaware of Frisch's experiment, looked for, and found, evidence of the fission fragments themselves. Because of some misunderstandings Frisch's finding was not immediately reported to Bohr, who became very concerned that Frisch and Lise Meitner might not be getting the recognition they deserved.

In retrospect, there is no doubt that Frisch was the first to show experimentally the existence of the fission fragments. However, this priority in time by a small margin should probably not be taken too seriously when several others did similar experiments shortly afterwards quite independently. There is no question about the importance of the note by Meitner and Frisch, with its clear discussion of the physics of the fission process in simple terms. It is an indication of the impact of this note that the term "fission" proposed in it immediately became accepted as the name of the new process.

During a subsequent visit of Lise Meitner to Copenhagen, she and Frisch performed experiments about the nature of the fission products, and showed the wide range of half–lives among them. Frisch proposed a simple empirical formula for the time dependence of the activity, resulting from a practically continuous distribution of half–lives.

---

[1]According to one account the experiment was started immediately after his return from Stockholm. In the book he says that this was only after Placzek had queried the Meitner–Frisch interpretation of the Hahn–Strassmann experiment and asked for a direct proof.

By now external events had again disturbed the peaceful and productive way of life in Copenhagen. Germany had annexed Austria, and Frisch had thereby become a German citizen. War was clearly imminent, and one could foresee that in a war Denmark was likely to be occupied by the German forces, and this would make it a very unsuitable place for a "Non–Aryan" emigré. Besides, the Danish police were beginning to make difficulties about renewing Frisch's permit, though he did not know this.

He had come to the conclusion that he would like to return to England, and made this desire clear to every visitor from England. In the end Oliphant responded with an invitation to Birmingham.

Frisch's trip to Birmingham was intended as a summer vacation, in which the possibility of staying there would be explored, and if it worked out he would return to Copenhagen for his belongings. However, during his stay the war started, and it would have been unwise to return, since, as a German citizen, he might not be allowed to enter Great Britain again. So he stayed, and his Copenhagen friends had to pack up his most essential belongings, vacate his flat and dispose of his half–paid–for piano. Meanwhile one of the Birmingham physics staff had left, so Frisch could be given a temporary appointment as a teaching assistant. The teaching duties did not occupy much of his time. The physics department was developing an interest in nuclear physics, but there was as yet little equipment. Oliphant and many of his collaborators were busy with radar research, in which foreigners were not allowed to participate.

His interest was still focused on the fission problem. By this time Niels Bohr had applied his ideas on nuclear reactions to the analysis of this problem, and had concluded that the observed slow–neutron–induced fission was entirely due to the rare isotope U235. This was comforting, because of the discovery that in the fission process secondary neutrons are emitted. If these can in turn cause fission the result may be a chain reaction. Because of the huge amount of energy this could release, it suggested the possibility of a new and devastating weapon.

Bohr's analysis, elaborated quantitatively in his paper with Wheeler implied that no explosive chain reaction was possible in natural uranium. Frisch was convinced by Bohr's arguments, but wanted to verify the assignment to the light isotope. For this purpose he tried to set up equipment which could separate the isotopes, or at least vary the composition. It appeared that the best method for this, and the only one that could be attempted with simple equipment, was thermal diffusion, amplified in a counter–flow column first used by Clusius and Dickel. By a coincidence, one of the most senior members of the Birmingham department, Dr. T. L. Ibbs, was one of the world's experts on thermal diffusion, but Frisch did not know this at the time. The experiment came to nothing because for UF_6, the only known gaseous uranium compound, the thermal diffusion coefficient, which depends on the intermolecular forces, happens to be practically zero. However, he remained preoccupied with thoughts about fission and its implications. These thoughts were further stimulated by his having to write an article about the subject.

The annual reports of the Chemical Society contained a section on nuclear physics, and Frisch and I had been asked to write this for the current volume. Naturally the part on the experimental situation, and on fission, fell to him. His report included the conclusion that there was no possibility of an explosive chain reaction.

But further reflection, stimulated by his interest in isotope separation, set him wondering what would happen if one could obtain a substantial quantity of pure U235. He raised this question in conversation with me, because I had recently worked out a formula giving the critical size for a chain reaction not involving the slowing–down of neutrons. I had been doubtful about the propriety of publishing such a calculation, but Frisch had convinced me that a weapon was not a practical possibility, and I sent the paper off for publication.

So now we tried to estimate the critical size of a sphere of U235. Bohr's analysis suggested that the fission cross section of this nucleus for fast neutrons should be about as large as the geometrical cross section, since fission would dominate over scattering and capture. This turned out later to be an overestimate, but since we had assumed a bare uranium sphere without a reflector that would scatter back some of the escaping neutrons, our estimate of "about a pound" was not too far out. This was much smaller than one would have guessed. Next we tried to estimate how much of the chain reaction would proceed before the heat generated would drive the uranium apart, and found that a substantial fraction of the available fission energy would be released.

This seemed an important enough finding to communicate to people in authority, and we wrote a memorandum (see page 187), divided into a technical and a general part, which, with Oliphant's help, was passed on to Sir Henry Tizard. This led to the work of the MAUD Committee, and later to the work of the Tube Alloys project. (Both names were just cover names for atomic energy.)

From then on Frisch spent most of his time on problems related to atomic energy. It was incongruous that, as a German citizen, and therefore an "enemy alien" he should be working on such an important and very secret project. In fact at first objections were raised, and it was suggested that he (and I, for that matter, though I had just become naturalized) should not be allowed any contact with the further development of the work, but this was abandoned after a short time.

One of the most urgent problems on the nuclear side was to check the guess about the fission cross section of U235. Since no sample enriched in this isotope was available, Frisch set up a source of photoneutrons from beryllium of an energy too low to cause fission in the abundant isotope U235. The rate of fission caused by these was low and hard to measure, but by counting continuously for 36 hours he did get an estimate which was lower than had been expected, showing that our guess had been somewhat high.

But carrying out these and other experiments in the Birmingham department was not easy, as the facilities and the technical staff were fully employed on urgent radar work. So Frisch moved to Liverpool in August 1940.

He was made welcome by Professor (later Sir James) Chadwick in his laboratory, and Frisch went to work on assorted problems. Some work was concerned with the nuclear cross sections relevant to the uranium chain reaction, but he also designed an instrument to measure the isotopic composition of a uranium sample by examining its α–ray spectrum. He also returned to the attempt to separate the isotopes by gaseous thermal diffusion, and confirmed that for uranium hexafluoride this cannot be done.

He made many new friends in the laboratory, including Joseph Rotblat, G. Pickavance and J. Holt. Life in wartime Liverpool was not very comfortable; there were frequent air raids. In addition, the city was in an area prohibited to enemy aliens, and he needed a special dispensation; another special permission was needed to ride a bicycle, and to be out of doors late at night. Even so he could not avoid breaking the rules occasionally, but no trouble resulted. He accepted all these complications with good humor, and his account of his Liverpool experiences reads like that of a hilarious adventure.

He remained in Liverpool until late in 1943, when it was decided to discontinue wartime work on atomic energy in Britain and to transfer many of the scientists engaged in this work, including Frisch, to the United States to work with the Manhattan District of the Army on the American atomic weapons project. He could not very well go there as an enemy alien, and steps were taken to make him a British subject. This would normally have been a protracted operation, but in this case bureaucracy proved its ability to move fast. He had to apply for naturalization, the application had to be granted by the Home Secretary, a certificate had to be issued, and with it he had to take the oath of allegiance, he had to register for National Service, and have his service deferred, a passport had to be issued, on which he needed a U.S. visa and an exit permit, and finally he had to be booked a passage on a boat and sail. All this was completed in 48 hours. Probably this could not have been achieved without Miss Vera Mayne, the secretary to the director of Tube Alloys, who took charge of the logistics, and dispatched a breathless Frisch to the various offices in the right order.

The Atlantic crossing, in the company of many other physicists about to join in the American work, was slow because the unescorted liner took a zig–zag course to evade the German submarines, but eventually the party reached the United States, and Frisch was assigned to the group which was to work in Los Alamos. Here he did not immediately join any particular project, but acted as a general adviser and trouble–shooter. With his experience and ingenuity in many techniques of nuclear physics, particularly concerning ionization chambers (to which he had made important contributions, including the "Frisch grid," which is still in use), he was able to give extremely valuable help in this way.

A very characteristic idea of his was to "mock up" the scattering of neutrons in a uranium block of complicated shape, a problem relevant to bomb design, by using the scattering of light in a partially absorbing mass of perspex of corresponding shape. This could have been used to check the theoreticians' calculations, which for general shapes were not too easy. In the end the procurement of the necessary transparent plastic with absorption took too long, and meanwhile enough confidence in the calculations had built up, so the project was abandoned.

As more U235 and plutonium became available, checks on neutron multiplication and determination of precise critical sizes became possible, and Frisch took charge of this work. It meant working with near–critical assemblies of fissile material, i.e., with an amount just enough not to sustain a chain reaction. This type of experiment required extreme care, because any unplanned movement of part of the fissile material might push the assembly past the critical point, and start a disastrous chain reaction. There were later to be two fatal accidents in this work, both caused by experts being carried away into making changes in the procedures on the spur of the moment, without adequate foresight. Even Frisch had a narrow escape when he was working with a bare mass of fissile material (appropriately code-named Godiva) without a surrounding scatterer. As he bent over the specimen, slow neutrons scattered back from his body added sufficiently to the multiplication factor to pass criticality. Only his alertness in observing the behavior of the indicator lights attached to the counters, and a quick reaction in pulling away part of the fissile material, stopped the chain reaction reaching a dangerous radiation level.

The most daring experiment of this kind was the "dragon," so-called because it was like tickling the tail of a dragon. The idea, proposed by Frisch, was to have a piece of fissile material which, because of a hole through its middle, was sub–critical. A plug, which fitted loosely in the hole, would make it supercritical. The plug was dropped through the hole from a suitable height, so that it passed through rapidly. The system would thus become supercritical for a brief interval, and during this time a chain reaction would develop but would not have time to build up to a dangerous level. Frisch said he was somewhat surprised to be given approval for this experiment, which was carried out safely and provided valuable information.

Work at Los Alamos was done with great urgency, as is natural for work in wartime on a weapon that could be decisive for the outcome of the war. Yet there was some spare time, and opportunities for meeting old friends and making new ones. In the strange scientists' town between the desert and the mountains of New Mexico, everything was a new experience to him: the climate and the beautiful landscape, the log houses remaining from the former boarding school and the temporary Army housing, the military policemen at the gates of the town and of the laboratories, the Indian maids and the Spanish–American technicians and laborers. But the feature which seems to have struck him most was the number of eminent scientists and intellectuals, some of them old friends and colleagues, which had gathered in this unusual community. It was not long before he had found congenial company for making music. He was certainly appreciated as a musician, and during one musical evening the remark was overheard: "This guy is wasting his time doing physics!"

He played the piano for the local radio station of the Los Alamos community, though for security reasons, as these broadcasts might be overheard outside, his name could not be mentioned and he was simply "our pianist."

There was time for walks in the hills (and for getting lost on the mesas where one was apt to be stopped by an unfamiliar canyon); there was time for resuming driving lessons, though not without some embarrassing incidents. In spite of the promising start mentioned earlier, he was never too comfortable driving a car with manual gear shift, because, as he commented later, he was too much aware of the mechanical operations, and this distracted him from the driving proper. This difficulty was lessened when automatic cars became available, but he never took much to driving.

Although the work of Los Alamos was done rather informally, a project of such size required meetings and discussions from time to time, and here, when others would idly "doddle" on the papers in front of them, Frisch indulged in his habit of drawing portraits of those around him. These sketches, of which some are reproduced in his autobiography, show considerably talent, perhaps inherited from, or inspired by, his father. Many of them bring out the personality of the subject very vividly, often by exaggerating the most characteristic features.

After the end of the war there was a general exodus from Los Alamos, as most of the scientists returned to their homes and to their normal occupations. Before leaving, the British contingent decided to give a large party in appreciation of the hospitality they had been shown. This included a play on local themes in the style of an English pantomime, in which Frisch was a great success in the part of an Indian maid.

He was invited to join the new Atomic Energy Research Establishment at Harwell. He returned to England early in 1946, to become the head of the Nuclear Physics Division of A.E.R.E., Harwell. While the buildings at the former airfield were being adapted for their new use, he visited the Chalk River Laboratory in Canada to acquaint himself with some of the problems of nuclear power, and then, for a while, had to work in the London office until Harwell became useable.

He occupied himself during that time with an ambitious calculation of the statistical fluctuations in chain reactions, which are important for accurate measurements of the multiplication factor, and for problems of reactor control. He describes this as "the most ambitious project in mathematical physics I have ever tackled" and comments that it "later got published with so many mistakes that I don't dare look at it and certainly never tell anybody where to find it." An internal Harwell report on this subject is indeed hard to find. Of a joint paper with D. J. Littler, published later, there exists a typed copy on which he wrote "Full of error." According to Sir Denys Wilkinson, these calculations about fluctuations turned out to be useful also in understanding the behavior of proportional counters, and were helpful in determining the optimum mode for using such a counter (Wilkinson 1950).

Frisch used his influence to create in the laboratory an informal atmosphere resembling that of a university laboratory. He recalls that he found that mahogany benches had been installed in the laboratories, and proceeded to drill holes in them at random, to encourage people to use them as real work benches.

He was not very interested in the administrative work of running the division, and says that he occasionally apologized to Dr. (now Sir Robert) Cockburn, his deputy, that he let him do all the administration. He adds, "But I think he enjoyed that." Cockburn certainly had greater ability as an administrator.

Cockburn describes this period in a letter to the author. He had to struggle with the problem of acquiring a background knowledge of nuclear physics after having concentrated for years on radar. Frisch, he says,

soon presented me with a golden key. "Just express energy in terms of the de Broglie wavelength, and all your radar equations are immediately applicable." This was typical of his approach. He would sit for hours apparently day–dreaming and then give a beautifully simple description of some complicated nuclear relationship. He once told me that he had solved the twelve–penny problem very easily by thinking in four dimensions!...There were endless competing working parties trying to establish a coherent program of research, and Otto was not alone in finding these tedious. I have known him to go to a meeting chaired by Cockcroft or Skinner, and decide after five minutes that it was of no interest to him and put his head on his arms and doze off. It fell to me to manage the division by default. It was not that I liked it; but that he didn't. It was not an onerous task.

Cockburn also comments about Frisch's autobiography: "...he never makes a harsh comment about anybody. He was completely impervious to the intellectual squabbles and power struggles rumbling all around him. He was a true scholar, completely at home in his subject. His particular talent was to ponder until he could present a problem in a form that admitted of a solution, the mark of a real physicist."

During the Harwell period Frisch also started to develop his talent for the popularization of science. Before leaving Los Alamos he had contributed an article on "The Tools of Nuclear Physics" to a popular account of atomic energy, and at Harwell he completed his book, *Meet the Atoms*, which he had started before the war.

He remained at Harwell less than two years. In 1947 he was offered the Jacksonian Professorship of Natural Philosophy in Cambridge. He took a little time to make up his mind, being a little awed by the status of the post, and by the names of the distinguished predecessors, among which he mistakenly included Rutherford.

Cambridge had once been one of the great centers for nuclear physics. After the death of Rutherford in 1937, and his succession by W. L. Bragg, the emphasis in the Cavendish Laboratory shifted to other fields, in which the laboratory later did work of great distinction. Nuclear physics went into a decline, although there were still a number of excellent nuclear physicists on the staff. One symptom of this decline was that after the war, when many other universities which were active in nuclear physics started to build large nuclear accelerators, the Cavendish was not able to attract financial support for such a development.

The appointment to the Jacksonian Chair of a distinguished nuclear physicist aroused in many minds the hope that he would rebuild nuclear physics in Cambridge, and become the leader of a new strong group in that field. This thought may even have been in the minds of the Electors who appointed Frisch to the chair. Such hopes were bound to be disappointed, because it was not in his character to be a leader. He was not the kind of person to fight for resources and grants, and to plan the provision of big machines, which in any case were not his favorite tools. To the nuclear physicists in the Cavendish it came as a surprise when, on one of his first days in the laboratory, he was observed queueing at the stores for parts for some gadget he wanted to build, instead of going around scattering brilliant ideas. No doubt he became aware of these expectations, and this imposed some strain on his enjoyment of the work in the Cavendish.

He did enjoy being there. He appreciated the high quality of the research and its breadth, and the presence of enthusiastic and bright students. For undergraduates his lectures seem to have been difficult, because he expected them to think too much for themselves. But for graduate students his ability to find simple pictures and analogies, and his insistence on simple explanations made his lectures attractive and inspiring. Dr. Alan Oxley remembers the verse: "If ever you're troubled by nuclear parity, Old Father Frisch will explain it with clarity."

Soon there was a group of research students who looked to him for inspiration, and to whom he communicated some of his attitude to physics and to experiments. His tendency was, once a student's problem was defined, to let him get on with it and not to interfere in the details. This meant sometimes that the student had to struggle on his own with obstacles, but that is a good way of learning.

He preferred to keep the number of collaborators modest. In a letter to Bragg (who was in hospital) in October 1962 he wrote: "This is a very busy term for me; I am giving for the first time a course Introduction to wave mechanics." For this course he had studied the history of the discovery of the electron. He goes on: "My group at the lab. has become too large: when we have tea I can no longer talk to all of them at once (more than a dozen)." Many of his students returned to Cambridge in 1974 to celebrate his seventieth birthday.

Most of his own work followed his interest in instrumentation. He had probably been the first to realize the importance of a pulse analyser (colloquially known as "kick–sorter"). This is a device which registers pulses from an ionization chamber, or later from a scintillation or solid-state detector, sorting them according to size, instead of the more primitive method of using a single recording device adjusted to accept only pulses of a certain size, which then has to be used many times to obtain a complete spectrum of pulse sizes. He had found the need for this in the Liverpool days in the course of developing his method of testing the isotopic composition of uranium samples by measuring the distribution of α–ray pulses.

At that time the problem was solved by an apparatus containing a few dozen valves, and this adventure into technology seemed quite appropriate in the context of the large wartime project. But for the everyday use of the nuclear physicist he preferred something simpler, and he designed, with the help of his collaborators, a mechanical device in which each pulse projected a steel ball diagonally up an inclined plane, with an impulse depending on the strength of the pulse. The parabolic path then followed by the ball brought it into one of 30 channels, so that each channel collected the balls corresponding to a particular pulse height. The pattern of balls therefore gave a histogram of pulse sizes.

G. G. Scarrott, who developed the electronics for the kick–sorter, comments: "I came to respect his formidable ingenuity. For example, he was responsible for the design of the parallel motion suspension system for the cue that projected the steel balls and the elegant technique for minimizing ball bounce by the use of a thin lamina damped by grease on the reverse side. He even wound the moving coil himself, so that he put its precise specification in the publication."

The device worked well, and was much admired. With its limited speed and accuracy it was, however, hardly more than a toy. Soon instruments serving the same purpose by purely electronic means became available, and came into general use. But it is likely that Frisch deserves the credit for recognizing the importance of such a device at an early stage.

Later, when the development of high–energy physics led to the extended use of bubble chambers, and therefore to the laborious work of analyzing tracks in bubble–chamber photographs, he became interested in the problem of devising semi–automatic methods for this purpose. One such design was described in a joint publication with A. J. Oxley in 1960.

Frisch was not satisfied with the speed that could be achieved with it, and decided that, by using a laser beam to scan the photographs, one could do better. The result was a much faster machine, to which he gave the name "Sweepnik" as its job was to sweep out the tracks, and its speed was reminiscent of the Sputnik. It was first described in a paper with J. Davies and B. S. B. Street in 1969. The success of this machine made it suitable for production on a commercial scale, and we shall return to that part of the story.

By this work on the design and use of such instruments he did, after all, contribute to making the Cavendish a center for modern particle physics. Today, as the front line of research shifts to ever higher energies, the necessary accelerators have become too large for any university laboratory, and research is done in large centers, such as C.E.R.N. or the Fermilab., by large teams comprising members of many laboratories. Much of this work involves bubble chambers, and the analysis of the photographs is shared between the various groups. The presence of track–measuring devices and of physicists experienced in supervising their use and understanding in simple terms the significance of the data is therefore essential for participation in this work. Later Frisch encouraged the development of a counter group for particle physics, though he never became closely involved himself in the physics of this field.

There were occasional excursions into other fields of physics, such as the work with D. N. Olsen during a sabbatical visit to Cornell University on coherent bremsstrahlung.

He also returned to his old field, the molecular beam, and under his guidance two students, H. Bellamy and K. Smith, set up apparatus with which to measure nuclear spins.

He was still intrigued by mechanical problems. According to K. F. Smith, "At one time he devised a method of supporting a horizontal platform three feet above the floor with six legs, all of them in compression, sticking out in various directions with ball and socket joints at each end to give good mechanical stability. He had a model made, which he used to keep on his desk in full view of any visitor who might be in the room waiting for him. Many times I have gone into the room with Robert to find the visitor on his knees hunting around on the floor for the ball bearings and the legs, for the slightest vertical displacement of the platform in the model resulted in a complete collapse of the system. The full–size system was built and used for a time, but we were never able to learn which legs to lengthen in order to move the platform in a particular direction—nor could Robert!"

Much of his energy went into the exposition and popularization of physics. His special talent for this had become obvious, and his early writings of this kind had been successful. He comments that none of his writings have become best sellers. "I enjoy writing and always get a little bit of fan mail which makes me feel good, and I tell myself that if just a few youngsters are attracted to physics by one of my books and become good scientists then it was well worth writing."

Besides his four popular books on physics, he wrote numerous review articles and comments intended for physicists and edited *Progress in Nuclear Physics*. He also wrote many articles for the intelligent layman. Perhaps the most often quoted, and certainly the most often reprinted of these, is the humorous article *On the Feasibility of Coal-Driven Power Stations* written originally for a small house journal circulating at Harwell, which makes fun of the opposition to nuclear power by pretending that the use of coal was to be introduced as an innovation.

The great variety of journals in which these articles appear illustrates how much he was in demand. The list, which is probably not complete, does not of course include lectures, broadcasts (except those printed) or television programs.

When the Sweepnik had proved itself, it was decided to set up a firm to manufacture and sell this machine. Laserscan Ltd. was incorporated in 1969, with Frisch as chairman of the board (he modestly says "as the oldest"). He remained chairman until his death. After his retirement from his university chair in 1972 he continued doing research in the firm, and continued contributing new ideas to the sophistication of their products, including the holographic "zone plate," which has since found other uses in addition to those in track scanning.

In the chair at meetings he was hardly the conventional type of business man; he is even known to have dropped off to sleep during particularly uninteresting business. But he was regarded with deep affection by everyone in the firm.

He maintained many other interests besides the work for Laserscan. One of the problems he worried about was the desirability of nuclear power. These thoughts were stimulated by the views of his old friend, Lew Kowarski, who had come to the conclusion that there should be a moratorium on building any further nuclear power stations. Frisch realized the importance of the problem, but found it difficult to form his judgment. In general, his feeling was that his ability lay in explaining the facts to people by way of background, rather than trying to influence their views. But he needed to make up his own mind, and he intended to get rid of some of his commitments so as to have more time to think about such basic problems. But the unfortunate accident which led to his death happened before he had implemented this resolution.

Cambridge, was, since Vienna, the first place where he could stay for more than five years, and apart from Copenhagen the longest stay anywhere was three years. So Cambridge became really the first place for him to settle, and he lived there for 32 years. In Cambridge professors are not automatically college fellows, but have to be elected. Frisch was elected by Trinity College. He recalls how a rather frivolous response to a request for a *curriculum vitae* had helped toward that decision. Unfortunately no copy of this document seems to have survived.

He was popular with everybody in the college as well as in the laboratory, and he was on very friendly terms with many, not only those in whose company he made, and enjoyed, music. But in spite of his cheerful friendliness he was rather reserved, and it is difficult to find people who, in those early Cambridge days, felt they had really come close to him.

In 1948 his parents came to live in Cambridge. His father had been taken to a concentration camp in Austria in 1938, but was released when offered a post in a publishing house in Stockholm and permission to enter Sweden. His parents stayed in Stockholm during the war, and moved to Cambridge on his father's retirement, but his father died shortly afterwards. His mother died two years later. Otto Robert had been an affectionate son, whom circumstances had kept from contact with his parents for many years, and he enjoyed their company at least for the short time they had left.

He had a great sense of family, and was said to have an aunt in every town. Indeed it was a frequent experience of his friends to tell him about a forthcoming trip to some out–of–the–way place, perhaps in South America, and get the response "If you have time, call on my aunt there."

The most distinguished aunt, Lise Meitner, came to live in Cambridge when, at the age of eighty-one, she retired from her research work in Stockholm, and Robert could, once again, share with her his thoughts about physics, until, near the age of ninety, her strength began to fail.

But by then he had become a family man in another sense. He married, in 1951, Ursula (Ulla) Blau, an artist, a Viennese like himself, and a charming person who shared his love for music and many other of his attitudes. She introduced him to the visual arts, for which he had not previously had much appreciation, in spite of his father's example, and through her he widened his circle of friends beyond the laboratory, the college and music. They had a daughter and a son, who has also become a physicist.

They were a happy family. There is an amusing story about the house in which they settled when their first, rented house ceased to be available. On a Sunday walk in the neighborhood they saw a house which they all agreed was the kind of house that would suit them. So Frisch rang the door bell and asked the lady who came to the door whether by any chance they were thinking of selling their house. It turned out they were thinking about it, but had not yet taken any steps to put it on the market, and before long the Frisch family was installed in the house.

He remained active and enjoyed life until the accidental fall which caused his death after a short period in hospital. He just missed his seventy–fifth birthday, to which he had been looking forward.

It is hard to do justice, in a brief memoir, to the charm of his personality, but fortunately he was able to complete his autobiography which can make up for this inadequacy.

In the writing of this memoir I have been assisted by information or comments from Mrs. L. Arnold, Professor E. H. Bellamy, Sir Robert Cockburn, Professor G. W. Hutchinson, Dr. A. J. Oxley, Sir Brian Pippard, Mr. G. G. Scarrott, Professor K. F. Smith, Mr. P. Woodford, Sir Denys Wilkinson and particularly Mrs. Ulla Frisch.

References

Bohr, N. and Wheeler, J. A., 1939 *Phys. Rev.* **55**, 426–450.
Wilkinson, D. H., 1950 *Ion chambers and counters*, p. 144. Cambridge: Cambridge University Press.

Paul Adrien Maurice Dirac, 1902–1984

Paul Adrien Maurice Dirac, who died on October 20, 1984, was one of the great physicists of the century. Born on August 8, 1902, he obtained an engineering degree at Bristol in 1921, but his outstanding mathematical ability was soon recognized, and after concentrating on mathematics at Bristol for two years he was offered a grant for graduate studies at Cambridge. There he worked under R. H. Fowler; five of his papers written during that time show his already firm command of relativity, quantum theory and statistical mechanics.

His first paper on quantum mechanics was written while he was still a student. He had obtained from Fowler a proof copy of Werner Heisenberg's paper and immediately saw the significance of the new ideas. In particular, he saw the fundamental role of noncommutative algebra and started extending the ideas even before Heisenberg's paper was published. From then on he developed the structure of quantum mechanics in his own way, introducing his own characteristic notation, much of which came into general use. He showed how to combine both Erwin Schrödinger's and Heisenberg's approaches into one common system by using what we now call transformation theory. In this context he introduced the bracket notation and later found it useful to split the brackets in two parts, which he called "bra" and "ket," further pieces of terminology that have come into general use. (I believe he was not aware at the time of the alternative meaning of bra.) The inclusion of continuous variables in the general transformation theory was made possible by the "delta function," another invention of Dirac, which horrified the purists until it was made respectable by Laurent Schwartz.

Dirac took many more important steps in completing the foundations of quantum mechanics, including the first discussion of the emission and absorption of radiation, and of dispersion. He introduced the density matrix, which besides being a convenient tool for quantum statistics is an important concept for the interpretation of quantum mechanics. He saw, independent of Heisenberg and Enrico Fermi, the connection between the symmetry of the wave function and the Pauli principle, leading to Fermi–Dirac statistics.

In 1928 he formulated what is now called the Dirac equation, the relativistic wave equation, which became the keystone of atomic quantum mechanics. Considered as a wave equation for one particle, it is beset by the difficulty of the negative–energy states—but by another stroke of genius he conceived the hole theory, in which all negative energy states are normally occupied. At first he thought the holes, or empty places in the negative–energy region, were protons, but this hypothesis could not be maintained, and they turned out to be the positrons.

In 1932 Dirac succeeded Sir Joseph Larmor as Lucasian Professor of Mathematics in Cambridge (where, traditionally, theoretical physics is regarded as part of mathematics) and he held this position until his retirement in 1969. In 1971 he became professor of physics at Florida State University.

By the early 1930s, quantum mechanics was essentially complete, and it owed an enormous debt to Dirac, who was recognized with an award of a Nobel prize in 1933.

His work would have secured him a position amongst the great names in physics, even if he had done nothing after 1933, but he continued without

Paul Adrien Maurice Dirac, Wolfgang Pauli and Sir Rudolf E. Peierls. (Photograph published courtesy of AIP Emilio Segrè Visual Archives.)

slackening his pace. Some of the later work involved applications of quantum mechanics, but most of it was trying new departures to advance into as yet uncharted areas of physics. His suggestion of a magnetic monopole ("one would be surprised if nature had not made use of [this possibility]") and of the change with time of the gravitational and other constants of physics have attracted much attention, but there is as yet no final verdict on their validity. Many other highly original and ingenious ideas have not made any impact on physics, but it would be rash to assert that we may not learn to make good use of them one day, or that we may not see their significance, as was the case with the brief aside in his book, *The Principles of Quantum Mechanics*, which turned out to contain the idea of Feynman's "integral–over–paths" method.

Dirac did not supervise many research students. perhaps because he did not like to discuss unsolved problems or speculate what the answers might be. But any students who asked questions would get clear and patient answers.

He received, of course, numerous honors and distinctions, including the Order of Merit—the highest and most selective British honor for great minds.

Dirac

T he first time I met Paul Dirac was when he visited Leipzig in 1928 to give a lecture. As a very young graduate student I was detailed to take him to the theatre. I do not remember what was the play we saw, or what was his reaction to it, but I remember being terribly worried by his refusal to part with his hat. It was then obligatory in German theatres to leave hats and coats in the cloakroom. It was summer, and we wore no overcoats, but Dirac kept his hat, as was the custom in England. I worried all through the evening about the disastrous consequences that would follow, but of course nothing happened.

His visit to Leipzig was one of his many journeys. He was an inveterate traveller. The story is told how he arrived at the Russian border with a visa valid for a different entry point, and had to wait some days in a miserable little border village for the paperwork to be sorted out. There was not always time to learn the language of the countries he was visiting, but the first words that he wanted to know in a new country were how to ask for a glass of water. He never took to alcoholic drinks. A colleague who noticed this, and the fact that he did not smoke, asked him whether he had any vices. "No obvious ones" was Paul's reply.

People called him very taciturn, but he could be very articulate when he had something to say—he just did not make idle conversation. He also did not share common prejudices. At a time when everything Russian was anathema, he questioned why each particular item was regarded as wrong, and this often caused raised eyebrows.

He could surprise us in other ways. One of the great surprises for his friends was his marriage to a lady who seemed so completely different from him in almost every way.

It was surprising to see the serious interest he took in domestic affairs, including the garden. He was a keen gardener, and tried to deal with horticultural problems from first principles, which did not always lead to good results.

The many topics outside quantum mechanics which attracted his interest included isotope separation. He not only thought about the theory, but carried out experiments. In fact he invented a method of isotope separation, and tried to make it work, using equipment put at his disposal by Kapitza. This work was abandoned when, in 1934, Kapitza failed to return from Moscow. The idea was revived during the war, when isotope separation acquired great practical importance, and a group in Oxford showed that the method was workable. They received much sound practical advice from Paul. The project was not pursued because other methods appeared to be more economical.

His general ideas about isotope separation became very useful to the work on atomic energy during the war. In unpublished notes he introduced the notions of "separating power" and "separation potential," which are now used generally, and help to simplify the discussion of plant design.

He also took an interest in other aspects of atomic energy work. In the course of time these contacts stopped. I believe this was because he was beginning to feel that atom bombs were not a matter he wanted to be associated with, and who would blame him?

When he was awarded the Nobel Prize he told Rutherford that he did not want to accept it because he disliked publicity. Rutherford said, "A refusal will get you much more publicity," and then he accepted.

But he would never compromise his principles and would stick absolutely to what he saw as right. No doubt the best comment on him was the remark by Niels Bohr: "Of all physicists, Dirac has the purest soul."

Dirac's Way

F riends and acquaintances of Paul Dirac were often struck by his surprising and sometimes "odd" reactions on topics arising in a conversation. Yet, when you had time to reflect, it became clear that his remark was the natural and logical response, and that it was only the automatic and unthinking associations of everybody else which made us expect something different. The same quality can be seen in his physics. The similarity is so close that, as I shall show, many of the famous anecdotes (some perhaps apocryphal, but nonetheless characteristic) can be put in parallel with some of his papers.

Take, for example, the well–known story, told to me by H. R. Hulme, of the pills in the bottle. Hulme apologized for the rattle in his pocket by explaining that a bottle of pills was no longer full, and therefore made a noise. Dirac's comment: "I suppose it makes the maximum noise when it is half full?" He had seized on the fact that the bottle was silent not only when empty, as is obvious, but also when completely filled. This thought is similar to the idea underlying his "hole theory." When I first heard this story, without the date, I thought it would have been very interesting if this conversation had preceded the hole theory, so that the phenomenon of the bottle might have led to the hole theory. However, it was much later, so Dirac merely repeated the train of thought that had given rise to the hole theory.

On another occasion, tea–time conversation concerned the fact that, of the children recently born to physicists in Cambridge, a surprising proportion had been girls. When someone airily remarked, "It must be something in the air!" Dirac added, after a pause, "or perhaps in the water." He had taken the phrase "in the air" not in its loose conventional meaning, but literally, seeing a possible application. This trend is reflected in much of his work, perhaps first in his picking up Heisenberg's observation that the quantum variables did not commute which, to Heisenberg, seemed an ugly feature of the formalism. Dirac showed instead that it had a very significant place in the new theory.

The importance of looking at the real meaning of a conventional remark extended even to comments on the weather. Jagdish Mehra reports how a visitor, sitting next to Dirac at college dinner and anxious to start a conversation, said, 'Very windy today.' Dirac got up and went to the door, so that the visitor began to fear he had somehow offended him. Dirac opened the door, looked out, and, resuming his seat at the table, said, 'Yes'.

Another characteristic is illustrated by the story that, during a visit to Copenhagen, it was decided that Pauli was putting on too much weight, and Dirac was asked to watch that Pauli did not eat too much. Pauli entered into the spirit of the game and asked Dirac how many lumps of sugar he was allowed in his coffee. "I think one is enough for you," said Dirac, adding after a moment: "I think one is enough for anybody." After some further reflection: "I think the lumps are made in such a way that one is enough for anybody."

Such faith in the orderliness of the world is reflected often in his writings, above all in the remark in the paper pointing out that a magnetic monopole would not contradict the known laws of quantum mechanics: "One would be surprised if nature had made no use of it."

The same idea is also reflected in his conviction that the true theory must have mathematical beauty. He says in his 1980 paper, "Why we believe in the Einstein theory,"[1] that the real basis for believing in general relativity does not lie in the experimental evidence. "It is the essential beauty of the theory which, I feel, is the real reason for believing in it."

It is said that he was once writing a paper at the dictation of Niels Bohr. (It was Bohr's usual habit of dictating a draft to one of his collaborators, while walking around the room.) At one point, Bohr interrupted himself and said, "Now I do not know how to finish this sentence." Dirac, so the story goes, put down his pen and said, "I was taught at school that you should never start a sentence without knowing the end of it."

He seems to have observed this injunction himself. At least, all his papers reflect very great tidiness of expression. Few drafts seem to have survived, but there were probably very few alterations or corrections. When he talked in a seminar about a recent paper, he explained his work almost verbatim in the same terms as in his published paper, the form he had chosen as the clearest expression.

One anecdote is instructive on how legends form. Tyabji, a retired lawyer from India, who was studying theoretical physics under Dirac, discovered

---

[1]*Symmetries in Science*, ed. B. Gruber and R. S. Millman. New York: Plenum Press, 1980.

that Dirac was very anxious to meet E. M. Forster, the novelist. He invited both to dinner, and when they were introduced, Dirac asked, so Tyabji told me, "What happened in the cave?" (referring, of course, to Forster's *Passage to India*), with Forster replying, "Nothing." This answer apparently satisfied Dirac, and he asked no more during the rest of the evening. I later heard another version, according to which Forster's reply was, "I do not know." I thought the earlier version, directly from the host, was likely to be right. Later, when I met Dirac, I asked if he remembered the occasion. He did remember it very well, and both versions were inaccurate. In fact, he had asked whether there could not have been a third person in the cave, which would have meant that neither of the protagonists had told an untruth, and Forster replied, "Absolutely not. There was no other person in the cave."

As usual, Dirac had thought of an interesting possibility which had occurred to nobody else. He did this, of course, in so many of his papers, that there would be no point in selecting specific examples.

These few stories might suffice to illustrate my point that unexpected and endearing turns of his conversation were typical of a mind which gave us so many fundamental contributions to physics.

Physics and Homi Bhabha

omi Bhabha was a highly talented scientist, and also a painter of distinction. But he will be remembered above all for the part he played in the development of science and atomic energy in India.

Bhabha came from a well–to–do and sophisticated Parsee family. After studying mechanical sciences at Cambridge, he started research in theoretical physics in 1930. This was an exciting time for physicists and Bhabha became attracted by the phenomenon of cosmic rays. He wrote his first paper on this subject, which remained his dominant interest thereafter. He developed the theory of cascade showers (simultaneously with Carlson and Oppenheimer) which explained many of the otherwise mysterious observations, and led to the conclusion that the penetrating component could be explained only by a breakdown of the quantum theory, or by a new particle, which indeed turned out to be the muon. His many papers on other subjects, such as electron–positron collisions ("Bhabha scattering"), new kinds of wave equations, and meson theory, all had potential applications to cosmic rays, which he kept clearly in mind.

By the late 1930s Bhabha was well established in Cambridge, while also visiting many other European institutions. We do not know how long he would have continued in Britain, but he happened to be in India on holiday when the Second World War broke out and prevented his return. He became a Reader, and later a Professor at the Indian Institute of Science in Bangalore, and built up a strong group, mainly on cosmic–ray research.

Bhabha, however, was also endowed with formidable administrative energies. He saw the need for a national institute devoted to fundamental research, and on his initiative the Tata Institute was founded, with him as direc-

This is a review of *Homi Jehangir Bhabha: Collected Scientific Papers*, edited by B. V. Sreekantan, Virendra Singh and B. M. Udgaonkar. Bombay, India: Tata Institute, 1986. For the complete reference, please see the Acknowledgments section.

tor. It became an institution of high standard and exerted a profound influence on science in India. Later he realized the importance for India of nuclear power and persuaded the government to set up the Atomic Energy Establishment in Bombay, of which he was put in charge. He also became the chairman of the Indian Atomic Energy Commission. As he was also chairman of the Scientific Advisory Committee to the Cabinet, devotees of Gilbert and Sullivan might well have thought of him as the "Poo-Bhabha!" But he managed to carry out all these functions efficiently.

He did not, however, regard science from a narrow, national point of view—he was also concerned with the international relations between scientists. For three years he was president of the International Union of Pure and Applied Physics, he chaired the Geneva Conference on the Peaceful Uses of Atomic Energy in 1955 and also served in the International Atomic Energy Agency. Bhabha's death at the age of 56 in a plane crash on Mont Blanc was thus a sad blow not only to India, but also to international science.

Two Mathematicians

J ohn von Neumann and Norbert Wiener were two mathematicians of outstanding ability. Their careers had a few features in common: both were infant prodigies; both started in the most abstract fields of pure mathematics, and derived some of their inspiration from David Hilbert in Göttingen. Both later extended their interest to topics that had practical applications, and their work had indeed a major impact on practical matters. Both were of Jewish origin.

Yet their lives, their attitudes, and their personalities were completely different. The contrasts, and the thoughts they induce about the function of scientists, no doubt provided the impetus for Steve Heims's embarking on this combined operation. It is unfortunate that from the very beginning he can be seen to be anything but neutral. While Wiener is described with great warmth, von Neumann is mentioned mostly in a tone of disapproval.

John von Neumann was Hungarian, the son of a banker, who had established himself in Budapest society and had acquired a title of nobility, which the son translated into the German "von." John graduated in mathematics in Budapest, and in chemical engineering in Zurich. He later worked in Göttingen, where his main interest was the foundations of mathematics. He followed Hilbert in a program to prove that the whole logical system of mathematics was free from contradiction. This was before Kurt Gödel proved that such a demonstration was impossible. He made many other important contributions to pure mathematics. In the mathematical circles of Göttingen, the late 1920s were also a time of great excitement over the new quantum mechanics, and von Neumann saw the challenge in this for a mathematician with a flexible

This is a review of *John von Neumann and Norbert Wiener: From Mathematics to the Technologies of Life and Death,* by Steve J. Heims. Cambridge: MIT Press, 1980. The book review is reprinted with permission from *The New York Review of Books.* Copyright © 1982 Nyrev, Inc. For the complete reference, please see the Acknowledgments section.

mind. He wrote several important papers and a book about the mathematical foundations of quantum mechanics, and some of his ideas are still fresh today.

He had a deep understanding of the practical physical side of the problems, yet his approach is very much that of the mathematician—some of his arguments turned out to be based on assumptions which to the physicist were not necessary or not reasonable. The impact of his work on the development of quantum mechanics was therefore not as crucial as Heims suggests. By this time, however, he had impressed everyone with whom he came into contact by the brilliance and the phenomenal speed of his reasoning, the width of his knowledge, and his memory.

He had no resemblance to the traditional image of a mathematics professor—he was always immaculate, with expensive tastes, sociable and gregarious, always courteous and reasonable, and willing to listen to an opposing point of view. He seemed cold and lacking in emotions, yet he was propelled by tensions among which ambition was certainly a strong one. Heims repeatedly claims that some of these characteristics derived from his origin as a banker's son in Budapest, but I have known many a banker's son with very different tastes and attitudes.

Heims also muses about the fact that there were, in von Neumann's generation, so many outstanding scientists from upper–middle–class backgrounds in Budapest. At least one other, the Nobel laureate Eugene Wigner, came from the same gymnasium and the same class as von Neumann. The school, says Heims, cannot be responsible for this phenomenon, because "many schools of this type existed in Europe." (One might have looked for the influence of one or a few outstanding teachers, rather than the "type" of school.) Instead Heims invokes the feeling of insecurity in Hungarian society as a driving force, quoting von Neumann on the necessity of "producing the unusual or facing extinction." Such tensions might indeed keep a young man on his toes, and drive him to great effort, but they are hardly likely to produce such outstanding abilities as von Neumann's or Wigner's.

In Berlin, where von Neumann spent some time after Göttingen, he wrote his first paper on the theory of games. This was a study of the logical structure of the decisions made by a player in a game such as poker, in which the aim is to make moves that will give the greatest chance of financial gain and the least risk of loss. The rules resulting from this theory depend on considering what the adversary may do, and ensuring that the player will not be left at a disadvantage by the most skillful play of his opponent. This paper and later work by von Neumann, and his book with Oskar Morgenstern, *The Theory of Games and Economic Behavior*, form the basis of game theory as a branch of mathematics. It is an interesting discipline, in clarifying the logical nature of the problem of choice, but it has not yet led any gambler into making a fortune, because in practice the job of enumerating all the situations that may arise from any strategy of the opponent is prohibitive except in the simplest examples, such as a game of tick–tack–toe on a small field.

In 1933, with the advent of the Nazi regime, von Neumann moved to Princeton, where he became, and remained, a member of the new Institute for Advanced Study.

Norbert Wiener was American, the son of a Polish Jew with a very strong personality who became a professor at Harvard. The father attempted to mold his son's personality according to his principles. Norbert entered college at the age of eleven, and obtained his Ph.D. in mathematical philosophy at Harvard when he was eighteen. During a postdoctoral year at Cambridge, England, and Göttingen he came under the influence of Bertrand Russell, G. H. Hardy, and Hilbert. While these contacts led him into very abstract fields of mathematics, he soon discovered the pleasure to be derived from applying mathematical tools to problems of practical interest. The first of these applications concerned Brownian motion, the irregular motion of tiny particles of dust immersed in a liquid, caused by the buffeting they receive from the motion of the molecules of the liquid. He found that the concept of Lebesgue integration, a highly esoteric mathematical method, was appropriate for handling this problem. Other abstract ideas that he picked up, such as functional integration—which he pioneered—proved unexpectedly to have very practical applications in statistical mechanics and quantum theory. One suspects that, without foreseeing these precise applications, he knew intuitively that such methods were likely to prove applicable.

Wiener was very much the eccentric professor: absent-minded, unsure of himself in spite of his early recognition, given to moods. Heims reports many anecdotes of his generosity to students, his concern—sometimes perhaps quixotic—with ethical principles, as when he resigned from the National Academy of Sciences in protest against its official powers and conformity with establishment views.

Wiener was not as great a wizard in mathematical reasoning as von Neumann, but he was more inventive. His best–known achievement is the foundation of a branch of science he called "cybernetics," which describes the functioning of control systems. Whether in the operation of a thermostat that keeps a constant temperature in a house or in a laboratory device, or in the coordination of the eye, brain, and hand of the driver who keeps his car following the road, we continuously meet examples of control mechanisms in which information received about a deviation from the intended object (temperature, or position and direction of the car) is fed back to cause changes which counteract the deviation. This principle of "negative feedback" is one of the essential ingredients of cybernetics, which explores the effectiveness and stability of such systems. Wiener's work did much to clarify the foundations of this field, which today has many applications through the growing use of automation. Attempts were also made by Wiener and others to use similar ideas in economics, and in psychology, in order to understand the functioning of the brain.

In some of these more general discussions von Neumann and Wiener, who got to know each other well, were collaborating. Indeed von Neumann's game theory also had some application to economics and to the problem of the brain. Heims devotes much effort to comparing and confronting the two approaches, but in my view he exaggerates the contradictions between them; they differ less in their fundamental assumptions than in their emphasis on different aspects. Neumann's approach seems deterministic, in that all possible moves in a game are to be foreseen and listed, whereas Wiener's treatment is statistical, allowing for unpredictable external influences. Yet Neumann's method must also allow for the unknown choices of the opponent in the game, and he also realized that in practice it is not usually feasible to have command of all possible situations; he is quoted as having stressed the impossibility of mechanizing decision-making.

Another difference is that in the theory of games there is an obvious single criterion: a move is good if it leads to an increase in the expectation of financial gain, and bad otherwise. Heims makes much of this single motive of greed. He does not seem to note that the theory applies equally directly to a situation in which other factors of merit apply, as long as one can define clearly the relative importance to be attached to each. There is a problem when there is no such clear principle to resolve the conflict between, say, moral and economic requirements. The same problem arises in Wiener's approach. A control mechanism can take account of diverse desiderata, but only if a decision is made about the weight to be given to each.

The lives of these two mathematicians, like those of many others, were profoundly affected by the Second World War, which Heims refers to as the "Watershed." He is fond of citing negative statements, about things which did not happen, with an air of surprise. Thus he records that von Neumann's and Wiener's attitudes toward formal logic, probabilities, time and process, limits and errors, did not undergo any abrupt changes. What did change were their activities. Von Neumann, in particular, became involved as a consultant in many military research projects including the US Army Manhattan District—the atom bomb project. He became a frequent visitor to Los Alamos, where the weapon design work was carried out, and his help with the many intricate mathematical problems arising from it soon became invaluable. He was more than a mathematical troubleshooter; with his clear mind and quick perception of issues he became increasingly useful on committees and in general administrative problems.

As a result he remained a trusted adviser to many government and private organizations after the war. This included serving on committees of the AEC and later on the commission itself. He had been interested in the early ideas about the hydrogen bomb, pressed by Edward Teller at Los Alamos, and he later favored the speedy completion of the hydrogen bomb program, and participated in its theoretical work. Regarding the political implications of all this, he was a hawk, favoring increases in the development of modern weapons, opposing disarmament, and regarding the Soviet Union as unavoidably the enemy of the United States. He is quoted as advocating preventive nuclear war against the Soviet Union, and while this is hearsay ("a number of...friends and associates") it is believable.

At Los Alamos he had become interested in the use of mechanical punch-card machines as aids to computing, and soon he was involved in the design of electronic computers, making important contributions to their development. High–speed computing was necessary for the work on the hydrogen bomb, but I do not believe Heims is right in seeing this as the major motivation for von Neumann's work on computers. He would have been fascinated by the challenge of the computing problem in any case.

His work on nuclear weapons and his views on weapons policy (with which indeed many of his colleagues disagree) are the basis of Heims's hostility. Heims does not like nuclear weapons (who does?), but surely there is no need to distort the facts relating to them. Historians of World War II will raise their eyebrows at his assertion that "Truman's deliberate tactic was to prevent Japan's surrender until the Americans had a chance to drop the atomic bomb"; radio biologists may be surprised by the statement that von Neumann "had increased his own chances of getting cancer by personally attending nuclear weapons tests and by staying at Los Alamos for long periods."

The reader senses the book's attitude to von Neumann as early as his school days. After mentioning the hated communist regime of Bela Kun (no doubt an important cause of the strong anticommunist and anti–Soviet emotions of so many Hungarian emigrés), Heims writes that "the available evidence indicates that [von Neumann] was not at all attracted to idealistic leftish regimes [another of the typical negative statements] and that his opinion was asked about how one should deal with his less reliable classmates." The evidence does not seem to tell us what was the nature of his answer, if any.

Von Neumann was fond of logical reasoning, and this is exaggerated by Heims into the statement that he tried to resolve all problems of life by formal logic. "This confidence in the power of logic, which after all is mechanical, to resolve the problems of human life is merely another form of the optimism and faith in technology that von Neumann inherited from his early years during Hungary's economic take–off." The same inclination to see connections between abstract epistemological considerations and practical matters shows up in a statement referring to von Neumann's theory of measurement in quantum mechanics, where he uses the phrase that the measuring instrument is merely an extension of the observer:

> During and after the war, when working with computers, von Neumann regarded these machines as extensions, supplements to his mental powers.... The continuity of this theme makes it possible to speculate that he also regarded nuclear weapons as means of self-aggrandizement; as extensions of himself.

What an impressive non sequitur! But it would take too long to summarize the chain of reasoning that leads to the obscure conclusion that "...von Neumann's interest in the bomb as an unusual tool he had helped produce reflected a desire for a long life, but focused it on the political scene in a tribal, parochial way."

Much is read into von Neumann's cooperation with the military and industrial establishment: "He seemed more like a talented nineteenth–century banker to some Central European court, aspiring to aristocracy, than a mid-twentieth–century American scientist accustomed to egalitarian and democratic values. Perhaps this was the mark that centuries of persecution of European Jewry had left on John von Neumann." He is blamed for the use of game theory by American strategy analysts, including those in the Rand Corporation. Without following up the many references, one has the impression that such conclusions are exaggerated. One even suspects there may be a confusion in Heims's mind between the theory of games and war games, the latter being an exercise that has gained in popularity since the availability of computers.

In criticizing the application of game theory to economics, Heims stresses that it rests on the traditional assumption of the scarcity of resources, and quotes Walter Weisskopf, who, he says, shows that this is an attitude characteristic of modern industrial society. However, the words he quotes from Weisskopf actually say, "There is...a sense in which the scarcity principle is universally valid because it is rooted in the conditions under which human beings exist."

The heading of one chapter "Von Neumann: Only Human, in Spite of Himself," adds to the flavor of the von Neumann part of the biography. But this chapter contains what is perhaps the best and one of the simplest sentences about von Neumann in the book: "Johnny von Neumann, who knew how to live so fully, did not know how to die."

Wiener had also been called in to help with military projects during the war, including work on directing anti–aircraft fire, which fitted in with his then developing ideas about control systems. He received the news about nuclear weapons and their use on Japan with horror and shock. His immediate reaction was to refuse from then on to provide information for military purposes. "I do not expect to publish any further work of mine which may do damage in the hands of irresponsible militarists." In many lectures, articles, and books he stressed the view that scientists carry a responsibility for the way in which their discoveries or inventions are used; he said, in Heims's paraphrase: "If an inventor regards his government as irresponsible in its military policies, then it is his moral responsibility to withhold his ideas on weaponry from that government."

Heims clearly approves of Wiener's attitude, and the parts of the book about him picture him as an attractive, though complicated, human being. This presentation is weakened only by the frequent cross–references to the deplorable attitudes of von Neumann, and by the tendency to seek motivations and connections between the mathematical concepts in Wiener's work and his attitudes to other matters. Heims makes a strange remark about Wiener's mathematical theories, "in which he perhaps incorporated his own image." Much is made of the pain he felt when confronting an unsolved problem, but this experience is surely common in mathematics, indeed in most creative activities. These, however, are minor points in a generally very acceptable presentation of Wiener's thoughts on general issues and his way of life.

In the epilogue Heims comes out as opposed to science in general. He refers to the "naive but conventional conviction...that scientific knowledge, confined to the accepted methodology of science, has some kind of absolute character as 'truth,'" and adds, "even as the only kind of truth acceptable." In this way he runs together a moderate and a provocatively extreme statement, and by quoting refutations of the latter, gives the impression of having disposed of the former. He quotes Einstein as opposing the university training of physicists, and recalls that the early members of the Royal Society in the seventeenth century were largely amateurs, without seeming to be aware of the quite exceptional ability of Einstein to teach himself modern physics, or of the shortage in our days of "gentlemen of leisure" who could command the complexities of modern science, let alone the necessary equipment.

Comments in this spirit are scattered throughout the book. "Rarely does a serious mathematician follow his muse as a freelance artist might, wholly outside the establishment. Mathematics is distinctly useful to the state and to industrial enterprises, and directly or indirectly they supply the mathematician's salary. Thus aesthetic pleasure and worldly practicality dovetail *all too neatly*" (my emphasis).

Heims's closing sentiment is that our civilization let its mode of existence "be determined by science and technology" but was awakened to realize that "what dominated it, after all, was people—play and affections, politics and passions, pleasures and pains." So science and technology did not dominate after all. May they still plead "Not Guilty"?

Some of the fifteen illustrations are excellent and include characteristic pictures of the subjects, many of them shown in assorted company and in not very interesting poses. One has no visible connection with either of the mathematicians. It is labeled "Quantum Theorists." Assuming it is meant to typify the species, I feel honored, as the photographer, by its selection.

Reading *John von Neumann and Norbert Wiener* takes time, not only because of its 414 pages of text and 115 pages of notes, but because so much of it, like some of the passages quoted above, expresses such far–fetched "explanations" and imagined connections that one casts about for some more plausible meaning. The convoluted style also holds up the reading. Some sentences have to be read at least twice before one can see what is intended. Typically: "I believe that generally a mathematician's predilection for mathematics related to empirics and applications expresses an impulse to have an impact on a wider world than that of the mathematical cognoscenti." Or, "...whereas von Neumann sought to explicate primarily through formal logical structure, Wiener sought an intellectually comprehensive synthesis as a context within which to examine a few concrete topics with mathematical rigor." A certain insensitivity to the sound of language is exemplified by: "Von Neumann's and Wiener's social theories...contained ideas about the nature of human nature, of what a human being is."

If, in studying his material, the author had picked up a little of von Neumann's clarity of thought and a little of Wiener's kindness, this would have been a better book.

Physicist Extraordinary

Mark Oliphant is one of the great personalities of the world of physics, whose career has had many very different facets. Original research in nuclear physics in the Cavendish Laboratory under, and in close collaboration with, Rutherford, was followed by leadership of the radar team at Birmingham which produced the cavity magnetron and by work with Lawrence at Berkeley on the electromagnetic isotope separation for the atom bomb. After the war came the idea of the synchrotron, conceived independently and probably before others, and the creation of a nuclear physics school at Birmingham, with the construction of a cyclotron and a proton synchrotron; then, after the return to Australia, a major share in the development of the National University and, after retirement from the laboratory, a term as Governor of his native state of South Australia.

But no list of this kind could bring out Oliphant's most characteristic quality, the fearlessness with which he speaks his mind, no matter whether his thoughts are popular or unpopular, never calculating their effect on his image. While therefore what he says and writes does not always please his listeners, they always respect his intrinsic honesty. This quality came out perhaps most strongly, and created the greatest surprise, in his last, most exalted position as Governor. Those chosen for such representative positions are often senior politicians or military officers, very conscious of protocol, and maybe stuffy. Stuffiness is a quality completely lacking in Oliphant, and his direct and outspoken ways went over well in a country where stuffiness is not welcomed.

He speaks with equal directness about his failures and errors of judgment, and this has made it easier for the authors to write this biography in his

This is a review of *Oliphant: The Life and Times of Sir Mark Oliphant,* by Stewart Cockburn and David Ellyard. 369 pages. Australia: Axiom Books, 1981. For the complete reference, please see the Acknowledgments section.

lifetime. Writing about living people who can answer back can be embarrassing, but evidently not in the case.

The authors have interviewed Oliphant and members of his family, as well as many colleagues in his different spheres, and they have had access to many letters and other documents. As a result the book brings out Oliphant's personality vividly. Perhaps the best part of the book is the account of his ten years in Cambridge. The authors do not give much detail of his papers on nuclear physics, but then they are not writing for physicists. What comes out so well is the atmosphere in the Cavendish, and Rutherford comes completely alive. Oliphant was perhaps Rutherford's favorite pupil, and we understand why. The characters of these two men had much in common in their directness, and in their enthusiasm for the work. A lecture by Rutherford had made a deep impression on Oliphant when still a student, and had led to his resolve to get to Cambridge and work under Rutherford; the great respect and awe gradually gave way to genuine affection.

Yet he left Cambridge, as other collaborators of Rutherford had done, to branch out on his own in Birmingham. He took up his new post just before Rutherford's death in 1937. His intention was to build up nuclear physics in Birmingham, but the process was soon interrupted by the war. The book describes the radar work, and his abortive trip to Australia to offer his services to the Australian government when the radar problems at Birmingham

Gerhard H. Dieke, Marcus L. E. Oliphant, and Ebbe Rasmussen at the 1934 Copenhagen Conference, Bohr Institute. (Photograph by Paul Ehrenfest, Jr. and published courtesy of AIP Emilio Segrè Visual Archives, Weisskopf Collection.)

had lost their urgency and the war in the Pacific was approaching his native country. Finding that no one was prepared to arrange for his services to be used, he returned to England, and soon afterwards joined E. O. Lawrence at Berkeley in work on the electromagnetic isotope separation for the atom bomb. Lawrence was another colleague to whom Oliphant had great affinity, and their relationship is also brought out very vividly here. The authors describe the shock Oliphant felt when he heard of the use of the atom bomb on Japan, and his determination to work for solutions that would make future atomic war impossible or at least unlikely.

Back to academic life, to complete the construction of the cyclotron started before the war, and to start the proton synchrotron, based on an idea proposed by him but published by others first. The construction of two machines in a very confined space, and in a Britain beset by post–War shortages, using do–it–yourself methods as far as possible, was a typical Oliphant act of courage and optimism. The less ambitious of the machines, the cyclotron, gave little trouble; the proton synchrotron took much longer to complete than had been hoped for, and was ready only long after the similar, though somewhat larger, "cosmotron" at Brookhaven.

In fact the machine was completed only after Oliphant had left for his next assignment, the Australian National University at Canberra. The book describes the part he played in building up the spirit of the new institution, with all the struggle that involved, and also his choice of a machine for the physics department, a 10 GeV synchrotron powered by a homopolar generator. Looking back, this was clearly an error of judgment. He was again over–optimistic about the prospect and the time scale. If the machine had been completed as planned, it would have been by some years the first to reach this energy range, though even then the slow repetition rate would have made it difficult to use. As it is, only the homopolar generator was completed, but it will not drive a synchrotron and has instead found application in work on plasma physics. In spite of this disappointment, he succeeded in creating a physics department with many activities in different fields, and a staff of high quality.

Last, but not least, comes his term as Governor, which I have already mentioned. This again is described with warmth, though without glossing over some errors and embarrassments.

Taken as a whole, the book is perhaps a little longer than necessary. But it is a pleasure to read, and it does justice to a great character, whose integrity, frankness and directness set an example to us all.

Landau in the 1930s

I first met Landau early in 1930 when he came to Zurich, where I was then Pauli's assistant. He came with a Soviet government fellowship, and he had trouble with the Swiss visa, because the Swiss and Soviet governments then had no diplomatic relations. His permit to stay was for a very short time, and even under pressure from Pauli and Scherrer was renewed for even shorter periods, until he had to leave. Landau was very proud: "Lenin lived here for several years and did not start a revolution, but they think I can!" He came again during the following academic year, and then there was no trouble. At that time he held a Rockefeller fellowship. If Mr. Rockefeller pays a Russian to travel, he must be all right.

I was immediately impressed by Landau's wide knowledge, deep understanding, and command of physics. His approach to a new theoretical paper that looked interesting was to glance at it to see what the problem was, and then to do the calculation himself. If the answer agreed with the author's it was a good paper.

He had a strong intuition and usually did not bother proving statements which were obvious to him. If you complained that they were not obvious to you he might say "In that case you should not be a physicist."

In that spirit he confronted me with the symbol $\sqrt{\Delta}$, the square root of the Laplacian, without explanation. It took me some time to work out that you could define this operation by a Fourier transform, and that its inverse, $1/\sqrt{\Delta}$, could be written as a respectable integral operator with the kernel $1/2R$. This was related to the paper in which we wrote the equations of quantum electrodynamics in the configuration space of photons. I realized later that this made no practical sense because the position of a photon is not an observable, but the idea of writing a Schrödinger–type equation in a number of coupled spaces (one for no photon, one for one photon, etc.) seemed to us new and interesting. Once the contents of the paper were agreed, he left the writing to me, as we have been told he did with later collaborators.

He was very critical, as was most of our generation of theoreticians, and the comment "falsch oder trivial" (wrong or trivial) about suspect papers, used often by Landau, was in common use. He was also fond of the term "pathologists" for people who wrote pathological papers, i.e. nonsense.

He had arrived in Zurich with his solution to the problem of the diamagnetism of an electron gas clear in his head, though the paper was written only later. His result was not easily accepted by many because it was known that it was easy to get a wrong answer for the corresponding classical problem, and it was therefore believed that the problem was very complicated, whereas Landau's solution was direct and simple. Pauli saw the point at once and respected Landau for it.

During Landau's second visit we became interested in whether in relativistic quantum mechanics there were limitations to the possible accuracy of measurement beyond those contained in the usual uncertainty relation, which might perhaps give a hint of how to find a better theory, which could avoid some of the infinities.

Landau knew already one such limitation, which links the accuracy of a momentum measurement and the time necessary to carry it out:

$$\delta p \; \delta t \geq \hbar/c$$

which we put in our paper, and which I believe to be correct, although it has been queried by Aharonov. But the major claim of the paper related to the measurement of one component of the electromagnetic field. In retrospect it is clear that this was of no great importance because it did not lead to a hint as to how to improve the theory, as we had hoped.

We later happened to be in Copenhagen at the same time, and showed our results to Niels Bohr, who disagreed violently. This led later to the paper by Bohr and Rosenfeld, which contained a proof that the measurement of electromagnetic fields is possible to the accuracy allowed by the conventional uncertainty relation. There is room for argument as to how far the device proposed there, which involves filling the volume element in question solidly with apparatus, constitutes a measurement of the field in that volume element.

We decided to publish our paper anyway but, because of Bohr's objections, we were not clear whether it was proper to acknowledge his help—this could be understood to imply that he approved of the paper. I consulted Bohr on this point, but he misunderstood and became very angry. He said he was trying his best to help people clear their minds, but if we even queried whether his help should be acknowledged, he insisted that his name not be mentioned. This was embarrassing, but it got straightened out (we did, of course, thank him in the paper). The episode does not seem to have upset the strong affection between Bohr and Landau.

Landau's approach to other matters was also very systematic. His classification of physicists has been mentioned. At the time he said he hoped ultimately to qualify for a position in the second class. He was convinced theoretical physicists did their best work early, and when a youngish man was mentioned, of whom he had not heard, he said, "What, so young and already so unknown?"

He could show strong personal disapproval of people; one of his favorite terms was "zanuda," which literally means "bore", but for him it carried stronger negative associations—perhaps Philistine, or bourgeois.

He was consciously trying to improve his response to social situations. He mentioned to me that in joining an unfamiliar group of people (which he was then doing frequently on his travels through Europe) one feels "disoriented," and he was working to get out of this expeditiously.

His analytic approach applied also to human relations, including "situations." A situation was a relation between a man and a woman, and it could be satisfactory or otherwise. If he saw in his acquaintance an unsatisfactory situation, it was of course his duty to inform the couple concerned, which did not always make him friends.

He also objected strongly to beards, particularly on young men; he regarded this as a survival from the nineteenth century. There was a young physicist in the laboratory, who did not have a beard, but had very pronounced sideboards. One day Landau phoned the man's wife (whom he had not met) and said "This is Landau; I am phoning to ask when you will get your husband to shave off his ridiculous sideboards." He claimed that there were more beards to be seen in Zurich than in Leningrad. I had been in Leningrad by then and doubted his claim, so we made a bet. We took samples in Zurich of the numbers of men with and without beards, and I was to repeat this on my next visit to Leningrad (Landau was still staying in the West). The count showed that I was right, and he paid up, but he said it was unfair—while he was abroad there had been the collectivization, which had driven many peasants into the towns, and everybody knew that peasants had beards.

His pronouncements often provoked negative reactions, particularly among people with traditional attitudes. A nice example of this occurred when he paid a short visit to Felix Bloch, who was then working in Haarlem, Netherlands, in an institution directed by Fokker, a very worthy citizen. Landau was invited, with Bloch, to dinner one evening to the Fokkers' house. On the next Sunday Bloch was there again for lunch, when Mrs. Fokker asked, "Is your Russian friend still here?" "No," said Bloch, "he left a few days ago." Whereupon the little daughter, who had never seen Landau, heaved a sigh: "Thank God!"

On the other hand many housewives tended to react to his very thin figure, which gave the impression that he was hungry, with a motherly urge to feed him.

He did not believe in any religion, though he was always clear about his Jewish origin (he used to refer later to the few non–Jewish members of his school as "the national minority"). I am reminded of an amusing confusion: a colleague, who was studying Landau's biography, asked me in a letter about Landau's "conversion to Christianity." When I refused to believe this he referred me to a passage in Livanova's biography, quoting Landau as saying, "I am a Christian now." However, the context showed clearly that this related to the old story of lions eating Christians in ancient Rome. He, and others, had called him a lion that eats people (his first name, Lev, means "lion"), and he was saying that he had now mellowed, he did not bite people any more.

I end with a conversation that took place in 1934 during an excursion in the Caucasus by Landau, myself and a friend of his, an engineer, Styrikovich. The latter asked one day: "What is this one hears about energy from the nucleus? Is this just science fiction?" Landau responded without a moment's hesitation: "It is a complicated problem. There are nuclear reactions which release more energy than they require, but they are mostly started by bombardment with charged particles. As we cannot aim at nuclei most of the particles go astray and lose their energy in passing through matter. So much more energy is wasted than we gain. Neutrons are different because they don't lose energy in passing through matter, and go on until they hit a nucleus, but so far we can produce neutrons only by charged–particle bombardment, so we are back with the same difficulty. But if someone discovers a reaction which can be initiated by neutrons, and produces secondary neutrons as a result, you are all set." This was barely two years after the discovery of the neutron, and it is yet another demonstration of Landau's vision and power of reasoning.

William George Penney, 1909–1991

William George Penney (Lord Penney), who died on March 3, 1991, was best known for his contributions to atomic energy, particularly for leading the development of the British atomic bombs.

Penney was born on June 24, 1909 in Gibraltar, where his father was stationed with the British army. He was a brilliant student at Imperial College, London, and published his first paper at age 20. He worked with Ralph de Laer Kronig, partly in London, partly in Groningen, the Netherlands, and together they developed the "Kronig–Penney model," a solvable model for the motion of electrons in periodic fields. After two years at the University of Wisconsin under John H. Van Vleck on a Commonwealth Fund fellowship he spent three years in Cambridge collaborating with J. E. Lennard Jones, under whose influence Penney's interests turned from crystalline fields to molecules. Penney made many important contributions to theoretical chemistry while at Cambridge, and he continued this work when he returned to Imperial College as an assistant professor.

On the outbreak of war in 1939, Penney started work for the British Admiralty and the Home Office on the structure and effect of blast waves. This work led to his being invited in 1944 to Los Alamos as a member of the British team participating in the Manhattan Project. Soon after his arrival he gave a talk about the effect of blast waves on people, including many gruesome details to which his American audience was not accustomed. He presented all this with his usual cheerful manner, and so acquired for a time the sobriquet "the smiling killer." Penney soon impressed people at Los Alamos with his expertise and ability. I remember the brilliantly simple method he used to estimate the intensity of the blast wave at the Trinity test at Alamogordo.

He had some wooden boxes prepared with circular holes of various sizes, covered with paper. The blast would puncture the paper covering the larger holes but not the smaller ones, and by noting the size of the largest holes still intact, one could find the strength of the wave.

On his return to England after the war, he was appointed chief superintendent of armaments research in the Ministry of Supply. When the government, under Clement Attlee, decided to start developing nuclear weapons, Penney was put in charge. He was the only member of his group who had experience in nuclear weapons work. Since the US McMahon Act of 1946 prohibited exchange of information on nuclear matters, Penney, with his Los Alamos experiences, carried the sole responsibility for the development of the British A-bombs—a responsibility he discharged brilliantly. In a short time he built up the Atomic Weapons Research Establishment at Aldermaston, which contained research and production facilities.

Leading this project required not only scientific brilliance but firmness of purpose and ability to handle people; Penney proved an outstanding success in these respects. Quiet and soft–spoken, patient and always cheerful, shunning publicity, and with an exceptional ability to explain complicated matters clearly and simply, he became popular with his staff. The first atomic bomb produced by his team was successfully tested in 1952. It led eventually to a hydrogen bomb in 1957.

Penney gradually acquired broader responsibilities for the United Kingdom Atomic Energy Authority, and became its chairman in 1964. He served in that post until 1967, when he accepted an appointment as head of the Imperial College. His open, informal manner gained him people's confidence also in the academic sphere, and in 1968, when most universities had violent student unrest, Penney's influence preserved peace at Imperial College. He retired in 1973.

At hearings of an Australian Royal Commission in 1985, it was suggested that the health of Australian aborigines had been damaged by British nuclear weapons tests. Though this allegation was not supported by the commission, Penney was deeply hurt, because he had taken great trouble to avoid any risks to the local population or the test staff.

Conservative Revolutionary

The title of John Heilbron's biography of the physicist Max Planck, *The Dilemmas of an Upright Man*, is well chosen. The chief characteristics of Planck were his integrity, his feeling for tradition, and his sense of duty. Yet throughout his adult life he faced conflicts, some of which involved reversing a stand he had taken on matters of principle; in other cases he pursued a futile struggle against inevitable disasters. Heilbron gives a very readable, and very balanced, account of the successes and disasters of this great physicist, without attempting to pass judgment. But the character of the man comes clearly through the narrative.

Planck's work was the starting point of the quantum theory, one of the two great revolutions in twentieth–century physics. Yet he was a most unlikely revolutionary. Descended from a long line of professors of theology and law, he was deeply religious, but without belief in a personal god. He was patriotic, but never chauvinistic, and he deeply regretted one lapse in the war fever of 1914. In his science he was driven by an urge for order and unity; he regarded the edifice of physics as it then was with awe, and with the wish to complete, and not to rebuild it. Once he had formed an opinion he was slow to change it. In his many administrative functions he would prefer to act through quiet diplomacy, never through public stands or angry confrontation. He had an iron self–discipline and sense of duty, which enabled him to cope with enormous administrative burdens, and to carry on in times of the most appalling personal tragedies.

The discovery that had such revolutionary consequence was the "Planck law of radiation," which described heat radiation. By the end of the nine-

This is a review of *The Dilemmas of an Upright Man: Max Planck as Spokesman for German Science,* by J. L. Heilbron. Berkeley: University of California Press, 1986. The book review is reprinted with permission from *The New York Review of Books.* Copyright © 1986 Nyrev, Inc. For the complete reference, please see the Acknowledgments section.

teenth century it was known from observations that the color and intensity of the radiation filling a cavity in a hot body did not depend on the nature of the walls of the cavity, but only on the temperature. It was a challenge to physicists to find and explain the law determining this dependence.

Planck succeeded in finding a formula that fitted the observations accurately. Now the problem was to derive this formula from the general laws of physics. He had previously worked on thermodynamics, the science of heat, and had clarified the important concept of entropy, a measure of disorder. The concept of entropy is vital to understanding the difference between a body falling under gravity, and heat passing from a hot to a cold body. The first is reversible; if the body is elastic, like a tennis ball, it will bounce back to where it came from, whereas heat will never flow back from the cold to the hot body. This is expressed by saying that the entropy increases as the heat spreads, and that it can never decrease.

Planck now had to find how to express the entropy of the radiating atoms and the radiation. He found what seemed a suitable expression for this, no doubt working backward from the radiation law he wanted to explain. He convinced himself that this was the correct expression for the entropy according to the known laws of physics. On the strength of this he published in 1900 his paper on what is now known as "Planck's law."

However, he was not satisfied with the derivation, and tried hard to find better reasons for the assumptions he had had to make. But gradually it became clear that the assumptions had no place in the established laws of physics. Finally, in 1908, the great Dutch theoretical physicist H. A. Lorentz proved that the Planck law could not be derived without a revolutionary change in the principles of physics.

Einstein had already recognized this and knew that Planck had implicitly introduced the concept of the light quantum, by which light is concentrated into packets, or quanta, each of which contains a definite amount of energy, the amount being linked to the color, or frequency, of the light by Planck's "quantum of action."

Einstein used this concept in his explanation of the photoelectric effect in 1905. Niels Bohr used the idea in his theory of the atom in 1912; the keystone of the quantum theory, quantum mechanics, was finally put in place by Heisenberg and Schrödinger in the 1920s.

Thus Planck had started the revolution in physics, but he himself was reluctant to accept that revolution. He struggled to reconcile what he had done with the traditional principles of physics. As late as 1910 he said, "The introduction of the quantum of action ... should be done as conservatively as possible, i.e., alterations should only be made that have shown themselves to be absolutely necessary."

Characteristically, when he proposed Einstein for election to the Prussian Academy in 1913, the recommendation by him and others said, after high praise: "That [Einstein] may sometimes have missed the target in his specu-lations, as, for example, in his hypothesis of light quanta, cannot really be held too much against him." Planck was almost like the sorcerer's appren-tice, having started the dramatic new developments in physics and being ·unhappy about it. His own comment in his scientific autobiography is:

> My futile attempts to fit the elementary quantum of action somehow into the classi-cal theory continued for a number of years, and they cost me a great deal of effort. Many of my colleagues saw in this something of a tragedy. But I feel differently about it. For the thorough enlightenment I thus received was all the more valuable. I now knew for a fact that the elementary quantum of action played a far more significant part in physics than I had originally been inclined to suspect.[1]

He made his peace with the new physics, but he never did any work to apply or extend it.

This was the greatest, but not the first dilemma in his scientific work. He had started his career with work on the laws of thermodynamics at a time when the kinetic theory (now called statistical mechanics) was developing. This explains heat as the random motion of atoms. Planck regarded it as an unnecessary hypothesis to introduce atoms, about which so little was known; his sense of order and simplicity required all physics to be expressed in terms of thermodynamics. This put him in opposition to Ludwig Boltzmann, the great pioneer of the atomistic theory of heat.

But he got shaken in this conviction when it became clear that no progress could be made about such phenomena as electrolysis and electric conduction without introducing atoms and ions. Finally his work on the radiation prob-lem demonstrated to him that he had to approach entropy in Boltzman's way, and he was converted to atomism. He now vigorously took Boltzmann's side against Ernst Mach, who denied the existence of atoms.

Another conflict, though perhaps not felt by Planck as a dilemma, was of a more philosophical nature: it was related to the problem of positivism in physics. The interpretation of quantum mechanics by Bohr and Heisenberg says that, when the velocity of an electron is known and, by the "uncertainty principle" its position cannot be determined, the question "Does the electron *really* have a position?" is meaningless. This attitude is described by some as positivism. Planck was always strongly opposed to positivism and inveighed against it in many lectures and essays, which again put him in opposition to Mach, a strong exponent of positivist views. When he had to accept the un-certainty principle of modern quantum mechanics, he argued his way out of

---

[1]*Scientific Autobiography and Other Papers* (Philosophical Library, 1949), p.44.

this dilemma[2] by saying essentially that it is the Schrödinger wave function that has reality, a view with which few physicists will agree.

Planck's career had a slow start, but he became an extraordinary (associate) professor in Berlin in 1889, at the age of thirty–one, and a full professor in 1892. He was a very conscientious teacher, and during most of his time in Berlin delivered a course on theoretical physics, which consisted of four lectures and one problem class a week, and covered the subject in a six–semester cycle. He maintained this lecture course at least until 1930, when he was seventy–two. I attended his lectures as a first–year undergraduate, and found them unhelpful—he was reading verbatim from one of his books, and there was nothing inspiring in his delivery. He had something in common with his predecessor, Kirchhoff, whose lectures he describes. "It would sound like a memorised text, dry and monotonous. We would admire him but not what he was saying."[3]

After his great success with the radiation law he became an established figure in science, and acquired many honors and duties. He was much in demand for his clarity of thought, his coolness of temper, and for what at a celebration at the Berlin Academy was called "the spotless purity of his conscience." He had responsibilities at the University of Berlin, the German Physical Society, and the Berlin Academy, and faithfully attended the scientific meetings of all these organizations.

His efforts did not slacken when his wife, to whom he had been very close in twenty–three years of marriage, died in 1909. He married again soon afterward. In Heilbron's words, "He needed another wife for his house and children, for companionship, and because a professor customarily had one."

At the outbreak of the First World War he was caught up in the patriotic fervor. He felt the war was an occasion for sacrifice in the service of the nation. He signed the famous Appeal of Ninety-three Intellectuals in October 1914, which supported the German Army and denied that Germany had committed atrocities in Belgium. But he opposed other, more aggressive, manifestoes. In the Berlin Academy he defeated a proposal that after the (of course, victorious) war German scientists would not cooperate with any foreign ones.

Later correspondence with Lorentz in Holland and an exchange of visits convinced him that many things had happened in Belgium "that do not conduce to the honor of Germans." In 1916 he sent an open letter to Lorentz, in which he tried to explain the Appeal of the Ninety–three Intellectuals as expresssing only support for the army, but he admitted that individual Germans might have done wrong. At Planck's request, Lorentz circulated this

---

[2]*Scientific Autobiography and Other Papers*, p. 121.
[3]*Scientific Autobiography and Other Papers*, p. 16.

letter to Allied scientists. It helped to raise Planck's reputation in the "enemy" countries, and above all to relieve his conscience.

His elder son, Karl, was fatally wounded at the front. In spite of his grief, the father felt that the sacrifice was necessary. "Everyone should be happy and proud to be able to sacrifice something for the whole." He had had a low opinion of Karl, who could not settle down to any worthwhile occupation; his war service had shown his real worth. "Without the war I would never have known his value, and now that I know it, I must lose him."

More tragedy not connected with the war came to the family. His twin daughters died, two years apart, each shortly after childbirth. Planck's sense of duty was too strong to let these events interrupt his activities, which now included much work for the Notgemeinschaft der Deutschen Wissenschaft (Emergency Association of German Science); this tried to secure the desperately needed funds for supporting research.

In the postwar period he worked to restore, or maintain, international contacts in science. Many international organizations had rules excluding German scientists, and he felt this as an injustice. When he was invited to the 1927 Solvay conference, he was at first inclined to refuse, because the physicist Arnold Sommerfeld was not invited. Only when the reasons for the choice were explained, and he was assured they did not involve any prejudice, did he agree to go.

His role was always that of a peacemaker. When Philipp Lenard started his ill-tempered anti-Semitic outbursts against Einstein and relativity, Planck called a meeting of scientists in Bad Nauheim in 1920, for a confrontation between Lenard and Einstein. As chairman he managed to conduct the debate in a civilized manner.

The greatest disaster that befell him—and Germany—was the coming to power of Hitler. He was shocked by the behavior of the Nazi regime, in particular by the dismissal of Jewish and liberal academics. But he made no public protest—that was not his way. He did consider resignation, but decided to remain and use his influence to work for a more reasonable course and to protect those under him as best he could. This required cooperating with the regime, at least in form; he had to give the Hitler salute at meetings.

His position is perhaps best characterized by a remark made by Robert Oppenheimer, when he was reminiscing about having remained too long in a post under a hostile authority. Oppenheimer referred to the attitude "As long as I ride on this train, it will not go to the wrong destination."

Planck had some successes. He prevented the appointment of some pro–Nazi nonentities to leading positions. As president of the Kaiser–Wilhelm–Gesellschaft, a government-sponsored and privately supported research organization, he could protect "non-Aryan" employees like Lise Meitner—for a time. When Lise Meitner finally had to go, posts had become much scarcer and it was much harder for her to find support abroad than it had been for earlier refugees. He succeeded in keeping the Kaiser Wilhelm Institute for Physics working on pure physics under Peter Debye, an excellent and independent-minded Dutch physicist for a time. A few years later Debye stayed abroad and the institute was taken over for work on atomic energy.

Planck even went to Hitler to tell him that the expulsion of Jews would weaken German science. Hitler replied that he had nothing against Jews, only against communists, and then flew into a rage. Planck was depressed, since this meant he had no basis for any further negotiation. Heisenberg, to whom Planck reported the conversation, took Hitler's words at face value and told Max Born and James Franck that they had nothing to fear, and could stay in their jobs. They left anyway.

Eventually Planck had to resign from the presidency of the Kaiser–Wilhelm–Gesellschaft in 1937 and from the secretaryship of the academy in 1938. He was now eighty, but he continued giving lectures and writing essays, mainly on philosophical and religious subjects, but always inserting remarks critical of the regime. In 1944, while he was on a lecture tour, his house was destroyed in an air raid, and with it his library and all his papers. Then his only surviving child, his son Erwin, was accused of taking part in the abortive assassination attempt against Hitler in 1944, and was executed in a particularly brutal way.

After this shattering blow his health broke down. The war zone came to his temporary home in the country, and he and his wife had to hide in the woods and sleep in haystacks. He was rescued by an American officer and taken to Göttingen, where he had relatives, and where he could get hospital treatment. After his release from the hospital he worked, in spite of his age and frailty, for the recovery of German science, and particularly for the re-establishment of the Kaiser–Wilhelm–Gesellschaft. The Allied occupying authorities were reluctant to allow this, because of the name, but to everybody's satisfaction the society now became the Max–Planck–Gesellschaft.

Heilbron's book is based on scholarly research, and the sources for every statement are given. To the relief of the reader, the footnotes appear at the bottom of each page, and not in a list at the end. There is a good collection of photographs. I suppose only a native German will be upset by the excessive use of the umlaut: Planck's house was in Grunewald, not Grünewald; his wartime home was Rogatz, not Rogätz. But this is the only complaint I have to make about this solid and readable book.

Reference

Planck, Max K. *Scientific Autobiography and Other Papers.* Gaynor, Frank, tr. 192 pages. Westport, Connecticut: Greenwood Press, 1968.

Recollections of James Chadwick

I first met Chadwick in the summer of 1933, when I was a visitor to Cambridge. He was then already famous for his discovery of the neutron. But he was the last person to become conceited as a result—his manner was, and always remained, matter–of–fact, if somewhat aloof. In physics he knew what he wanted to do and how to get it done. This is perhaps illustrated by a story which went around in Cambridge at the time. An experiment he was doing required observations late at night when it was quiet. But the gates which gave access to the Cavendish Laboratory were locked at night, so if he stayed late he could not get out. He knew that any idea of a gate being kept open late, or of his being given a key, would be regarded as a shocking breach of tradition, and he did not try. Instead he had a camp bed brought to his office and spent the rest of the night there.

During 1933 to 35, when I was in Manchester, I often came to Cambridge for a day to find out what was going on in physics, and I found Chadwick always ready to talk and to discuss. It took a little effort to get used to his manner. On the first such visit I saw him chatting with some students in the entrance hall, when he recognized me and asked what I was doing in Cambridge. I explained that I was there for the day, and if he would spare me the time I would much like to talk to him. He looked at me over the top of his glasses and said, "Yes? What for?" This was very intimidating, but I was not easily dashed, and explained what I wanted; we had then, and many times later, a very useful discussion.

On one occasion both Hans Bethe and I were talking with him, when he suddenly said, "bet you could not calculate the cross section for the photodisintegration of the deutron." (He probably said "the diplon", as that was the name then used in Cambridge.) He did not tell us that he already knew the answer, having carried out the experiment. This went back to a suggestion of Maurice Goldhaber, then a young graduate student. Chadwick had not only

Sir James Chadwick. (Photograph published courtesy of AIP Emilio Segrè Visual Archives, Numeroff Collection.)

accepted the suggestion, but invited Goldhaber to take part in the experiment and published the work as a joint paper.

To Bethe and me this was a challenge, and it led us to construct an approximate theory, which worked well, though it failed for the inverse process of radiative capture of neutrons by protons, because we did not yet know about the strong spin dependence of the nuclear forces.

After 1935, when Chadwick moved to Liverpool, I did not see much of him until 1940 when interest started in atomic energy. Otto Robert Frisch and I had shown that the critical mass of uranium 235 was not as large as one might have guessed, and that a chain reaction in it would not stop before an appreciable part of the available fission energy had been released. In other words, a new and terrible weapon could be made, though with a considerable effort. Such a violent reaction was possible only with fast neutrons, and a fast chain reaction was possible only in the rare isotope uranium 235.

Chadwick had been thinking along similar lines. He also stressed the importance of fast neutrons. For a time he even thought that one could use natural uranium, but it followed from Niels Bohr's theory that in the predominant isotope, uranium 238, only neutrons with energy above a certain threshold will cause fission, and it was doubtful whether enough of the secondary neutrons produced in fission had energies above the threshold to keep the chain reaction going. In addition he realized that there would be inelastic collisions, in which the energy even of a neutron above threshold would be reduced below it. He reserved his judgment about the fission cross section of U235 and hence on the critical size: as an experimentalist he was less inclined than we were to accept the prediction of Niels Bohr's theory on the matter.

He accepted, however, Bohr's argument about the threshold, which shows as a steep rise in the fission cross section of uranium at an energy of 1 MeV. According to Bohr, the small fission cross section at energies below this threshold is due to U235. Chadwick set up measurements of the variation of the fission cross section of natural uranium with energy, and from this could derive a reasonable estimate of the cross section for the 235 isotope, using only the minimum of theory. Such experiments had been done by Tuve and others in Washington, yielding a very small fission cross section for 235, which would have made the critical mass prohibitively large, but Chadwick was rather sceptical of these experiments.

These and other nuclear experiments on uranium required much effort. In this he had a few collaborators including Joseph Rotblat, John Holt and Frisch. By this time Frisch had moved from Birmingham to Liverpool, where he could be more useful.

As the nuclear physics problems expanded, Chadwick also directed the work in Cambridge by a group under Norman Feather and Egon Bretscher. At that time Rotblat and Bretscher independently concluded that the result of U238 capturing a neutron (now known as plutonium) would also be capable of sustaining an explosion, but this did not affect the program in the UK.

Chadwick also took an active part in the committee overseeing the atomic–weapons work, which later became the "Maud Commitee." I suppose the origin of the name is by now well know. Early after the Nazi occupation of Denmark, a cable arrived from Lise Meitner, who had been visiting Copenhagen, but had managed to get back to Sweden. It reported that Bohr and his family were well, and ended "Inform Cockcroft and Maud Ray Kent." This wording seemed mysterious. If the recipients knew Maud Ray, why say "Kent,"? If they did not (nobody did) this was hardly an adequate address. It was suggested that there might be a hidden message, perhaps an anagram. But there were too many possibilities. This was discussed at a time when it seemed useful to have an innocent–sounding cover name for the committee, and so the name Maud suggested itself. (In spite of this, many people were convinced that MAUD stood for Military Applications of the Uranium Disintegration.) The mystery was cleared after the war: Maud Ray was a former governess of the Bohr family, and the telegram had been mangled in transmission; between "Ray" and "Kent" there had been a full address.

Chadwick's influence on the committee was very important. When the memorandum by Frisch and myself was received, he backed the recommendation to go ahead with research on the physics of fission and on large–scale isotope separation. In the work of the committee he would always stress the basic issues, and was never afraid of supporting unusual solutions for unusual problems. At an early stage it became clear that someone must take charge of research on isotope separation. I felt sure that the right person for this was Francis Simon. He was, like Frisch, a former "enemy alien," but by then most first–rate British physicists were busy with other important war research. The committee chairman, Sir George Thomson, put forward another name, of a charming but not very effective man. I was not then a member of the committee and worried that the chairman might proceed with the appointment; so I appealed to Chadwick. He saw the point at once, and after some words with Thomson, Simon was brought in.

I remember this visit to Chadwick. We discussed also other matters concerning atomic energy, and for this he wanted to look at some correspondence. He opened a drawer in his desk, which was fairly full of letters, each in its envelope. Yet after shuffling a little, he pulled out the papers he wanted. A far cry from the steel cabinets and safes that became compulsory later, but it worked.

In 1940 Hans von Halban and Lev Kowarski escaped from France, where they had been working on a thermal–neutron chain reaction with heavy water as moderator, and continued this project with a group they built up in Cambridge. In discussing the future of this group the possibility was considered that they might work in the United States, or in Canada in collaboration with U.S. teams. The American physicists, who had great respect for Chadwick's judgment, asked that he review the work of this group. He did so, and reported that he had confidence in the standard of their work. Thereby he paved the way for the creation of a joint project in Canada.

By the summer of 1941 the work sponsored by the Maud Committee had made enough progress to show that a nuclear weapon was indeed a strong possibility, and to give some indication of the effort involved. A first draft of its report was ready at the end of June. Some members felt that a document of this importance should be considered at leisure, but others, including Chadwick, urged speed. But the draft needed to be shortened and made clearer.

The rewriting fell largely to Chadwick. The result was two reports, one on the bomb, the other on the power reactor, or "boiler." They were completed and approved during July. As the chairman of the Maud Committee, Sir George Thomson, was unavailable, Chadwick presented the reports to a panel of the scientific advisory committee of the cabinet.

The panel, and subsequently the committee, accepted the Maud reports' recommendations that the power reactor could not make a contribution to the war effort, but the bomb project should be developed urgently. As a result a new organization was formed, called the Directorate of Tube Alloys, a meaningless, but plausible sounding cover–name. It was to sponsor the next stage of development. Chadwick continued in charge of the nuclear physics work.

During this time exchange of information with scientists in the U.S. was regarded as important. A copy of the Maud Committee reports was passed to America, and two Americans, Harold Urey, the discoverer of heavy hydrogen, and George Pegram, a nuclear physicist, both from Columbia University, attended the first meeting of the new Tube Alloys committee. During their visit Chadwick told them: "I wish I could tell you that the bomb is not going to work, but I am 90 percent certain it will." This remark no doubt reflected his attitude, and that of all of us, that the world would be better off without the bomb, but that, if it could be made, and in particular by Nazi Germany, we could not afford to neglect it.

To reinforce the exchange of information, a group of British physicists visited the United States early in 1942. The Americans had expressed the hope that this would include Chadwick, but he declined to go because of his health. It is possible that, if he had been able to go, some of the later difficulties in the relations between the countries could have been avoided.

Indeed relations over atomic energy became difficult later in 1942. By this time the United States had launched their full–scale project, operated by the Army Corps of Engineers as the "Manhattan District" under General Leslie Groves. Negotiations over collaboration with the UK failed, and eventually all exchange of information ceased. This is not the place to discuss the reasons for this failure, but they involved errors of judgment at high administrative levels on both sides of the Atlantic.

Tube Alloys work continued though there was uncertainty whether full–scale isotope separation could be carried out in Britain under wartime shortages, or whether relations with the U.S. could be restored. This uncertainty continued until at the Quebec Conference in August 1943 Churchill and Roosevelt discussed the matter and decided to restore collaboration, though only to a limited extent. Again some of the British scientists went to the U.S., this time including Chadwick, though his health was still giving trouble, and he arrived a day after the rest of us.

His presence was of the utmost importance; not only was he greatly respected by American scientists, but there developed immediately a close personal rapport between him and General Groves. It was now decided that all those Tube Alloys scientists who could make useful contributions to the American project would join it, and that work in Britain would be discontinued, except for a small number of groups whose facilities could solve problems that were essential for the Americans. This was a serious decision, because it meant abandoning nuclear physics in Britain for the time being. Chadwick realized that the United Kingdom would want to develop atomic energy after the war, and that the present move would make this more difficult. But he backed the decision strongly, because he saw that, in the interests of winning the war, all priority must go toward assisting the Americans in their project. He also understood that this would improve American attitudes to collaboration.

A Combined Policy Committee representing the U.S., the UK and Canada was set up, and Chadwick was appointed technical adviser to its British members. He was also invited to join the Los Alamos Laboratory, where the work on bomb design, including the relevant nuclear physics, was concentrated. In addition he was to take charge of the atomic–energy section of the British office in Washington. Wearing these three hats kept him extremely busy, but he coped with all his duties, in spite of his indifferent state of health.

At first he made his home at Los Alamos, but later found that he was needed more urgently in Washington. In Los Alamos he was welcomed by the scientists and soon made familiar with the state of the work and the outstanding problems. He was the leader of the British group, and that meant looking after their personal, as well as their scientific problems, as I discovered when I deputized for him after his move. One anecdote is more typical of life in Los Alamos than of Chadwick. For security reasons we were not allowed to use our names outside the settlement; for example, local drivers' licenses for Los Alamos staff had numbers instead of names. Chadwick wanted to get a fishing license, and was taken by an American colleague to the appropriate state office. When the official asked his name, Chadwick hesitated and turned to his colleague for guidance. The official long afterwards muttered about the dumb foreigner who did not know his own name.

While still resident at Los Alamos, but paying frequent visits to Washington, Chadwick was involved in many delicate negotiations. One problem that took long to settle was the future of the Montreal Laboratory, which under Halban was studying the possibility of a heavy–water moderated reactor with natural uranium. The Americans were critical of the running of this laboratory, and would support the building of a full–scale reactor only if a scientist of acknowledged reputation was brought in to take charge. John

Cockcroft was approached, but would accept only if the support for a large reactor was assured. After long negotiations, Groves and Chadwick were asked to make a recommendation and, while Groves was originally opposed, Chadwick's persuasion prevailed in the end. Cockcroft took over as director of the Montreal Laboratory (later Chalk River), and its work toward a large reactor was given American help, particularly with supplies of heavy water and uranium.

There were many other problems demanding Chadwick's attention and delicate negotiations, particularly with Groves. In addition he was involved in the discussion of the future of atomic–energy work in Britain. Some of his colleagues argued that, as the American project approached completion, the British scientists were no longer required and should return home to start work toward producing fissile materials (uranium 235 or plutonium) in Britain. Chadwick resolutely opposed this, because he expected it to damage relations with the Americans. The prospects of close collaboration with the U.S. after the war were most valuable, and he did not want them damaged for the sake of short–term advantages. Even making definite plans for post–war production plants was undesirable.

However, planning for a research establishment, he felt, would not upset the Americans, and he vigorously urged that the government should go ahead with this. He took part in many detailed discussions about the requirements for such a laboratory. It was important to find the best person to become the director of the laboratory. Chadwick made it known that he was not willing to be considered, as he wanted to return to Liverpool when the war was over, though he was willing to continue as a consultant, and he endorsed the choice of Cockcroft. However, the process of decision–making was disappointingly slow, and nothing concrete about the new research establishment at Harwell had been done by the time the war ended.

As Chadwick watched the test explosion at Alamagordo, he shared the general feeling of awe at the terrific power that had been unleashed. He had no regrets about his part in bringing this about, but he had doubts about the morality of the decision to use the bomb on two large towns without prior warning. He also shared Niels Bohr's concern about the future.

Bohr had been in Copenhagen under the German occupation. Early in 1943 Chadwick sent him a message, through intelligence channels, inviting him to come to England; the intelligence services would arrange to smuggle him out. But Bohr felt obliged to stay and do what he could to protect the members of his institute, which included many refugees from Germany. But later, when the Nazis decided to round up all Jews and liberals in Denmark, he escaped to Sweden and was taken to England, from where he went to the U.S., attached to the British Tube Alloys group. Besides taking an interest in the physics of the project, and indeed making a contribution to it, he thought deeply about the likely effect of nuclear weapons on international relations after the war. He foresaw the arms race and the cold war, and tried to persuade the authorities to be frank with the Soviet Union about the existence of the weapons, in the hope that this would reduce the tension. In this he did not succeed, and we do not know whether relations would have been better if his advice had been followed. Chadwick had many conversations with Bohr about these problems, and supported his view.

When, after the destruction of Hiroshima and Nagasaki, the Japanese had surrendered, and Chadwick had time to wind up his activities in Washington, he returned to Liverpool, to devote himself to building up his department there, as he had always intended.

So ended the period in which the great physicist had proved himself to be a great diplomat and statesman, carrying many burdens of which I have mentioned only a selection. To most of us, each of these burdens would have seemed a full–time load. In this his main asset, besides his great energy, was the confidence in his integrity and good sense which he inspired in all who came into contact with him.

Bell's Early Work

J ohn Bell came to my department in Birmingham in October 1953 on a year's leave from the Atomic Energy Research Establishment, Harwell, UK. Technically his status was that of a graduate student, but he was evidently much more mature than his 25 years. He also had already had substantial research experience for in his four years at Harwell he had worked on accelerator design, particularly on aspects of particle orbits and focusing. It may well be that this experience, of approaching physics through work on concrete problems relating to hardware, influenced his later style. When dealing with abstract problems he would always find some simple, tangible example to test his ideas.

He quickly became popular in the department, and it did not take long before we were impressed with his ability and the clarity of his thoughts. We became accustomed to his way of speaking, which at first may have sounded pedantic but on closer acquaintance revealed care to get the essential points across.

He had come to Birmingham to learn about modern theoretical physics. He started studying field theory and in a short time acquired an up–to–date knowledge of the subject. At the time we heard of experiments which seemed to reveal evidence for a negatively charged particle which was stable, but with a mass less than that of the proton. The experimenters asked us whether this could possibly be the antiproton. This seemed unlikely, but could it be firmly ruled out? Everybody expected particle and antiparticle to have the same mass, but was this strictly necessary?

This was a problem after his heart. He did not like to take commonly held views for granted, but tended to ask "How do you know?" In due course he came up with the "CPT theorem," that the results of any field theory must remain unchanged if one reverses the sign of the space coordinates and of time, and interchanges particles and antiparticles. (He said cautiously, "in any theory of the present form," but nobody has yet given an example of a sensible theory in which the theorem would not hold.) The theorem ensures, in particular, that any particle and antiparticle must have the same mass. Any evidence contradicting the theorem would be very hard to reconcile with our present basic physics; so far no such evidence has been found. Indeed, the experiment which had raised the question was not confirmed.

The proof of the theorem formed the basis of John's Ph.D. thesis completed after his return to Harwell. Before he had completed writing it the same result was published by Lüders. So John lost priority, but this did not diminish the merit of his insight.

After returning to Harwell, he retained his interest in problems relating to time reversal. He showed that time–reversal arguments cannot be strictly applied to β decay because the inverse reaction is not in practice observable, but that useful conclusions can be drawn provided first–order perturbation theory is applicable, which of course it is to high accuracy. He also continued to think about field theory. He developed a formalism proposed by Skyrme and together with him applied it to an attempt to calculate the magnetic moments of nucleons.

But his main effort went into problems of nuclear physics which came up in the work at Harwell. Here again he was never satisfied with routine applications of standard methods, but always went back to foundations. For example, he showed how the spin–orbit term in the shell–model potential could be derived from the spin–orbit force in the two–nucleon interaction.

He discussed how far β decay would be influenced by taking place in a many–body situation, and with Blin–Stoyle he considered the effect of virtual mesons in the nucleus of β decay. Two papers with Mandl discuss the identity relating polarization and asymmetry in scattering, and show that this is valid if longitudinal polarization plays no part, but needs correcting if there is longitudinal polarization, which is always possible because of parity violation.

Another fundamental problem is bremsstrahlung in multiple scattering. A characteristic approach adopted by John Bell was to solve a many–body problem in a one–body potential, because this can serve as a model for genuine many–body problems.

Charge conjugation in the shell model yields simple rules which lead to simple deviations of results which had been obtained beforehand using more cumbersome methods.

This is only a short selection from the many important and original contributions he made between 1954 and 1960 when he moved to CERN, where after a short period of adjustment he started contributing to particle physics. It may be worth noting, however, that his command of nuclear physics put him in a strong position to deal with problems straddling the borderline be-tween nuclear and particle physics, such as muon capture in heavy nuclei, or the nuclear optical modes for pions.

In recent years he has spoken out strongly against the usual interpretation of quantum mechanics, and this was also expressed in his last paper. Some, including myself, do not agree with his views, but we respect his arguments as raising and clarifying important issues and provoking serious thought. The issue he has defined will be debated for many years.

II. ATOMIC ENERGY
AND ARMS CONTROL

The Frisch-Peierls Memorandum

T he attached detailed report concerns the possibility of constructing a "super–bomb" which utilises the energy stored in atomic nuclei as a source of energy. The energy liberated in the explosion of such a super–bomb is about the same as that produced by the explosion of 1,000 tons of dynamite. This energy is liberated in a small volume, in which it will, for an instant, produce a temperature comparable to that in the interior of the sun. The blast from such an explosion would destroy life in a wide area. The size of this area is difficult to estimate, but it will probably cover the center of a big city.

In addition, some part of the energy set free by the bomb goes to produce radioactive substances, and these will emit very powerful and dangerous radiations. The effects of these radiations is greatest immediately after the explosion, but it decays only gradually and even for days after the explosion any person entering the affected area will be killed.

Some of this radioactivity will be carried along with the wind and will spread the contamination; several miles downwind this may kill people.

In order to produce such a bomb it is necessary to treat a substantial amount of uranium by a process which will separate from the uranium its light isotope (U_{235}) of which it contains about 0.7 percent. Methods for the separation of such isotopes have recently been developed. They are slow and they have not until now been applied to uranium, whose chemical properties give rise to technical difficulties. But these difficulties are by no means insuperable. We have not sufficient experience with large–scale chemical plant to give a reliable estimate of the cost, but it is certainly not prohibitive.

It is a property of these super–bombs that there exists a "critical size" of about one pound. A quantity of the separated uranium isotope that exceeds the critical amount is explosive; yet a quantity less than the critical amount is absolutely safe. The bomb would therefore be manufactured in two (or more) parts, each being less than the critical size, and in transport all danger of a premature explosion would be avoided if these parts were kept at a distance of a few inches from each other. The bomb would be provided with a mechanism that brings the two parts together when the bomb is intended to go off. Once the parts are joined to form a block which exceeds the critical amount, the effect of the penetrating radiation always present in the atmosphere will initiate the explosion within a second or so.

The mechanism which brings the parts of the bomb together must be arranged to work fairly rapidly because of the possibility of the bomb exploding when the critical conditions have just only been reached. In this case the explosion will be far less powerful. It is never possible to exclude this altogether, but one can easily ensure that only, say, one bomb out of 100 will fail in this way, and since in any case the explosion is strong enough to destroy the bomb itself, this point is not serious.

We do not feel competent to discuss the strategic value of such a bomb, but the following conclusions seem certain:

1. As a weapon, the super–bomb would be practically irresistible. There is no material or structure that could be expected to resist the force of the explosion. If one thinks of using the bomb for breaking through a line of fortifcafions, it should be kept in mind that the radioactive radiations will prevent anyone from approaching the affected territory for several days; they will equally prevent defenders from reoccupying the affected positions. The advantage would lie with the side which can determine most accurately just when it is safe to re–enter the area; this is likely to be the aggressor, who knows the location of the bomb in advance.

2. Owing to the spread of radioactive substances with the wind, the bomb could probably not be used without killing large numbers of civilians, and this may make it unsuitable as a weapon for use by this country. (Use as a depth charge near a naval base suggests itself, but even there it is likely that it would cause great loss of civilian life by flooding and by the radioactive radiations.)

3. We have no information that the same idea has also occurred to other scientists but since all the theoretical data bearing on this problem are published, it is quite conceivable that Germany is, in fact, developing this weapon. Whether this is the case is difficult to find out, since the plant for the separation of isotopes need not be of such a size as to attract attention. Information that could be helpful in this respect would be data about the exploitation of the uranium mines under German control (mainly in Czechoslovakia) and about any recent German purchases of uranium abroad. It is likely that the plant would be controlled by Dr. K. Clusius (Professor of Physical Chemistry in Munich University), the inventor of the best method for separating isotopes, and therefore information as to his whereabouts and status might also give an important clue. At the same time it is quite possible that nobody in Germany has yet realized that the separation of the uranium isotopes would make the construction of a super–bomb possible. Hence it is of ex-

treme importance to keep this report secret since any rumour about the connection between uranium separation and a super-bomb may set a German scientist thinking along the right lines.

4. If one works on the assumption that Germany is, or will be, in the possession of this weapon, it must be realized that no shelters are available that would be effective and that could be used on a large scale. The most effective reply would be a counter-threat with a similar bomb. Therefore it seems to us important to start production as soon and as rapidly as possible, even if it is not intended to use the bomb as a means of attack. Since the separation of the necessary amount of uranium is, in the most favourable circumstances, a matter of several months, it would obviously be too late to start production when such a bomb is known to be in the hands of Germany, and the matter seems, therefore, very urgent.

5. As a measure of precaution, it is important to have detection squads available in order to deal with the radioactive effects of such a bomb. Their task would be to approach the danger zone with measuring instruments, to determine the extent and probable duration of the danger and to prevent people from entering the danger zone. This is vital since the radiations kill instantly only in very strong doses whereas weaker doses produce delayed effects and hence near the edges of the danger zone people would have no warning until it was too late. For their own protection, the detection squads would enter the danger zone in motor-cars or airplanes which would be armoured with lead plates, which absorb most of the dangerous radiation. The cabin would have to be hermetically sealed and oxygen carried in cylinders because of the danger from contaminated air. The detection staff would have to know exactly the greatest dose of radiation to which a human being can safely be exposed for a short time. This safety limit is not at present known with sufficient accuracy and further biological research for this purpose is urgently required.

As regards the reliability of the conclusions outlined above, it may be said that they are not based on direct experiments, since nobody has ever built a super-bomb yet, but they are mostly based on facts which, by recent research in nuclear physics, have been very safely established. The only uncertainty concerns the critical size for the bomb. We are fairly confident that the critical size is roughly a pound or so, but for this estimate we have to rely on certain theoretical ideas which have not been positively confirmed. If the critical size were appreciably larger than we believe it to be, the technical difficulties in the way of constructing the bomb would be enhanced. The point can be definitely settled as soon as a small amount of uranium has been separated, and we think that in view of the importance of the matter immediate steps should be taken to reach at least this stage; meanwhile it is also possible to carry out certain experiments which, while they cannot settle the question with absolute finality, could, if their result were positive, give strong support to our conclusions.

On the Construction of a "Super–bomb"; Based on a Nuclear Chain Reaction in Uranium

The possible construction of "super–bombs" based on a nuclear chain re-action in uranium has been discussed a great deal and arguments have been brought forward which seemed to exclude this possibility. We wish here to point out and discuss a possibility which seems to have been overlooked in these earlier discussions.

Uranium consists essentially of two isotopes, uranium 238 (99.3 percent) and uranium 235 (0.7 percent). If a uranium nucleus is hit by a neutron, three processes are possible: (1) scattering, whereby the neutron changes direction and, if its energy is above about 0.1 MeV, loses energy; (2) capture, when the neutron is taken up by the nucleus; and (3) fission, i.e., the nucleus breaks up into two nuclei of comparable size, with the liberation of an energy of about 200 MeV.

The possibility of a chain reaction is given by the fact that neutrons are emitted in the fission and that the number of these neutrons per fission is greater than 1. The most probable value for this figure seems to be 2.3, from two independent determinations.

However, it has been shown that even in a large block of ordinary uranium no chain reaction would take place since too many neutrons would be slowed down by inelastic scattering into the energy region where they are strongly absorbed by uranium 238.

Several people have tried to make chain reaction possible by mixing the uranium with water, which reduces the energy of the neutrons still further and thereby increases their efficiency again. It seems fairly certain, however, that even then it is impossible to sustain a chain reaction.

In any case, no arrangement containing hydrogen and based on the action of slow neutrons could act as an effective super–bomb, because the reaction would be too slow. The time required to slow down a neutron is about 10^{-5} sec and the average time lost before a neutron hits a uranium nucleus is even 10^{-4} sec. In the reaction, the number of neutrons would increase exponen-tially, like $e^{t/\tau}$ where τ would be at least $10^{10^{-4}}$ sec. When the temperature reaches several thousand degrees the container of the bomb will break and within 10^{-4} sec the uranium would have expanded sufficiently to let the neu-trons escape and so to stop the reaction. The energy liberated would, there-fore, be only a few times the energy required to break the container, i.e., of the same order of magnitude as with ordinary high explosives.

Bohr has put forward strong arguments for the suggestion that the fission observed with slow neutrons is to be ascribed to the rare isotope uranium 235, and that this isotope has, on the whole, a much greater fission probability than the common isotope uranium 238. Effective methods for the separation of isotopes have been developed recently, of which the method of thermal diffusion is simple enough to permit separation on a fairly large scale.

This permits, in principle, the use of nearly pure uranium 235 in such a bomb, a possibility which apparently has not so far been seriously considered. We have discussed this possibility and come to the conclusion that a moderate amount of uranium 235 would indeed constitute an extremely efficient explosive.

The behavior of uranium 235 under bombardment with fast neutrons is not known experimentally, but from rather simple theoretical arguments it can be concluded that almost every collision produces fission and that neutrons of any energy are effective. Therefore it is not necessary to add hydrogen, and the reaction, depending on the action of fast neutrons, develops with very great rapidity so that a considerable part of the total energy is liberated before the reaction gets stopped on account of the expansion of the material.

The critical radius r_0—i.e., the radius of a sphere in which the surplus of neutrons created by the fission is just equal to the loss of neutrons by escape through the surface—is, for a material with a given composition, in a fixed ratio to the mean free path of the neutrons, and this in turn is inversely proportional to the density. It therefore pays to bring the material into the densest possible form, i.e., the metallic state, probably sintered or hammered. If we assume, for uranium 235, no appreciable scattering, and 2.3 neutrons emitted per fission, then the critical radius is found to be 0.8 times the mean free path. In the metallic state (density 15), and assuming a fission cross–section of 10^{-23} cm^2,[1] the mean free path would be 2.6 cm and r_0 would be 2.1 cm, corresponding to a mass of 600 grams. A sphere of metallic uranium 235 of a radius greater than r_0 would be explosive, and one might think of about 1 kg as a suitable size for the bomb.

The speed of the reaction is easy to estimate. The neutrons emitted in the fission have velocities of about 10^9 cm/sec and they have to travel 2.6 cm before hitting a uranium nucleus. For a sphere well above the critical size the loss through neutron escape would be small, so we may assume that each neutron, after a life of 2.6×10^{-9} sec, produces fission, giving birth to two neutrons. In the expression $e^{t/\tau}$ for the increase of neutron density with time, τ would be about 4×10^{-9} sec, very much shorter than in the case of a chain reaction depending on slow neutrons.

If the reactions proceed until most of the uranium is used up, temperatures of the order of 10^{10} degrees and pressures of about 10^{13} atmospheres are produced. It is difficult to predict accurately the behavior of matter under these

[1] Note added March 1994: This was a surprising slip. The correct figure is about 3 times smaller, and the estimate of critical mass is therefore 3^3, or about 30 times too small.

extreme conditions, and the mathematical difficulties of the problem are considerable. By a rough calculation we get the following expression for the energy liberated before the mass expands so much that the reaction is interrupted:

$$E = 0.2M(r^2 / \tau^2)\sqrt{(r / r_0)-1} \tag{1}$$

(M, total mass of uranium; r, radius of sphere; r_0, critical radius; τ, time required for neutron density to multiply by a factor e). For a sphere of diameter 4.2 cm ($r = 2.1$ cm), $M = 4700$ grams, $\tau = 4 \times 10^{-9}$ sec, we find $E = 4 \times 10^{22}$ ergs, which is about one-tenth of the total fission energy. For a radius of about 8 cm ($M = 32$ kg) the whole fission energy is liberated, according to formula (1). For small radii the efficiency falls off even faster than indicated by formula (1) because τ goes up as r approaches r_0. The energy liberated by a 5 kg bomb would be equivalent to that of several thousand tons of dynamite, while that of a 1 kg bomb, though about 500 times less, would still be formidable.

It is necessary that such a sphere should be made in two (or more) parts which are brought together first when the explosion is wanted. Once assembled, the bomb would explode within a second or less, since one neutron is sufficient to start the reaction and there are several neutrons passing through the bomb in every second, from the cosmic radiation. (Neutrons originating from the action of uranium alpha rays on light–element impurities would be negligible provided the uranium is reasonably pure.) A sphere with a radius of less than about 3 cm could be made up in two hemispheres, which are pulled together by springs and kept separated by a suitable structure which is removed at the desired moment. A larger sphere would have to be composed of more than two parts, if the parts, taken separately, are to be stable.

It is important that the assembling of the parts should be done as rapidly as possible, in order to minimize the chance of a reaction getting started at a moment when the critical conditions have only just been reached. If this happened, the reaction rate would be much slower and the energy liberation would be considerably reduced; it would, however, always be sufficient to destroy the bomb.

It may be well to emphasize that a sphere only slightly below the critical size is entirely safe and harmless. By experimenting with spheres of gradually increasing size and measuring the number of neutrons emerging from them under a known neutron bombardment, one could accurately determine the critical size, without any danger of a premature explosion.

For the separation of the uranium 235, the method of thermal diffusion, developed by Clusius and others, seems to be the only one which can cope with the large amounts required. A gaseous uranium compound, for example, uranium hexafluoride, is placed between two vertical surfaces which are kept at a different temperature. The light isotope tends to get more concentrated near the hot surface, where it is carried upwards by the convection current. Exchange with the current moving downwards along the cold surface produces a fractionating effect, and after some time a state of equilibrium is reached when the gas near the upper end contains markedly more of the light isotope than near the lower end.

For example, a system of two concentric tubes, of 2 mm separation and 3 cm diameter, 150 cm long, would produce a difference of about 40 percent in the concentration of the rare isotope between its ends, and about 1 gram per day could be drawn from the upper end without unduly upsetting the equilibrium.

In order to produce large amounts of highly concentrated uranium 235 a great number of these separating units will have to be used, being arranged in parallel as well as in series. For a daily production of 100 grams of uranium 235 of 90 percent purity, we estimate that about 100,000 of these tubes would be required. This seems a large number, but it would undoubtedly be possible to design some kind of a system which would have the same effective area in a more compact and less expensive form.

In addition to the destructive effect of the explosion itself, the whole material of the bomb would be transformed into a highly radioactive state. The energy radiated by these active substances amount to about 20 percent of the energy liberated in the explosion, and the radiation would be fatal to living beings even a long time after the explosion.

The fission of uranium results in the formation of a great number of active bodies with periods between, roughly speaking, a second and a year. The resulting radiation is found to decay in such a way that the intensity is about inversely proportional to the time. Even one day after the explosion the radiation will correspond to a power expenditure of the order of 1000 kW, or to the radiation of a hundred tons of radium.

Any estimates of the effects of this radiation on humans must be rather uncertain because it is difficult to tell what will happen to the radioactive material after the explosion. Most of it will probably be blown into the air and carried away by the wind. This cloud of radioactive material will kill everybody within a strip estimated to be several miles long. If it rained the danger would be even worse because active material would be carried down to the ground and stick to it, and persons entering the contaminated area would be subjected to dangerous radiation even after days. If one percent of the active material sticks to the debris in the vicinity of the explosion and if the debris is spread over an area of, say, a square mile, any person entering this area would be in serious danger, even several days after the explosion.

In these estimates, the lethal dose of penetrating radiation was assumed to be 1,000 roentgens; consultation of a medical specialist on X–ray treatment and perhaps further biological research may enable one to fix the danger limit more accurately. The main source of uncertainty is our lack of knowledge as to the behavior of materials in such a super–explosion, and an expert on high explosives may be able to clarify some of these problems.

Effective protection is hardly possible. Houses would offer protection only at the margins of the danger zone. Deep cellars or tunnels may be comparatively safe from the effects of radiation, provided air can be supplied from an uncontaminated area (some of the active substances would be noble gases which are not stopped by ordinary filters).

The irradiation is not felt until hours later when it may be too late. Therefore it would be very important to have an organization which determines the exact extent of the danger area, by means of ionization measurements, so that people can be warned from entering it.

Defence Against the Atom Bomb

In a recent article[1], D. G. Christopherson gives an impression that the disastrous effects of atomic weapons have been grossly overrated. His main points are that the atomic bomb is effective mostly against large cities, and that dispersal would leave a density of population against which atomic bombs would be too costly to use, provided there were adequate shelters and an efficient warning system.

It seems to us that Mr. Christopherson's conclusions cannot be maintained if one remembers the following facts:

(1) The power of atomic bombs against cities has been demonstrated and is not questioned in Mr. Christopherson's note. It is true that shelters would reduce casualties from collapse of buildings, injuries from glass and other fragments and flash–burn. There remains even then a smaller area in which the blast intensity and the intensity of penetrating radiations are so high that adequate protection would not be practicable. Present knowledge of the mechanism of radiation injuries leaves little room for the hope, expressed by Mr. Christopherson, that an effective treatment after exposure could be found, except in marginal cases.

(2) Even if one accepts the statement that large cities are undesirable, their dispersion in a manner governed not by the natural growth of other units, but by the fear of atomic war, would represent a most drastic dislocation of the economic life of Great Britain. To start dispersal when war is imminent would clearly be too late. Indeed, a group of American scientific workers have seriously discussed a plan[2] of dispersing all major cities of the United States. The data given in their article make it clear what a formidable undertaking this would represent.

---

[1]*Nature*, **158**, 151 (1946).
[2]Marshak, J., and Teller, E., *L.R.K. Bull. Assoc. Atomic Sci. Chicago*, April 15, 1946.

Great Britain is, of all big powers, the least favourably placed in this respect, because of the limitations in resources and building materials, high centralization, dependence on imports, and because of the relatively high population density that would remain even after perfect dispersal. It must be remembered that, for protection against atomic bombs, one must disperse not only dwellings, but also all other important installations, such as ports, transport centres and warehouses, power stations and factories, unless they can be rebuilt underground. A comparison of the amount of rebuilding required for this with the number of houses involved in the present building program gives an idea of the staggering magnitude of the job.

(3) Dispersal of all large cities would still be insufficient if it turned out that atomic bombs could be used without prohibitive effort against the largest remaining units. Thus one's plan for dispersal must depend on the cost of a bomb. This should not be estimated by comparison with the published figures for expenditure on the American project, which deliberately put speed and certainty of operation before economy. It is known from the Smyth Report that the effort was spread over three independent lines of attack, of which each is now known to have been successful. The facts published in the Smyth Report alone are sufficient to save anyone embarking now on a similar project much of the expense of development work.

In particular, Mr. Christopherson's argument based on power–consumption is a fallacy. The figure of 1,000 kilowatts which Smyth quotes as being associated with a production–rate of one gram of plutonium per day refers to power *produced*, not *consumed*. It should therefore appear on the opposite side of the balance sheet, particularly when once the remaining engineering problems in the way of utilization of atomic power have been solved.

(4) In discussing the protection by shelters and a warning system, it must be borne in mind that the warning, to be effective, must come into action on the approach of a single aeroplane or rocket. In the case of the latter, if any warning can be given in advance the available time must be exceedingly short and at best sufficient for people to drop what they are doing and to run for shelter. An enemy can therefore cripple the life of a country by sending at frequent intervals aeroplanes, rockets or other missiles of which only a few need carry atomic bombs. The experience of the war has shown that in these circumstances the warning system becomes useless, since people will soon refuse to take cover.

(5) The difficulties connected with a warning system are enhanced if it becomes necessary to guard against a surprise attack of the Pearl Harbour type. In that case the warning system would have to be alerted in peace–time and might have to come into action on the approach of single unidentified aeroplanes and other objects. To take this step in peace–time could in itself have bad effects on international relations.

(6) Lastly, we must remember that the very idea of an atomic bomb is only six or seven years old. To base one's ideas, or even a complete reorganization of a country, on the assumption that atomic bombs will never be more effective or less costly than they are now, would be as shortsighted as it would have been in 1938 to assume that atomic weapons are impossible.

We do not wish to overstate the effect of such bombs. We do not wish to assert that all civilized life will stop if there is a major atomic war. But it seems clear to us that in all probability the effects of such a war would be such a serious blow to civilization that the problem of ruling out atomic warfare (and also, of course, warfare using other new means of mass destruction) should be regarded as the foremost political problem of our time. Any attempt to blind ourselves to the seriousness of the dangers is liable to diminish the sense of urgency that alone will ensure a determined and sustained attack on the problem.

Atomic Energy:
Threat and Promise

In public discussion of the prospects of atomic energy, the threat to the future of civilization which the atomic bomb represents tends to overshadow the promise of benefits from the constructive applications of atomic power. After the first public announcements on the bomb, exaggerated claims for the possibilities of an impending "atomic age" appeared, particularly in the popular press, and the reaction to them discredited the whole subject.

Serious thought about the prospects is, in any case, made difficult by the lack of information, due partly to secrecy restrictions and partly to real uncertainties in the physical, chemical, and engineering data.

I shall endeavor to appraise the positive possibilities as they appear today, and as far as this is possible within the limits of available information. In many instances it will not be possible to give definite answers, but merely to indicate the factors on which the answers depend and the types of problems that have to be solved before the answers can be found.

The Atomic Bomb

In discussing the potentialities of atomic energy it would be futile to ignore atomic weapons, not only because they were the first and, up to now, the only important practical application, but because the threat which they represent must always override all imaginable advantages. If there is to be another world war fought with atomic weapons and, as we have to expect, with other new weapons of mass destruction, prospects are so grim that the constructive applications of atomic energy would give us poor comfort. If, on the other hand, international politics were to progress to a point where we felt reasonably confident that there would not be another major war, this would represent such an enormous step forward that the direct benefits would for a long time appear insignificant in comparison.

This problem is essentially political, but it may be appropriate to summarize here the features of the atomic bomb that make it play such an outstanding part. We have often been reminded that the atomic bomb raids were not the most destructive single operations in the past war, nor the ones involving the greatest number of casualties. The essential new feature lies in the comparative cheapness of these weapons, once the necessary scientific and development work has been done, and in their small bulk, which makes it possible to deliver them by a variety of means against which an effective defense and even an adequate warning system seem quite impossible. As a result of this, once stocks of atomic bombs are available, surprise action becomes much easier than with other forms of attack, which require large operational forces.

It is probably not true that atomic bombs in themselves would make a future war much more destructive. Other destructive techniques were perfected during the recent war, and there is no doubt that such weapons as rockets of the V_2 type were only just beginning to play their part. Other new developments, like chemical and biological warfare, about which very little has been said in public, share many of the features of atomic bombs. Attention has focused on the atomic bomb because its effect has been demonstrated.

One of the essential features of atomic warfare is that it is bound to result in heavy damage and casualties to all parties (assuming that both sides possess these weapons) and that, while it may be possible to deliver a crippling blow in a very short time, this will not prevent retaliation striking even the country whose stocks of atomic bombs are superior. Hand in hand with that goes the fact that, while they could destroy a country, atomic bombs could not be used to occupy and hold it. Hence they lend themselves to irresponsible destructive action but not to any calculated aggressive war that is likely to benefit the aggressor.

It is important to stress this feature because it indicates that the destructive nature of atomic weapons may itself be of ultimate benefit if it helps to bring home to everyone the lesson that the use of military force for aggression does not pay.

Constructive Uses of Atomic Bombs

Supposing the threat of atomic warfare could be eliminated, are there any other uses to which atomic bombs could be put? No definite suggestion has so far been made that can bear examination, but the problem is so new that the possibility should not be ruled out altogether. The use for mining and other similar engineering operations would seem extremely difficult, for three reasons. One is the enormous intensity of the blast action in the immediate neighborhood of the explosion center, which leads to a much more thorough destruction than is necessary or desirable. This immediately rules out the use in all those mining operations where one is dealing with a narrow vein of ores. One would probably then find the ores thoroughly mixed up with the surrounding matter, and their recovery would be more difficult than before.

The second difficulty lies in the fact that it is very difficult to use atomic explosions in a controlled way and to limit their effects, since in any use underground the range of the blasting action will depend greatly on the nature of the geological structures, which are hard to explore over a region as large as the likely range of the explosion. In addition, the power of atomic bombs is not exactly predictable and will vary somewhat from one instance to another, depending in most types of design on whether or not a stray neutron happens to initiate a chain reaction before the mechanical assembly of the components is complete. Even if the power of the explosion could be made precisely reproducible there would still remain the difficulty that the precise range in different types of rock or soil could be ascertained only by experimentation (as it is with ordinary mining explosives), and such experimentation covering the necessary variety of conditions would be a prohibitive task.

The third difficulty is that in an underground explosion the greater part of the fission products, with their intense radioactivity, would probably remain on or below the ground and therefore the affected area would become inaccessible for a very long period. This difficulty did not arise in the past because, for explosions in air, most of the radioactivity is carried up into the upper atmosphere, while in underwater explosions it is reduced fairly rapidly by dilution.

In spite of these difficulties, one could imagine cases in which it is not necessary to limit the range precisely. This applies, for example, to breaking a gap in a mountain range so as to alter the course of a river, or to breaking up an ice–pack that threatens to cause serious obstruction. The danger of radioactivity may not be serious where speedy access to the area is not needed or where one can make sure that rising air will disperse the products.

Atomic Power

Leaving aside the very speculative possibility of utilizing atomic explosions, we are left with atomic chain reactions conducted in a controlled way. Their most obvious use lies in the power they generate. From the point of view of basic physics this problem is completely solved, and there is no difficulty in operating an atomic pile in such a way as to generate heat at whatever rate it can be removed from the pile by a suitable cooling medium. Considerable engineering problems remain, however, of which the most serious is the question of reaching a temperature at which the generation of power becomes economical.

The chief reason for this difficulty is that in all piles constructed with ordinary uranium the reproduction factor, i.e., the number of secondary neutrons available to cause further fission for each primary neutron resulting from the previous fission process, is only slightly larger than unity. The chain reaction can, of course, proceed only if this reproduction factor exceeds unity, and any cause that will reduce it by more than the slight available margin will prevent the pile from operating. Such a cause is represented by the absorption of neutrons in impurities, in the cooling medium, or in materials introduced for structural reasons. Hence if we think of piles made of uranium metal and graphite, like the piles now operating at Hanford in North America, we must not only use uranium and graphite of an extremely high degree of purity, with tolerances for certain impurities of the order of a few parts in a million or better, but the amount of water or other cooling medium used in the pile and the amount of other materials introduced for structural purposes or for preventing corrosion of the uranium metal must be extremely limited. Because of such limitations the Hanford piles do not permit a temperature–rise that would allow the generation of steam at even a modest pressure. There seems no doubt, however, that a determined attack on the remaining engineering problems will lead to a solution in which the temperature of the pile can be raised considerably. As far as the basic nature of the reaction is concerned, there is no practical limitation to the temperature. There is, in principle, no reason why it should not be taken up to a point where the utilization of the heat by means of steam engines or turbines or any other methods of generating mechanical energy is most favourable. How far this can be carried in practice is as yet unknown, and the higher the temperature the greater the difficulties in protecting the structural components against corrosion or against mechanical distortion. Both problems are accentuated by the presence of intense radiation, which accelerates chemical action because of the high degree of ionization and dislocation of atomic structures that it causes, and which may even reduce mechanical strength.

Some of the problems in the way of high–power piles would be eased if the piles were constructed not of ordinary uranium but of enriched material, obtained either by separation of isotopes or as a result of nuclear transformations carrried on in piles of the Hanford type or in the power piles themselves. In this case the active content of the piles is increased, and with it the margin in the reproduction factor. Piles constructed in this way have much more latitude in the use of structural materials and cooling medium, but the corrosion problems remain, and at the same time the piles burn a very much more expensive fuel.

Another difficulty in the operation of all piles is represented by the dangerous radiations, contributed in about equal parts by the neutrons themselves and by the gamma rays resulting from neutron–capture in the pile and from the radioactive fission products. This necessitates heavy protective shields around the piles, and arrangements for doing any adjustments on them by remote control. It also requires arrangements to prevent the escape of radioactive materials by way of the cooling medium into the machinery driven by the pile. Otherwise this machinery would become radioactive and would have to operate inside the radiation shield; it could then not be serviced except by remote control.

Similarly, refuelling of the pile and removal and treatment of depleted fuel have to be carried out by remote control. These obstacles have been overcome in the piles now in operation, but they represent severe limitations in the engineering design.

Advantages of Atomic Power

We have seen that, even if not all engineering problems in the way of the production of atomic power have been solved, they at least appear soluble. What will be gained by making atomic power available on a large scale? We shall first consider large stationary power units, used to generate electricity.

The cost of electricity from such stations will depend on the cost of the fuel, the capital charges and running expenses of the pile itself and its auxiliaries, and the cost of generating machinery and distribution. The last item is much the same as for conventional power stations, except that the factors governing the location of atomic power stations are different; this may affect the distribution cost. The pile itself probably represents a much greater capital investment than a boiler plant, though not necessarily a higher running cost. The actual cost is hard to estimate when the design is as yet unknown. Since the pure materials (uranium, graphite, etc.) required for the pile represent a major item, the cost must also depend on the scale on which these are being manufactured, and hence on the overall scale of the atomic power program.

The cost of the raw material at pre–war market prices is very small compared to that of a corresponding amount of coal. No doubt the applications of atomic energy make uranium a more valuable material, and indeed, because of its military importance, there is presumably no free market for it at all. Even so, quite a considerable rise in the cost of raw material would hardly affect the cost of electricity from such a plant.

These considerations show that, since at present the generating and distribution costs are a large item in the price of electricity, the cost of atomic power is likely to be about the same as for electricity from coal and that, even if we assume the pile to cost no more than a boiler plant and the fuel to cost nothing, the price of electricity would fall by only some 20 or 30 per cent.

A recent official statement by the American authorities confirms these rough estimates, and arrives at a price for electricity that is slightly higher than the current one. A slight increase in the cost of coal at the power station would equalize the difference. No details were quoted, but it seems likely that a similar estimate for conditions in Great Britain would show a slight balance in favour of atomic power. These figures are not meant to be precise, and further experience may well show a somewhat greater margin on either side.

It follows that we can hardly expect a revolutionary change in the price of electricity. One specific advantage of atomic power is the small bulk of the fuel consumed, which makes the power stations independent of access to railways and waterways. This allows the planning of power stations close to the consumer, and reduces the difficulties of distribution; it also gives more freedom in placing the stations in positions where the low–grade heat that is always a by–product of the generation of power can be used for heating or for industrial purposes.

This specific advantage is of outstanding importance in the case of countries, such as Italy, in which there are no adequate resources of power and where the transportation of coal becomes an important factor. It may even make possible new projects, such as the irrigation of desert areas in which the supply of power would today present prohibitive difficulties.

In addition, atomic power may add to our resources if and when the depletion of known coal deposits, or lack of people who are attracted to mining as a career, limits the supply of coal. At equal cost, it would surely be preferable to employ fewer men in coal mines. It should not be assumed, however, that atomic power stations may in the immediate future reduce the demand for coal. Even disregarding the time still needed for the necessary development work, it will take a long time to build atomic power stations on a sufficient scale to have any effect on world economy. It will take an even longer time to convert to electricity such things as railways, domestic fires, and so on, the conversion of which would become worthwhile only if the price of coal were to increase relative to that of electricity. In any case it must, of course, be remembered that there will always remain some need for coal for metallurgical, chemical, and other purposes.

All these considerations are based on the assumption that sufficient amounts of atomic fuel can be found. The next section will deal with the question of their availability.

Supply of Raw Materials

From the previous section it is clear that the availability of atomic fuel is a vital question for the success of any atomic power project. It would seem an easy task to estimate the accessible deposits of suitable ores and compare them with the expected demand, but there are a number of factors complicating this issue. Firstly, the only large atomic piles that have so far been operated work with ordinary uranium. Since the major constituent of this is uranium 238, which does not undergo fission with the slow neutrons on which these piles operate, it is only the rare constituent uranium 235 which is consumed in this operation. This represents only 0.7 percent of the uranium found in nature. In the course of this operation, however, some of the neutrons are captured in uranium 238 and convert it (through the intermediate nucleus, neptunium 239) into plutonium 239. It is well known that this conversion was, in fact, the main object of the Hanford piles. It has been stated that the number υ of secondary neutrons obtained from each fission process is greater than one (otherwise no chain reaction would be possible) but less than three. Of these υ secondary neutrons from each fission, one must in a steady chain reaction be captured in another fissile nucleus so as to maintain the chain reaction, and the remainder (υ-1) will either escape from the surface of the pile or be absorbed in one or the other of its constituents. Escape from the surface can be made small by increasing the size of the pile, but there are unavoidable losses due to absorption in the "moderator," i.e., the substance used for slowing down neutrons (which in the Hanford piles is graphite), in the cooling medium, and in the necessary structural elements of the pile. The remainder may be available for absorption in uranium 238 or in other suitable materials mixed into the pile for this purpose. This number is less than υ-1, and from the published figures is therefore certainly less than two. We thus know with certainty that at most two atoms of plutonium will be made for every atom of 235 burned in the pile, and it is also clear that this is a generous upper limit.

If it were possible to run the pile in such a way that this number were greater than one, the amount of plutonium generated would be greater than the amount of 235 burnt up. If the properties of plutonium were similar to those of 235, this would mean that by leaving in the pile an appropriate fraction of the plutonium so made we could in this way make up for the uranium 235 that was burnt, and such a pile could therefore continue operating until the depletion of uranium 238 reduced the rate of production of plutonium. In these conditions it would therefore be possible to burn not only the uranium 235 but a substantial fraction of the uranium 238.

With our assumptions, the condition for this process is that υ-1 should exceed one, i.e., that υ should exceed two. There is no guidance from published figures as to whether or not this is the case. Actually we have made the assumption that plutonium is about as efficient a fuel as uranium 235; if it differs appreciably, the critical limit for υ may be somewhat different from two. It will, in any case, be increased by the unavoidable loss of neutrons through absorption in other parts of the pile.

Since, as this process continues, we shall be burning plutonium rather than uranium 235, its success will also depend on the value of υ for the fission of plutonium.

Lastly, it is known that a new fissile material can be made by the bombardment of thorium with neutrons, and that this new material, which is an isotope of uranium of weight 233, may also be suitable for a chain reaction. No evidence is available on the merits of this isotope as an atomic fuel compared with uranium 235 or plutonium, or on the number of neutrons emitted in its fission.

It is, of course, conceivable that there may be similar processes as yet undiscovered starting from other raw materials, but this does not appear very probable.

All the above uncertainties in the data make the position about raw materials quite obscure. If it should turn out that in a pile starting with ordinary uranium the plutonium produced will more than offset the consumption of uranium 235, and also that as the pile turns over to consuming plutonium the amount of plutonium produced is greater than that consumed, it will be possible to utilize most of the atoms of ordinary uranium. On the other hand, if this is not possible the material will become almost useless when a substantial fraction of the 235 has been burnt. There is, therefore, an uncertainty of about a factor 100 in the amount of uranium required to maintain a given power output. Alternatively, the reaction can be conducted in such a way that some of the excess neutrons are not absorbed in uranium 238 but in thorium, and if it should turn out that the production of 233 more than offsets the initial consumption of 235 or plutonium and later on the consumption of 233, when this becomes the major fuel in the pile, it may be possible to run the power stations in a way in which ultimately the fuel is only thorium without a continuing consumption of uranium. This would at once add substantially to the supply of raw materials upon which we can draw.

Quite apart from this uncertainty, the amount of uranium that we can expect to be available is hard to estimate. There are at present only a few known deposits containing uranium ore in sufficient concentration to make its extraction profitable, but even if further surveys should not lead to the discovery of new important deposits, it is well to remember that uranium is a common element contained in many minerals (such as ordinary granite) in very low concentration. No methods have as yet been worked out that would make the extraction of uranium from such poor ores a feasible operation. The concentration of uranium in some poor ores is, however, larger than, for example, the concentration of gold in most ores that are commercially worked for gold, and it is possible that a chemical process could be found which would effectively make this uranium available. This would open up sufficient quantities of uranium for any power program that is likely to be needed in the foreseeable future.

Evidently the question whether any particular extraction process will make low–grade ores economically interesting depends on the price at which the fuel would make the production of atomic power compete successfully with those from other fuels. At present the price of the raw uranium is quite a small fraction of the estimated cost of atomic power, and a considerable increase in the cost of mining and extraction could be tolerated without altering the picture substantially. The limit depends on the place in which power plants are contemplated, because of the cost of transportation of coal, and on the likelihood of an increase in the price of coal owing to the exhaustion of suitable deposits, and other factors.

We must conclude that, while the uncertainty in all these facts certainly does not allow us at this stage to feel assured that a sufficient supply of atomic fuel can be obtained, it is equally impossible at present to exclude this possibility. It is certainly of sufficient promise to justify intensive study and development work.

Mobile Power and Other Specific Applications

Can atomic piles be made suitable for small units to be used in ships, trains, etc.? There are a number of factors which make the operation of small plants uneconomical. The first is the critical size. It is known that for each reacting system there is a least size at which chain reaction becomes possible, since for any system of smaller size the number of neutrons lost by leakage from the surface reduces the reproduction factor below the critical limit. In the ordinary pile made of uranium metal and graphite we have already seen that the margin in the reproduction factor is very small. Hence only a small leakage of neutrons can be tolerated, and the least size of plant becomes very substantial. If one desires a power unit with a small power output, the plant will still have to be at least of the critical size and will therefore be very bulky. In addition, the amount of atomic fuel locked up in such a power unit is then very substantial, and the unit therefore represents a relatively large capital investment.

Things are more favorable if one can use concentrated atomic fuel, i.e., either uranium 235 obtained from a separation plant, or plutonium, or possibly uranium 233 from thorium. In the case of these substances the reproduction factor is considerably larger, and this brings the critical size down to quite a convenient magnitude. In fact, it is likely that in practice the size of such a unit would be determined by the requirement of getting in the necessary amount of internal surfaces for heat–extraction purposes. On the other hand, such concentrated fuels are very much more expensive. In the case of uranium 235 this would have to be produced by an isotope separation plant of the type of the plants that produced the material for one variety of atomic bomb, although the degree of separation need not be the same as that suitable for the bomb. Plutonium and uranium 233 are produced in piles, and their cost depends essentially on the considerations referred to in the previous section. If it turns out that a suitable pile can not only regenerate the fissile material necessary to keep the pile in operation, but also leaves something to spare, it might be comparatively easy to get fissile materials as byproducts of power production.

Even for piles with enriched fuel, however, there remains the need for protective shielding, and it has been estimated that the minimum weight of such a shield is about a hundred tons. This clearly at once rules out use in cars or trains, but a weight of this order is quite modest if one thinks of its use, for example, in a large ship. In fact, it would appear possible in the case of a ship to use the surrounding water as part of the shield and thereby further reduce the deadweight that has to be carried. Whether atomic ships are a practicable proposition is therefore merely an economic question. In comparing the cost of atomic power with coal or oil one has to bear in mind that ships would have to use concentrated atomic fuel, as it does not seem likely that a power unit of the size, of the Hanford piles would be suitable for a ship. It is impossible to estimate at the moment the cost of the concentrated fuel. On the other hand, it is certainly an advantage that the weight of the fuel consumed during the voyage would be quite negligible, and this factor would be particularly important for long journeys without refuelling.

As regards to airplanes, the need for a heavy shield is certainly a most serious handicap, and atomic planes cannot be practical unless one increases by a large factor the overall size of the plane, and with it the size of the runways, etc. Even if one is prepared to do this, one would probably not want to operate the power unit of a plane on low–grade heat, which would necessitate the use of steam engines. The solution of this problem is therefore dependent on raising the temperature in the pile to a level where more modern forms of conversion to mechanical power become possible.

On the other hand, the low weight of the consumed fuel is in this case a most important consideration, and if an atomic plane is possible at all it will have a practically unlimited range without reduction of the carrying capacity.

A particularly interesting problem is that of using atomic energy in rockets for interplanetary travel. If one looks at this problem purely from the point of view of the balance of energy, it is clear that atomic power overcomes the major handicap of other such rockets. The energy necessary to carry a certain amount of ordinary fuel beyond the range of attraction of the earth is greater than the energy obtained by burning the fuel, whereas for atomic fuel it forms a negligible fraction. However, rockets operate on the recoil produced by ejecting the combustion product backwards, and, since the mass of the combustion product is negligible in the case of atomic power such rockets would have to carry material specially for the purpose of ejecting it. Thereby one loses all the advantage of the low fuel weight unless it is possible to eject materials at a speed in excess of that of the combustion products in ordinary material. This means that one would have to use atomic power for heating a suitable substance to a temperature higher than the flame temperature in ordinary rockets. Though there is nothing to prevent this in the basic process of a chain reaction, the engineering difficulties are clearly formidable.

Radioactive Elements

To return from speculation to well–established fact, we next consider the uses of radioactive elements produced in piles, either as fission fragments or by the neutron bombardment of almost any substance. These can be used for most of the purposes for which naturally radioactive elements have been used in the past, such as medical therapy and industrial radiology, but they are becoming available in practically unlimited amounts, and with a much greater variety of properties. A doctor can, for example, choose for his treatment a radioactive element of a suitable lifetime and can thus inject it close to the affected spot. The decay of the activity will then limit the total dose. At the same time there is more freedom in the choice of the hardness of the radiation. By choosing a radioactive isotope of a suitable element it may be possible to cause the radioactivity to accumulate in a suitable place in the human body, and this principle has been successfully applied in the United States to the treatment of a rare form of thyroid disease by means of radioactive iodine.

In radiology, gamma rays may for certain purposes be preferable to X–rays because of their greater penetrating power, and because they allow the use of a very compact source which may, for example, be placed inside a small opening insufficient to house an X–ray tube. Atomic piles can produce such sources with intensities greatly in excess of those practicable with radium.

A much more important application, however, lies in the use of radioactive tracers for chemical or biological research. This gives the possibility of labelling certain atoms and recognizing them afterwards, while allowing them to take part in a chemical or biological process in which their history will be precisely the same as that of other atoms of the same element. This allows us to recognize, for example, the exchange of two atoms of the same element between two different molecules, and it allows us to identify the origin of substances in a living body—to see, for example, whether they are taken up as part of the current food intake, or whether they come from deposits of the same substance in other parts of the body. Such tracer methods were already being employed before the war, using radioactive isotopes made in the laboratory by means of cyclotrons or other machines. Atomic piles provide both greater quantities and greater variety.

A further advantage of such tracers is that their radioactivity may make it possible to observe them in the living body, while keeping the intensity of radiation well below the safe tolerance limit. It is immediately evident that this method of studying chemical reactions in the living body is an extremely powerful tool.

Similarly, there are applications of such tracer elements in other fields, notably in engineering research. These possibilities have not yet been adequately explored.

The amounts required for such experiments are usually quite small, so that a pile which, as a power station, has a very modest output can keep a number of laboratories supplied. In the United States such radioactive tracers are already freely available, but their export is prohibited by Act of Congress. In England work is in hand on the construction of a pile for this purpose.

Other Possibilities

All the applications discussed above amount really to doing better, or more cheaply, or more conveniently, things that could also be done by other means. Nobody has so far shown enough foresight to say what completely new uses may be found for atomic energy. Similar reasoning applied to the steam engine or to electricity or to the internal combustion engine, when they were first introduced, would evidently have given a very incomplete picture of their promise. It may well be that the really important applications will not be thought of until the world has got more used to the new possibilities.

However that may be, I hope to have made it clear in this necessarily vague and incomplete report that a great wealth of new possibilities is opened up by atomic power, and that, if we show enough enterprise and determination in exploring them, they are bound to add materially to the resources of the world.

Britain in the Atomic Age

I n this general review of the past 25 years, I will comment on events in Great Britain and how they were affected by the release of nuclear power. Since there is hardly any aspect of life which has not in some measure been influenced by the new technology, this might involve writing a complete history of postwar Britain. Therefore I shall have to confine myself to a few general comments. My selection will necessarily reflect the knowledge and interests of a scientist who participated in work on atomic energy during the war, but has since had only peripheral contact with this field.

Great Britain has now, in the United Kingdom Atomic Energy Authority (AEA), a sizeable organization concerned with atomic energy, which was built up after the war. In 1945 there was only a very small amount of work going on here. In the earlier part of the war there had been a number of active research groups and some industrial development work, but in 1943 it was decided to discontinue this work. All those members of these teams who could be used in the United States and Canada were transferred there to give what help they could. When the war was over, nearly all these people returned, and while some went back to their normal occupations in the universities and in industry, others were recruited to the new atomic energy organization, to form the core of the new laboratories. The motive for starting this organization was to develop the experience necessary to exploit the applications of atomic energy. Clearly, one needed work that would lead to the construction of atomic power stations, as well as facilities for research applications such as the use of radioisotopes. It was important to explore the long-term possibilities, such as thermonuclear power, and to keep open the option of producing nuclear weapons.

The basic research was concentrated in the Atomic Energy Research Establishment at Harwell. Under its first director, Sir John Cockcroft, the laboratory built up activities ranging from the directly project–based ones to some

The Peierls Group pictured here in Birmingham in Spring 1949. (Photograph published courtesy of AIP Emilio Segrè Visual Archives.)

of mainly academic interest, so as to preserve breadth and to attract and keep some of the best scientists of the country. When the growth of numbers and equipment threatened to exceed the capacity of the site, the reactor research was moved to a separate laboratory at Winfrith in Dorset, and the fusion work to the Culham Laboratory not far from Harwell.

A production organization was set up with headquarters at Risley. Of the production plants, the chemical plant at Springfields, Lancashire, was the first to come into operation, first for the processing of uranium, and later also for the treatment of used fuel elements. Later came the plutonium-producing reactors at Windscale, Cumberland; the isotope separation plant at Capenhurst, Cheshire; and the first power-producing reactors at Calder Hall. A fast fission reactor is being developed at Dounreay in Scotland. The Radio–Chemical Center at Amersham deals with the processing and distribution of radioisotopes.

Meanwhile the decision was taken to design and manufacture nuclear, and later thermonuclear, weapons, and for this purpose the Atomic Weapons Research Establishment at Aldermaston, Berks, was set up.

One of the most substantial results of all this was, of course, nuclear power development. This was a welcome new resource to a country in which one could see the coal deposits diminishing—at least the richer or more easily worked seams—and in which the use of oil for power meant a burden on the trade balance. Once feasibility of nuclear power production had been proved by the success of the Calder Hall station, plans developed for very rapid expansion. Calder Hall had dual–purpose reactors, which were intended to add to plutonium production as well as producing electricity. For this reason they had to be built by and for the AEA. For the same reason they could not easily serve to forecast the economics of later, more advanced, reactors built primarily for power. It was clear that for some time to come nuclear power would be more expensive than that from coal–fired stations, but improvements in nuclear technology and the increasing cost of coal–mining would reduce the gap. Even today the comparison between nuclear and conventional power is difficult to make, since for nuclear reactors a larger fraction of the cost is in the capital charges, and therefore sensitive to the expectation of life of the reactor, on which there has not yet been time to gain experience.

Such uncertainties are no doubt unavoidable in any new technology, and meanwhile the construction of large new nuclear power stations is going ahead. After the first few stations were built and operated by the AEA, the design and construction of such stations has been carried out by private industry, and they are ordered and operated by the Central Electricity Generating Board, the nationalized successor to the former private and municipal power companies. The AEA continues research and development and advises industry. At first there were four groups of firms to compete for the orders for new stations which were expected to be placed rapidly, and to seek orders for stations abroad. However, the program had to be slowed down considerably for several reasons. One was the fear of unemployment in the coal-mining industry. Where coal mining is the major industry, too rapid a drop in coal consumption—already accelerated by conversion to other sources of power of many industrial processes and by the discovery of natural gas in the North Sea—could create serious social problems. It was therefore decided to continue building some coal–fired power stations. The other factor was the economic crisis, which forced a reduction in the rate of investment and brought an increase in the interest rate. As a result, the market was not large enough for the number of firms in the industry, and different groups merged, to leave only two independent groups. However, even the reduced scale of construction is still considerable. The design of the latest stations shows considerable advance over the early ones, and their generating capacity is increasing. A substantial fraction of the country's electric power already comes from nuclear energy.

Compared to the power program, the production and application of radioisotopes for research and for medical purposes has attracted much less publicity. It is much less expensive and causes little controversy, yet its benefits, which are harder to quantify, are by no means negligible.

To most people, however, atomic energy does not mean power stations or isotopes, but nuclear and thermonuclear weapons. It is not necessary to spell out the thoughts and fears raised by the existence of these weapons. One point to make here is, however, that although Britain is a nuclear power, and has her own weapons, both of the fission and fusion variety, the dominant fear is of the possibility of global conflict, which would involve American and Soviet weapons. There are few people here who would not enthusiastically welcome a real possibility of disarmament (and the few are mostly people who question its feasibility). But the major problem is that of American and Soviet arms and, before long, perhaps, Chinese. Britain's nuclear weapons, if perhaps not very useful, (I shall return to this question later), do not seem very dangerous in the presence of the larger arsenals, and would, of course, be regulated as a by–product of any disarmament treaty.

For this reason the danger of world war looks from here almost as it looks from the "have-not" viewpoint of the nonnuclear countries. The dangers inherent in the arms race are serious, and we wish that something sensible could be done about them, but there is not much that we can do about them ourselves. This is one reason why the groups of scientists and others who are acutely worried about these problems in this country found it more difficult to enlist active support than has been the case in the United States. The prospect of giving abstract thought to possible solutions which might commend themselves to others does not attract wide support.

Some feel they know the answers. To those who are enthusiastic for unilateral disarmament there is scope for trying to exert pressure even on a domestic level, or to demonstrate against, for example, British bases for American atomic submarines. For those who do not see the problem in quite such simple terms there is less possibility of appropriate action.

Public action could serve, at best, to exert pressure on our government to urge the U.S. government (and, more remotely, the Soviet government) to do the right thing, or what we regard as right.

Shortly after World War II, an Atomic Scientists Association was formed, for the purpose of drawing attention to the dangers and working for solutions. Its aims were similar to those of the Federation of American Scientists. It did some useful work, wrote some statements which were commented on sympathetically both by U.S. and Soviet statesmen (not bad going at the height of the cold war) but failed to attract enough support, particularly among the younger scientists, mainly for the reasons sketched above. Eventually the work proved too much for the limited number of active member, and the ASA terminated.

About this time the Pugwash Conferences were started, with a good deal of the initiative originating in England, where the central office is still located, and many of the people formerly active in the ASA are now active supporters of Pugwash, strengthened, fortunately, by a number of younger scientists, though even there the average age is greater than it ought to be.

Last year a new society with related interests was started, the Society for Social Responsibility in Science. Its aims are broad, extending far beyond the problems of world war and of the arms race, and its membership is much more numerous than that of the bodies mentioned before, though by no means all its members are scientists in the normal sense of the term. It is still too early to form a clear picture of its attitudes, of its way of working or of the impact it will have.

These are not the only—or even necessarily the most important—groups concerned with world problems. The United Nations Association and numerous political, religious and educational groups are making serious studies, and sometimes pronouncements, on these problems, but I have referred in detail to those in which the dominant influence is that of scientists.

Our serious concern is with the nuclear arsenals of the superpowers. But Britain does have some nuclear weapons, and means for delivering them, and their existence is causing some controversy. Their military function is an extremely subtle problem. They might conceivably be an effective deterrent against a nonnuclear enemy if it could be believed that the fear of involving one of the superpowers would not firmly rule out their use. This would be of value only if it could also be believed that a serious attack on this country would be tolerated by the United States without a threat of nuclear action by that nation. They might even be regarded as a deterrent against a major nuclear power, if one could regard the surely suicidal use of such weapons by Britain as credible.

To a layman in military matters, not brought up on abstract war games, such speculations seem somehow to be lacking in reality. Yet one should remember that some of this unreality applies even to consideration of the real function of the deterrent effect of the large nuclear stock piles of the superpowers. The deterrent effect against direct nuclear attack by the other superpower is clear enough, but if this were the whole story, the power game might happily proceed as if nobody had any nuclear weapons. A more general deterrent effect may consist in inducing each of the superpowers to act with caution lest the other be provoked into unreasonable action. It is thus desirable for each to have a government which is considered by the other to be easy on the nuclear trigger, while in the interests of national survival it is of course vital that the government be absolutely safe with the trigger.

Such thoughts suggest that some of the feeling of lack of reality attached to the British deterrent is really shared with anybody else's deterrent. There remains, however, the important difference that the complete abolition of the

nuclear stockpile of either of the superpowers would create a dangerous un-
balance in the world situation, whereas the disappearance of the British deterrent would not do so.

One cannot help wondering whether the decision by this country to produce nuclear weapons, and the present policy of retaining them, were not
also influenced by other motives. One possible motive is prestige, or more
precisely the idea that the possession of nuclear weapons is the fee one pays
for membership in the "nuclear club" whose voice is listened to in the councils that shape world affairs. In the discussions on the Non–Proliferation Treaty,
when it was a question of persuading nonnuclear powers to forego the option of
joining the club, we were assured that this "top–table" argument was not valid.
But one wonders whether perhaps the membership was felt to be a valuable insurance even though the privileges of membership were open to some doubt.

Another line of thought is that, compared to large armies with conventional weapons, nuclear weapons look attractive in terms of a cost–effectiveness calculation, particularly if the assessment of cost includes the political
price of an alternative that might necessitate conscription. The trouble with
this argument is, of course, that the assessment of cost is relatively easy,
whereas the estimate of effectiveness is vitally sensitive to one's picture of
the actual or potential military situations that may arise, and the way in which
nuclear weapons would be used or threatened. At the present time public
debate on this point, in relation to the contingency planning of NATO, is
reviving, with highly–respected experts on both sides of the argument.

But my comments on the possible reasons for the British deterrent are
only irreverent musings, not based on adequate evidence.

In recent years concern has grown here, as elsewhere, regarding the danger from chemical and bacteriological weapons. As regards to biological
weapons, most people in this country accept the assurance of the government
that there are no stocks of materials for these, and no facilities for their production, except on a research scale. Some feel strongly that such research
should be entirely open, though others accept that there may be arguments
for continued secrecy. Some university departments have come under criticism from student groups and others for accepting research contracts from
the Porton Down research station, which is working on possible biological
warfare agents. Their reply has been that these contracts relate only to basic
problems which, while their solution may be of interest to Porton Down, are
also of general value.

Public opinion has generally welcomed Britain's initiative in the United
Nations toward a treaty banning biological weapons, and the statement by
President Nixon announcing the abandonment of American stockpiles and
production plants.

With regard to chemical warfare agents, nobody questions the intention of the government to respect the prohibition in the Geneva Protocol of gas warfare in the strict sense, i.e., of chemical agents intended to injure or kill. However, a major controversy is developing over the recent statement by the government which claims that the military use of certain "nonlethal" gases, in particular of the British–developed chemical irritant CS, would not contravene the Geneva Protocol. This apparent change of policy, from previous statements affirming an interpretation of the Protocol as banning tear gases, is explained by the greater safety of CS, which, it is claimed, can cause damage to health only in the most exceptional circumstances.

It is not clear how far the purpose of this statement is connected with our relations with the United States, which reportedly used CS in Vietnam, and perhaps reflects an intention to facilitate U.S. adherence to the Geneva Protocol, and how far it may be to forestall domestic agitation against the use of CS in riot control (which is not excluded by the Geneva Protocol).

Another development giving rise to some controversy is the proposal for a joint project by Great Britain, West Germany and the Netherlands for a plant to separate uranium isotopes in gas centrifuges. It is believed that this might now become economically superior to the present gaseous diffusion method, although no quantitative comparison has been published. The primary purpose of such a project would be to produce slightly enriched material for use in reactors. However, since such a centrifuge plant would presumably use in each stage large numbers of units in parallel, it could relatively easily be adapted to working in series, and thus to produce smaller quantities of highly enriched, weapons grade material. The view is now put forward that the existence of centrifuge plants, with access to them by nonnuclear nations, would be contrary to the intentions of the Non–Proliferation Treaty, which we all hope to see implemented in the near future, unless provision was made to bring these plants within the safeguards system of the Treaty.

The release of atomic energy has caused important changes in other fields, particularly in the position of scientists, and of science. Attributing cause to effect is always speculative, and we can only guess how things would have gone in the absence of atomic energy.

The first immediate result was a marked enhancement of the prestige of scientists, and the amount of support they could command. To this other developments contributed, such as the success of radar, and medical advances such as penicillin. The attitude of the general public showed, and still maintains, a marked ambivalence. Scientists are wizards who can confer great benefits on the community, including the promised benefits of the Atomic Age, e.g., getting electric power for practically nothing, or having cars powered by pellets of uranium. At the same time scientists were the villains responsible for the annihilation of Hiroshima and Nagasaki, and for worse horrors to come. Which of these reactions dominates depends on the person or group, but also on the context. Lately the anti–scientific view seems to have been gaining in strength. In either view, however, scientists are more important, and more newsworthy, than they were.

The first practical result was greatly increased financial support for science. Apart from the government–owned research laboratories, including those of the AEA, increased government support became available to academic science. Throughout the period there was a considerable expansion in the size of universities, and the later part of the period saw the foundation of a number of new universities. This growth was not confined to science, and was necessitated by a shortage of university graduates in many fields; the size of the student population in Britain had been very small by comparison with other industrial countries. But within this expansion, great weight was given to science, and science departments grew more rapidly in numbers and resources than others.

Financial support comes both through a general government grant to the universities, in the use of which the universities are almost completely autonomous, and through grants for research equipment and other research expenditures directly to departments. Such grants are made by the Research Councils, government agencies with functions somewhat comparable to the American National Science Foundation. Nuclear physics was particularly dependent on this kind of support, because of its increasing need for expensive installations. It was perhaps a fortunate accident that the time of increased support for nuclear physics happened to coincide more or less with the time when the nature of the subject forced a change from the traditional string–and–sealing-wax methods to "big science," with accelerators and highly sophisticated detection devices. This statement is not invalidated by the fact that some of the important discoveries of the period, notably those made by C. F. Powell and his colleagues with photographic emulsions, required only a relatively modest expenditure.

Government support for nuclear physics was deliberate policy. There is a temptation for nuclear physicists in claiming support, or for a government in granting it, to pretend, or at least to encourage the belief, that such work will directly serve the development of atomic energy in leading either to better bombs or to better reactors. Neither the physicists nor the administrators yielded to this temptation: the case for support was always discussed in terms of real, but less direct, arguments. Active participation in the discussion is important as an intellectual challenge which attracts first–rate minds to the subject, thus raising the educational standard. It produces trained professionals of high ability who can be useful in many activities, including atomic energy, and it keeps the university departments up–to–date with developments which—who knows?—some day may lead to practical applications with tangible benefits.

Drawing a clear distinction between atomic energy and academic research in nuclear physics has the advantage of avoiding the impression that there were secrets of potential military significance lying around in the universities, and this shielded the universities from misconceived security restrictions. Occasionally a newspaper might get confused about the distinction, and might come up with stories about possible risks from the presence in university departments of scientists suspected of undesirable political views, but this never had any serious consequences, even at the time when the trials of Allan Nunn May and Klaus Fuchs aroused great public interest in the question of loyalty of nuclear physicists. At that time the mature and unemotional reaction of the British public to these disclosures and the complete absence of hysteria were most impressive.

The provision of better financial support was by no means all that was necessary to buildup nuclear physics research. In the immediate post–war period materials were in short supply, buildings were desperately hard to get and, above all, there was a shortage of experienced physicists, since during the six years of war, with total mobilization of resources, practically no students were trained in research. (Of those who had completed their training before the war most had had to switch to work of direct relevance to the war effort.)

As a result there was a serious gap in an important age group, and of the remaining people with the right experience some were absorbed by the new AEA. The development of accelerators and other equipment therefore represented years of hard work by too few, and a preoccupation with hardware, largely to the exclusion of the other important task of following the rapidly growing new ideas and principles in basic physics.

The post–war growth in the prestige of science also attracted an increased number of science students to fill the enlarged departments; at times there was a shortage of places. Part of this trend took the form of schools (high schools, in American terminology) which traditionally had encouraged their ablest pupils to study the arts, now regarding it as their—sometimes painful—duty to encourage them to enter science courses.

The past ten years or so have shown a reversal, a swing away from science. The reasons are still a matter of debate: One factor may have been that the new bright image of science became a little worn; the expected immediate benefits of science had not materialized. Another was a serious shortage of good and inspiring teachers of science and of mathematics in the schools. In addition, prospective students began to regard university science courses as more demanding than others. Stories about the shortage of places were exaggerated, leaving the impression that it required exceptional ability and application even to be accepted for a science course. If the last factor played a major part it would be expected to correct itself because of the publicity now given to the unfilled places in science courses, and there are indeed signs that the swing from science is diminishing.

Meanwhile the expansion of the government support of university science (and of universities in general) continued though the rate of growth was never enough to satisfy the scientists. In nuclear physics in particular, where the front line of research had moved to higher and higher energies, from nuclear structure to elementary particles, the needs could no longer be met by equipment belonging to individual universities. It was doubtful even whether any countries other than the United States and the Soviet Union could afford up–to–date national accelerators. This led to the creation of CERN (The European Council for Nuclear Research) as a cooperative venture for Western Europe, and this has been a brilliant success. Britain first decided to stay out of CERN, but a few years later she became a full member, thus allowing a number of British physicists to participate in research work of the most fundamental kind.

In spite of the success of CERN, a need was felt for national facilities as well. Two national laboratories with accelerators were built, in which most of the experiments are done by universities.

The past few years have seen a substantial change in climate. An economic crisis developed, with a dangerously adverse trade balance. It had always been clear that a densely populated country which had to rely on substantial imports depended for its solvency on an effective export trade. Therefore, it was vital for industry to be competitive in the world market. Britain needed quality products, a high level of technical innovation, and efficient organization of production. These aims had encouraged the expansion of higher education and an increased output of trained scientists. But as the economic crisis developed, the view gained ground that too much emphasis had been placed on pure rather than applied science, that one needed engineers and technologists more than scientists, and that too few scientists were moving into industry to contribute to its earning and exporting capacity. It is certainly true that during the rapid expansion of the universities in the 1950s most of the ablest research workers were absorbed to staff the new or growing universities, and that their successors came to regard an academic rather than industrial career as the normal outcome of their research training. In fact, some branches of industry never had the experience of the work of enthusiastic scientists of high quality, and are not even sure that they want them.

Study of these problems by numerous official commissions led to recommendations urging a diversion of support from academic to more useful training and research. Some of these new measures have already been implemented, but there are still many changes to come, and the pattern for the next few years is not yet clear. It is certainly going to be a period of tightened belts in the universities and in university science.

During this period of readjustment a decision was required whether Britain was willing to join the new European project for an accelerator to reach 300 GeV (Giga-electron volt), a further big step from the 30 GeV machine now operating at CERN. The nuclear physicists in Britain worked out a plan to reduce the national facilities, in order to fit the cost of participation into an existing budget, but the decision was negative. Perhaps they can draw some slight comfort from the history of our participation in CERN, and hope for a similar change of heart.

The current agonizing reappraisal concerns not only the universities, but also government laboratories, including those of the AEA. For example, the Harwell laboratory uses only a fraction of its strength on problems of atomic energy, now that the power program is well under way, and much of its work is in the hands of industry or other branches of the Authority. Its pursuits of basic academic studies are regarded as a luxury which must be kept within bounds. A plan has therefore been developed to devote some of its research potential to projects of direct value to industry. This has made a promising start and is being extended.

A rather different fate has come to the Culham Laboratory, which is devoted to the problems of thermonuclear power. Since the realization of this project does not seem to be around the corner, authority has decreed that the laboratory is on too grand a scale, and should be reduced by one–half over a five–year period, cutting its manpower and its budget by 10 percent each year.

The last parts of my sketch of the past quarter–century in Great Britain are hardly a characterization of the period, but rather a description of its end, and of the beginning of a new period. Thus 25 years is not just a convenient round number, but an appropriate period to form a chapter in history.

Limited Nuclear War?

E urope is full of "tactical" or "battle field" nuclear weapons. Under
the doctrine of flexible response, it appears that the NATO strategy
is to use these weapons in the event of a conventional attack by Warsaw Pact forces that NATO conventional forces were unable to stop. No sane
person would contemplate a strategy which could lead to a global nuclear
war, so the present strategy must be based on confidence that the dividing
line between tactical and strategic weapons can be held.

This confidence might conceivably be justified if there were a clear distinction between the two kinds of weapons, and if there were a consensus
about this between the Warsaw Pact and NATO nations. But I am not aware
of any statement by the Soviet Union or its allies which recognizes that there
is a distinction in quality, nor does there seem to be any clear definition of
tactical weapons. They come in all sorts of sizes and ranges, and from the
smallest and shortest–range ones there is a continuous spectrum leading to
very powerful medium–range missiles.

Perhaps we should look for the distinction in the spirit in which weapons
are used. Perhaps the use of weapons against troop concentrations, tanks,
ammunition stores and so on would be "tactical"? But here one must distinguish between the intention and the result. In World War II no air force admitted bombing civilians; the communiqué always described attacks on "military objectives." And quite possibly it was military targets that were aimed at
in many cases, but the result was indistinguishable from a strategic attack on
the civilian population.

I do not see any likelihood of an attack from the East in the near future, but
this is what all the planning is about. And if NATO conventional forces failed
to contain it, the NATO command would face the choice between accepting
defeat and starting a tactical nuclear exchange which could escalate to an
unpredictable extent and, very probably, leave at least Europe in ruins.

There is, however, another limitation which would be clearly defined and would have some credibility. It is the possibility that both superpowers would refrain from attacking each other's territory. The United States would realize that a nuclear attack on Soviet territory would invite retaliation in kind, and that there would then be serious danger of escalation. This might indeed induce the U.S. command to observe this limitation, and the converse argument might induce a similar caution in the Soviet Union.

In this sense, limited nuclear war is a credible scenario, one which, for Europeans, is totally unacceptable. The reasons for the recent antinuclear demonstrations in Europe have not been precisely articulated, but probably the fear of the superpowers fighting each other to the last European played a part in their motivation. If this fear became more widespread it could mean the end of the NATO alliance. This prospect may, in turn, prompt a reassessment of NATO doctrine and strategy.

In any case, the need for such reassessment is becoming increasingly urgent. This is not a very original idea; it was eloquently expressed by the late Lord Mountbatten. It is about time his words were heeded, even if this should lead to the politically unpopular step of increasing NATO's conventional strength.

Agonising Misappraisal

A common human weakness is a refusal to think about unpleasant facts. In the world situation today we have to face a fact of truly gigantic unpleasantness, namely the existence of atomic and thermonuclear weapons, and means for delivering them, which provide the technical possibility of East and West inflicting on each other damage and casualties on a terrifying scale. It is not surprising that we should experience a degree of reluctance to think straight corresponding to the degree of horror of the possibilities, and an equal temptation to indulge in wishful (or fearful) thinking, or to clutch at simple solutions.

One is therefore hopeful of enlightenment when one picks up the book by Dr. Herman Kahn,[1] a member of the RAND corporation, that strange research organization which conducts studies for the U.S. Air Force and other government agencies. Here indeed we are going to face the facts, for 651 pages. It will, of course, take some effort to study a book of this size, particularly since it consists of the text of a series of lectures, and to read lecture notes verbatim is usually almost as hard as to listen to the recital of a written text.

We shall, in the words of the foreword by Klaus Knorr, "boggle at the unfamiliar idiom," with its 75 tables listing items usually labelled by new names and then explained (even if the term is not used again), but with no index to help us track down a definition we have forgotten.

We shall boggle even more at the endless repetitions, which only in some cases are marked by an emphatic "I repeat," which assists us in skipping a

---

[1] *On Thermonuclear War*, by Herman Kahn, (Princeton University Press and Oxford University Press).

Editor's Note: Kahn, Herman. *On Thermonuclear War.* 668 pages, reprint. Westport, Connecticut: Greenwood Press, 1978.

few pages. But mostly the repetition contains some variations, so that we may miss a point if we omit it.

But we shall not be deterred and shall follow the book to the end in the hope of learning the answers to the important questions.

What are the issues with which one is concerned? It is not easy to summarise a presentation of this length and as amorphous in structure, but the central theme is the question of deterrence.

It is usually believed that as long as both East and West have the power to attack each other with "modern" weapons the danger of a deliberate aggression from either side is negligible, because the unavoidable retaliation would involve too high a price for whatever advantages the attack might be supposed to bring. This is quite apart from the question whether either side is likely to be in the mood for a deliberate unprovoked attack if it could do so with impunity. Kahn is quite satisfied that the U.S. would not strike Russia "out of the blue" (p. 210). He evidently is not willing to discount the opposite possibility. In a different context (p. 33) he remarks, "I have not asked any Soviet citizen" in a tone which implies that it is a completely fantastic notion to talk on any such matters with a Russian. If he had had an opportunity of doing so, he would no doubt have discovered that they are willing, with equal or greater conviction, to discard the possibility of an attack out of the blue from their side.

But as far as the possibility of an unprovoked attack exists, it is obviously a good thing for the prospects of peace that there exists the certainty of retaliation. There would be doubt about this certainty if the aggressor had a chance of knocking out the retaliatory potential of the other side with the first blow. At present probably a good deal of American strength lies in manned bombers or "soft" missile bases (i.e., bases vulnerable to attack by modern weapons) and it is therefore technically feasible that a Russian attack might destroy or severely reduce America's power of retaliation unless the bombers got away and the missiles were fired before the Russian missiles or planes reached their target. This need for "split-second" retaliation increases the danger of mistakes, and therefore frightens both sides. The development of "hard" bases and such developments as the virtually invulnerable Polaris missiles reduce the need for automatic retaliation, and therefore tend to lessen tensions.

A deliberate unprovoked attack against a country capable of retaliating would seem so crazy that we need not take it seriously (except that this may be straining our confidence in the sanity of statesmen; one cannot imagine the consequences of Hitler possessing thermonuclear weapons), but there remains the possibility of accidents and of pre–emptive war. The thought of an accidental firing of a missile, or of an attack by a plane because of a fault in communications, is most disturbing, however small the likelihood of such errors may be. An accident need not automatically lead to fullscale retaliation, if the country responsible for it realises what has happened and notifies the other side of the fact before the process has gone too far. But there are evidently terrible risks in this, and in any case even the effects of a single weapon are so serious that the risk of accidents is, in my opinion, by itself an extremely powerful motive for getting on with the problem of disarmament.

The possibility of pre–emptive war, i.e., of an attack to forestall an apparently impending attack from the other side which would lead to greater disaster, depends very much on the magnitude of the disaster which retaliation will bring to the aggressor. If the result of full–scale action by both sides is in any case mutual annihilation or, as is sometimes claimed, the total destruction of life on the earth, there would be no sense in any pre-emptive action, since it would be substitute certain annihilation for a probable annihilation of perhaps a more drastic kind. This reasoning therefore leads to the question. 'How complete is the disaster caused by thermonuclear war?'

The same question arises in connection with what Kahn calls "Type II Deterrent," i.e., the use of the threat of nuclear weapons to stop different kinds of aggressive behaviour. Can the fear of thermonuclear attack prevent or limit crises of the type of Korea or Laos (or Cuba)? If any use of thermonuclear weapons is bound to lead to the complete mutual annihilation of both sides, and this is known, it is evident that no sane government would resort to the use of these weapons no matter how serious the provocation.

Kahn rightly warns us against sloppy reasoning on this question. He reminds us that the populations of English and German cities survived during the Second World War air raids of an intensity which, it was predicted, would break morale and make life in the city impossible. (I remember hearing during the Abyssinian crisis the argument that Britain was not in a position to take a stronger line because the Italian Air Force had a few planes ready to drop bombs on London.) It is necessary to form an idea of the consequences of thermonuclear war in order to know not only whether we would ourselves be willing to accept these consequences to avoid military or political defeat, but also whether it is likely that (rightly or wrongly) some government will be willing to accept them.

The study of these consequences, and of the capacity of the United States to recover from them, forms the most original part of the book. It reports in many pages, and many tables, the results of calculations which are no doubt correct. But the same can hardly be said for the assumptions and hypotheses on which these calculations are based.

In the first place it is assumed that the attacker cannot afford to waste his weapons on cities, but must concentrate on attacking the missile and bomber bases of the other side to minimise the retaliation to which he will be exposed. The defender's bases will have been damaged and his retaliation capacity somewhat limited. Perhaps this is right if the aggressor makes a rational choice about his tactics. But are we justified in relying on rational choices being made at this particular point?

Kahn's assumptions about the post-attack problems are based on severe simplifications of which I can give only a few examples, by way of illustration. The productive capacity of the country is assumed to continue at a rate reduced in proportion with the surviving resources. The possibility of bottlenecks arising from specific shortages is mentioned, but not taken very seriously. It is assumed that normal processes of government continue, sufficient to get the best distribution of remaining food resources and to ration the remaining stocks and supplies of petrol and the remaining vehicles. At this point one is staggered to read on p. 92 the admission that the study has so far dealt with the effect of the various kinds of damage, radioactivity, physical destruction, casualties, etc., as if each occurred by itself. "But if all these things happened together, and all the other effects were added at the same time, one cannot help but have some doubts." A poor layman would, of course, try to picture the situation with all these terrible things happening together. Since the mutual interplay of the different causes is less easily expressed in terms of numbers, the experts can only look at one at a time.

Elsewhere one discusses the question of surviving military capacity. There will be some bombers and some usable airfields left; one just has to find out where they are and what crews and servicing they need (p. 108). "This calls for the survival of at least a minimum communications monitoring, dataprocessing, and computing capability... ." The emphasis on dataprocessing and computing in the situation described seems to me incredibly misplaced.

One's doubts about the basis of such studies are enhanced by the reference (p. 119) to a technological breakthrough at the RAND corporation in military studies. This was the discovery that it is not right to assume a set of circumstances and then determine one's plans to give the best possible result in such circumstances, if the same plan works out badly in a different set of circumstances. Hence one must study a range of possible situations and choose a plan which is reasonably good in all of them. Dr. Kahn does not state how many man–hours and computer hours were used on military studies before this great discovery was made.

By these studies the conclusion is reached that the consequences of an attack leading to the loss of several, or all, major cities, and to five or perhaps even twenty million casualties in the United States might in some circumstances be tolerable, given adequate advance planning and civil defense. As I have tried to show, the basis of the arguments is so unrealistic that the conclusion is most unconvincing.

Dr. Kahn advocates making these preparations in order to make the deterrent "credible," and to make it also a "Type II Deterrent," i.e., use it to discourage provocative action short of direct atomic aggression. He does not seem to make sufficient allowance for the effect in the present situation of uncertainty. Either side is restrained from too provocative adventures by the fear that too much tension may result in a major thermonuclear war, even if from any rational point of view it was crazy to resort to this. (How far are we willing to gamble on other people's sanity?)

It is hard in a short summary to mention all the interesting points thrown up on some of the 651 pages. While disarmament is dismissed quite early (p. 6) as not realistic at the moment, arms control is discussed later (p. 241) with the suggestion that one should accept the Russian proposal of a ban on the use of atomic weapons, except in direct retaliation against atomic attack, and perhaps excepting use in aerial defence, or against naval units at sea. He suggests coupling this with the U.S. giving up their overseas bases, in return for an opening-up of the USSR. It is strange to see this followed in the same breath, still under the heading of arms control, by mention of a proposal to make a firm commitment that any attack on one's own territory (suitably defined for the purpose) would be met by atomic retaliation, while one reserved one's freedom of action in the event of conflict elsewhere. Nor can one mention all the little asides which suggest that the sources of information on the facts on which the studies were based must have had their weaknesses, such as the remark (p. 99) that in Russia "most current peasant housing has earth walls two or three feet thick."

I am afraid I have not gained from Dr. Kahn's book any confidence that we can learn much about the future, let alone improve it, unless we succeed in negotiating a substantial measure of disarmament.

Reflections of a
British Participant

The story of the development of atomic energy started for me, a British scientist, one day in March 1940, when O. R. Frisch asked me, "Suppose one had a large quantity of separated uranium 235, what would happen?" We both knew the Bohr–Wheeler theory of fission and could make a rough guess at the fission cross–section of the light uranium isotope. I had derived a formula for determining the critical size to achieve a self–sustaining chain reaction, in terms of the cross–section and the number of neutrons emitted per fission, which was known approximately. We were surprised that the critical size came out quite small, in pounds instead of the tons one might have guessed.

Next was the question of how far a chain reaction would go before the developing heat would drive the uranium apart. My formula also gave the time scale of the chain reaction, and though we could not get more than a rough estimate, that estimate indicated that a substantial fraction of the uranium would undergo fission.

We were staggered by these findings, because an atom bomb now seemed a practical possibility. We knew that isotope separation on a large scale posed big problems, but the power of the weapon would justify a large effort. In what turned out to be a classic understatement we said to ourselves: "Even if the isotope separation plant costs as much as a battleship, it would be worth it."

Our immediate fear was that the Germans might also have seen the point and might be working on an atom bomb. The thought of this weapon exclusively in Hitler's hands was a nightmare. We wrote a memorandum (see pages 187–194) with our arguments and conclusions, urgently recommending that research on the weapon be started, and that, if it proved feasible, it be developed as a deterrent against a possible German weapon.

The technical part of the memorandum was accompanied by a nontechnical note describing the likely effects, including radioactivity and fallout. We pointed out that since the bomb could probably not be used without killing large numbers of civilians, "this may make it unsuitable as a weapon for use by this country." It might conceivably be used against fleet concentrations, though in a harbor it would still be likely to involve many civilians. We did not spell out the obvious fact that, even in the absence of a German atom bomb, British possession of such a weapon would drastically alter the military balance. It was too early to argue what use of this terrible new weapon would make military sense and be morally justified.

As a result, research work was started under the MAUD Committee (a code name), with teams including a group under James Chadwick, in Liverpool, who seemed to have come to conclusions similar to ours, and Francis Simon's group in Oxford starting to study the problem of isotope separation. The committee's report in 1941 led to a government decision to increase the scale of the work, and a new administrative structure, with the meaningless, dull–sounding cover name of "Directorate of Tube Alloys," was set up to support it. Harold Urey and G. B. Pegram from Columbia University attended the first meeting of the new committee and reported back in the United States that the British government was taking the project very seriously. This is believed to have influenced the U.S. decision to give their own project high priority.

There was little doubt that the atom bomb could be made, but it was very doubtful that it could be made within a reasonable time in wartime Britain. The possibility of a joint project with the United States was raised very early, but for a long time did not get anywhere; it is clear now that this was due to political short–sightedness on both sides. Eventually the political problems were settled in conversations between Roosevelt and Churchill and incorporated into the Quebec Agreement of 1943.

After this a number of British scientists were sent to join in the Manhattan District of the U.S. Army Engineers, and work in the United Kingdom was closed down, except for some research that could prove useful to the U.S. project. I worked for the first half of 1944 in New York, in contact with the Kellex Corporation, the designers of the Oak Ridge isotope separation plant. From mid–1944 until the end of the war I worked in Los Alamos.

By 1944 our fear of the Germans developing the atom bomb first had abated somewhat. I conducted my own intelligence operation, based on published information. I knew, for example, that each semester the *Physikalische Zeitschrift* printed the physics lecture lists of all German universities. These lists, which would have been difficult to fake, showed that most of the German physicists were in their normal places, lecturing on their normal subjects. There were some exceptions, and these were probably people involved with the atom bomb or other wartime projects. The picture suggested that some research was going on, but no crash program. All the same, one dared not rely on this conclusion, however plausible.

But when the war in Europe ended with the defeat of Nazi Germany, the fear of a German atom bomb was also over. I have been asked many times why I continued working for the project when the bomb was no longer needed as a deterrent, and whether I felt happy about developing a weapon that was going to be used to cause unprecedented destruction and suffering. My answer to this question may differ from that of other scientists, but I believe it is not untypical. The war in the Pacific was still raging, with many people being killed or wounded. The possession of the atom bomb would obviously strengthen the Allies' position enormously. Its use on cities was likely to kill and hurt large numbers of civilians, as Frisch and I had pointed out in the very beginning. At that time we felt that a British government would be willing to use such a weapon only in special circumstances, and I felt that the U.S. military and political leaders likewise would not resort lightly to its use.

It was of course essential that the scientists who understood the physical consequences of an atom bomb explosion and had thought about the implications should explain these ideas to the authorities. This could not be done by every scientist on the project, but we knew that people like Robert Oppenheimer, Enrico Fermi, and Arthur Compton were in touch with the men who would have to make the decisions. I had great faith in my colleagues' depth of understanding and their capacity for simple and clear explanation. The leaders, I felt, were also intelligent men of good will and would try to make wise and humane decisions. In retrospect I have to admit that these views were a little naive.

The thought that the bomb would be used on a city without warning never occurred to me. While an explosion in an empty place would not have been enough—the scene after the Trinity test in Alamogordo did not look impressive—it could have been dropped on a sparsely inhabited place, perhaps some island, to show its power. Such a demonstration could have been accompanied by the threat to use the bomb in earnest, unless Japan surrendered. It would have killed some people and wrecked some buildings, but this was war, and people were being killed all the time.

This idea, which seemed obvious to me, was apparently never considered by the high–level committees. They did discuss an announced demonstration, in the presence of international observers, but this idea was rejected. The reliability of the mechanical and electronic devices in the bomb had not yet been proved, and failure of the demonstration would have been counterproductive.

I do not want to imply that I regarded any use of an atom bomb in itself as immoral. The number of casualties in Hiroshima was less than in one fire raid on Tokyo or Dresden. It was therefore not the scale of the results that put the atom bomb in a new category, but the ease with which it could be used. Instead of the great effort of staging a massive air raid, a single plane was enough to carry the weapon, and one man's decision to press the release button over one city or another would decide the fate of many thousands of people. It was this ease which created the temptation to use it irresponsibly. One can debate the merits of the decision to drop the first atom bomb on Hiroshima, but few would dispute that the second one on Nagasaki was unnecessary and therefore irresponsible.

The raid on Hiroshima was no more and no less unethical than the raids on Tokyo or on Dresden. It is remarkable how attitudes to strategic air raids on cities changed in the course of World War II. The attitude of outrage at the raids on Guernica and Rotterdam changed to acceptance, and even approval, of large–scale raids on cities. Hiroshima would not have been attacked without the climate of opinion that made the Tokyo and Dresden raids acceptable.

Apart from the effect of the bomb on the war in progress, we were of course aware of the problems of its future consequences. In thinking about the future, however, I had no idea of the tension between East and West which would develop and be a major factor in the atomic–weapons problem. I had no illusions about the nature of the Soviet regime but expected that, as we had managed to fight as allies in the war, we would also manage to coexist in peace.

But even without a clear idea of who would be potential adversaries, we felt the nature of war had changed. With generous supplies of nuclear weapons, war would acquire such grotesque proportions that any nation would be too frightened to start a war if it was likely to bring in nuclear weapons. That expectation has so far been borne out.

It seemed an attractive idea to create international machinery that would institutionalize the fear of nuclear war and thus prevent it. With many other scientists, I welcomed the ideas on international control of atomic energy put forward in the Acheson–Lilienthal report of March 1946. In the United Kingdom we founded the Atomic Scientists Association (ASA), which backed the idea of international control.

The ASA was dissolved in 1959 mainly because of weaknesses in its structure, and because the activities of the Pugwash Conferences appeared more promising. Instead of international control, which probably never had a chance, the aim of these conferences was arms control and disarmament.

In spite of some important but minor achievements, there has been no success with disarmament so far, and an insane arms race continues. This arms race is one of the developments I did not foresee at the early stages.

Perhaps the scientist who is used to rational arguments makes the mistake of expecting politicians to act rationally.

Everybody accepts the fact that a war cannot be fought with nuclear weapons if both sides possess them, and therefore their only purpose is to act as a deterrent against a nuclear attack by an adversary. This requires an invulnerable retaliatory capacity, sufficient to inflict unacceptable damage on a nuclear aggressor. One can argue about the number of weapons necessary for this, but there is no question that the present stockpiles of the superpowers are vastly greater than needed for this purpose. Why, then, do both sides continue to increase their arsenals and worry if the numbers or the power of their weapons are in some respect inferior to those of the other side? I have heard it said that the weapons count might affect a nation's political will; that is, that they might be more frightened by the threat of a nuclear attack if the enemy had a greater number of weapons, even if it would be suicidal to use them. I find it very hard to believe in the reality of this proposition.

From the point of view of the danger of nuclear war, British nuclear weapons do not seem very important. Since the early 1950s, when the decision was made to create a British nuclear weapon, the purpose it was supposed to serve has varied a great deal, but the policy has not changed. It has always been denied that a substantial role was played by the prestige argument—that is, that nuclear nations could expect to be listened to in international policy debates. One could not of course admit the validity of this argument and at the same time favor nonproliferation, which would dissuade others from becoming nuclear. Yet one cannot help feeling that something like this played a part in the British decision.

The British deterrent is supposed to be independent and available in the event of a conflict in which the United States turned isolationist. The scenario of the Soviet Union, or some other nuclear power, threatening a nuclear attack on the United Kingdom without involving the United States seems hardly likely enough to justify a heavy insurance premium against it. If such insurance is considered worthwhile, it is obviously important to make the deterrent force as invulnerable as possible. This would favor the maximum number of submarines—not necessarily carrying weapons of very great destructive power—rather than the planned Trident system which is very sophisticated and therefore so expensive that the country will be able to afford only a small number of vessels.

The U.S. cruise missiles now being deployed are an addition to the existing excessive overkill capacity and therefore unnecessary and wasteful. I do not, however, believe that their presence in Britain is harmful; in particular I do not believe that the danger of a nuclear attack on this country is increased by their presence. They are a small factor within the overall insanity of the arms race.

Energy from Heaven and Earth

T his is a readable, thought-provoking, but in part irritating book. Most readers will regard it primarily as a source of information and of arguments for the current energy debate, and to this most of the book is indeed relevant. It reviews traditional energy sources and their potential, energy consumption, the nature and safety of nuclear energy, the prospects of fusion and other novel sources, such as solar, wind and geothermal energy, and the possibilities of conservation. Finally, alternative strategies and policies are reviewed.

The presentation of the facts and figures is mostly clear, if somewhat repetitious. The "quad," a very large unit of power, is introduced at least three times, probably because much of the book originates from lecture series (in Buffalo, N.Y. and Israel), but in lectures one cannot give a reference to a previous page. Presumably the cartoons by George Bing were originally slides for the lectures; in the book they do not appear very informative or funny, at least to my sense of humor. I do not know why Figure 4-2 uses a jet airplane engine to illustrate the functioning of a stationary gas turbine (it reads almost like the traditional explanation of the wireless telegraph in terms of a dog) unless it is there to prepare for the joke in Figure 14-1 about an owl, which I cannot quite follow either.

The facts taken from physics are evidently reliably stated; one would expect nothing less from an author who is a distinguished and original physicist. I would hope that the statements from other subjects, such as geology, meteorology and economics, are equally reliables but as I am no expert in these areas, and few references are given, I have to take them on trust.

Few serious readers will disagree with some of the main conclusions: that energy is short and will get shorter if nothing is done about it; that conserva-

This is a review of *Energy from Heaven and Earth,* by Edward Teller. New York: W. H. Freeman & Son, 1979. For the complete reference, please see the Acknowledgments section.

tion is valuable but will not suffice to balance the energy budget, particularly if there is to be a rise in the standard of living in the underdeveloped world; that more oil can be found at a price and that the output of coal can be increased; that environmental problems should be discussed rationally, so that damage done by installations can be weighed against the disadvantage of not having them; that novel resources, including fusion, solar and wind power, cannot make more than a marginal contribution during the rest of this century.

When it comes to nuclear power, the author is on more controversial ground. I happen to share his view that it would be hard to get through the next 20 years or so without more nuclear power, and that its risks are not unacceptable, provided the safety problems are handled with sufficient determination and responsibility. I am not sure that those holding opposing views will be convinced by Teller's arguments. He presents the problem of accident prevention as a technical matter which can be handled by effective engineering research. He says of the Brown's Ferry accident that it was "a demonstration of the health safety inherent in reactors." But it could also be viewed as a demonstration of how close one can get to breaching all technical defenses by unforeseen human behavior. In thinking about industrial accidents it is good to remember the old adage: "No man is wise enough to think of all the ideas that may occur to a fool." The story of the Three Mile Island event (which occurred after the book was written) emphasizes the further great problem of insuring that safety rules are observed. Such problems are not insoluble, but they must be tackled, and they are not technical problems.

On the question of low–level radiation in normal operation, I would agree that the doses are minimal, but the anti–nuclear reader may not be convinced. Nor does it help that the arithmetic is hard to follow. On page 186 the dose received by a "bystander" at a nuclear reactor is estimated as less than 0.0017 rems per year, and the dose for someone living within 50 miles of the reactor as 0.001! Here again it is not stressed that vigilance is essential; there have been reports that variations in wind and currents have increased the effect of routine effluents well beyond what was foreseen.

On waste disposal, Teller insists that several safe methods have already been proved, but goes into detail only about vitrification. His position on this seems a little complicated: On page 180 he assures us that such a glass could safely contain 30 percent fission products, but a footnote refers to a snag, which calls for further dilution, and mentions an alternative method now being developed.

Many opponents of nuclear power want it stopped because its wide spread use may accelerate proliferation of nuclear weapons. Teller argues that the connection is slight, and that in particular the abandonment of nuclear power in the United States would not have much effect on slowing down proliferation. With this I agree.

By contrast to the extensive discussion of technical factors, political and economic questions are treated rather cursorily. The reasoning about the extent to which government regulations were responsible for aggravating the oil shortage in the United States occupies just one and a half pages.

Chapter 13 discusses four alternative, not very sharply defined, energy policies, accompanied by a table (13-1) presenting detailed figures for the expenditure on energy for each of these. The table is preceded by the candid statement: "Of course, all of the projections are guesses." But if they are, what use is the table? In spite of the warning, the nonscientific reader is likely to attach too much weight to such figures. As described, and quite apart from the figures, the "international" option sounds the most attractive; it is also the one preferred by Teller. One could argue long what other policies, or what variations about the ones suggested, should be considered.

Similar comments can be made about another table (14-1), which relates to a model for the more distant future, but which is also rather hard to interpret, if only because a line labelled "Total Exports" lists mainly imports, with its one export figure carrying a minus sign. In introducing this table the author says "I am less interested in being right than in stimulating ideas on the subject."

The introductory chapters on the history of the universe, the sun and the Earth, while not essential to the energy debate, are lively and, as far as I can judge, intelligible to the non–scientific reader. Chapter 8 on the history of the atom bomb is a very personal account, giving details only of those events in which Teller himself was involved. But this makes it an entertaining story. Anecdotes outside his direct experience are not always convincing. He attributes a lecture by Ernest Rutherford, in which he severely denounced people who believed in the release of nuclear energy, to Rutherford's having heard Leo Szilard propose this idea shortly before. According to Szilard's recollections, however, Rutherford spoke in that spirit at a lecture (possibly even the one that Teller heard) before the idea of a nuclear chain reaction occurred to Szilard. It is also hard to believe that the data on nuclear masses in the literature, around 1935, were so wrong as to cause Enrico Fermi to conclude that the potential barrier for fission was prohibitively high.

On the use of the atom bombs, Teller believes it would have been better to demonstrate their power first, and here I strongly agree. A more unusual suggestion is that it would have been better not to maintain any secrecy about atomic weapons after World War II.

Throughout, the story is interspersed with many asides, comments and ideas, often very interesting. Some pages of five paragraphs contain as many distinct thoughts; no reviewer could possibly do justice to this variety. Teller says of his friend Szilard that he "never said anything that was expected of him." Would it be unkind to suggest that one of the author's strong ambitions is to emulate Szilard?

Memories of the Secret City

L os Alamos was the laboratory set up, during the Second World War, to design and produce atomic bombs from the fissile materials, uranium 235 and plutonium, provided by the gigantic installations of the Manhattan District of the U.S. Army. A remote location in the hills of New Mexico was chosen, partly because of the high degree of secrecy required, and partly because the work might involve hazardous experiments. Conceived originally as a settlement for 265 people—six scientists with their support staff and families—it had grown to accommodate many thousands by the end of the War, largely because the design of a plutonium bomb proved much more difficult than had been foreseen.

Much has been written about the achievement of the laboratory in solving the difficult technical problems so quickly, and about the life in this unusual community. The present collection of lectures, given at Santa Barbara in 1975, is a most interesting addition to this literature, containing the recollections of scientists, of scientists' wives and of an army officer.

Imagine a team of scientists, many of them at the top of their professions, working in improvised buildings, living in temporary housing, depending for their professional and domestic needs on the army, and subject to security restrictions they did not always see as necessary or consistent; imagine a collection of army personnel supporting the activities of scientists whose objectives they were not supposed to know, and whose methods and manners were quite foreign to the army's style of operation. There were bound to be conflicts, irritations and grievances. The recollections indeed include many examples of such friction. The fact that all major problems were somehow resolved, and the minor nuisances accepted, is a tribute to the good sense of

This is a review of *Reminiscences of Los Alamos, 1943–1945*, edited by Lawrence Badash, Joseph O. Hirschfelder, and Herbert P. Broida. Dordrecht, Netherlands: Reidel/Kluwer Academic, 1980. For the complete reference, please see the Acknowledgments section.

the people involved, their feeling of urgency and in no small measure to the personality of the director, Robert Oppenheimer.

There is already some folklore about the place. The first lecture, by General Dudley, reports four versions of how Los Alamos came to be selected as the site of the laboratory—his own and three others—of which at least two are quite incompatible with his and with each other. The recollections of the residents do not show any such disagreements on facts, but they differ much in emphasis. Some seem to remember above all the irritations, the shortages, the beautiful school building which for a time had no qualified teachers, the strange facets of the voluntary mail censorship. Others stress the attractions of the landscape and climate, the contacts with the local Indian culture, and above all the proximity of a large number of unusually gifted and interesting people. But all would probably agree with Laura Fermi's remark:

> One interesting thing to me is that during the war I was in Los Alamos only a year and a half, and *still* it seems such a big portion of my life... It was so different, it was such intense living, that it seems impossible that in my case it actually lasted hardly eighteen months.

Her account and those of Bernice Brode and Elsie MacMillan give a lively picture of the domestic scene. Feynman's "Los Alamos from Below" is as entertaining as one has come to expect from him. He reveals the secret of his notorious skill in opening combination locks, and how he accidentally stumbled on a safety hazard in a complicated plant design. There are hilarious accounts of his brushes with the censor and other bureaucratic elements. In stressing his "low" status he has his tongue in his cheek. He was recognized at once as a man of outstanding promise. The opening statement that he did not even have a degree, has to be withdrawn a little later when he admits that he did complete his Ph.D. before going to Los Alamos.

Kistiakowsky's reminiscences form the most substantial part of the book. He describes, in non–technical language, the work of his division in designing and making the "implosion" device. He and his collaborators, whose contributions are individually acknowledged, had to fight on two fronts: against professional engineers who insisted on using traditional methods for unprecedented problems, and against physicists who wanted to verify the functioning of the devices by clever, but ambitious and time–consuming, and not always reliable, tests. His lecture ends with a disarmingly frank account of how he had a little too much drink in celebrating VJ Day, and how he fired 21 high–explosive bursts in place of a 21–gun salute, only it came out to be 22.

John Manley and Edwin MacMillan describe the early stages in the creation of the laboratory. Joseph Hirschfelder contributes a brief outline of the work of the laboratory with thumbnail sketches of some of its most famous members. Norris Bradbury, director from 1945 to 1970, deals with the transition to a peacetime institution. There is a historical introduction by Lawrence Badash.

A few questions after each talk give the impression that the audience (or at least its more vocal members) followed the lectures less intelligently than the reader will, but they still elicited interesting answers.

All in all, this mixture of styles and approaches adds up to a lively picture of an unusual, and maybe unique, experience.

Counting Weapons

N uclear weapons, and the knowledge of the horrors they are capable of producing, have been with us for 35 years. We might be tempted to let familiarity blunt the impact of these facts on our mind, were it not so frequently refreshed by news of ever more powerful weapons, ever–increasing numbers in the stockpiles, and ever more efficient means of delivering them to their targets. Any future nuclear attack could be, and probably will be, enormously more devastating than those experienced in the unfortunate cities of Hiroshima and Nagasaki. Fortunately, the danger of this happening is very much reduced by the "balance of terror": by the fact that the United States and the Soviet Union both possess these weapons in profusion, so that any nuclear attack on either country or on their allies must expect a reply in kind. In this kind of nuclear war there can be no victors.

This prospect of retribution, the deterrent effect, is a positive factor in a gloomy situation. Related to it is the fear of a global nuclear war, which introduces a certain amount of caution into international politics, and makes statesmen think of arms control as a serious possibility. The achievements of arms–control negotiations—the ban on atmospheric testing, the first SALT treaty and the unratified SALT II—are by no means impressive, but arms-control talks in pre–nuclear days did not even get that far.

So the present purpose of nuclear weapons is essentially to prevent nuclear war. If deterrence fails, the result will be global disaster. Experts who discuss military strategy often overlook the fact that the effectiveness of the deterrent does not depend on the actual damage the weapons would do, or on the cer-

---

This is a review of the books *Britain and Nuclear Weapons,* by Lawrence Freedman. New York: Papermac/St. Martin's Press, 1980; *Countdown: Britain's Strategic Forces,* by Stewart Menaul. London: Robert Hale Ltd., 1980; *The War Machine,* by James Avery Joyce. London: Quartet Books Ltd., 1980; *Protest and Survive,* edited by E. P. Thompson and Dan Smith. New York: Penguin, 1980.

tainty that they would be used in retaliation against an attack, but on how the potential enemy *estimates* the damage, and what he *thinks* is the likelihood of retaliation. He does not have to be sure that the damage will be unacceptable, he does not have to be certain that the trigger will be pulled by the country he is about to attack, as long as these are serious possibilities. Responsible leaders do not gamble at long odds on the survival of their country. "Responsible" is an important qualification: heaven preserve us from a Hitler in charge of a superpower.

Why is there such an intense race in nuclear weaponry, damped only slightly by SALT and similar agreements? Each of the superpowers wants to have more, bigger and better weapons than the other, and any evidence of a "gap," of a superiority on the other side, produces dismay and an increase in the weapons effort. The result is that each side has the physical capacity to kill all the other's citizens many times over. The arms race would make some sense if the potential enemy could use weapons in such a way as to avoid retribution. But even if it were possible to mount an operation which was sure to put all the other side's land–based missiles and planes out of action simultaneously—and there are plenty of doubts about that—there would remain the missiles carried by submarines, powerful and numerous enough to inflict terrible damage. The art of locating submarines does not seem to have got very far. Surely these facts are known to the superpowers' statesmen and military establishments. Why do they insist that their arms must be bigger and better than the other side's? Is it that tradition and habit make them continue to count and weigh weapons? Is it pressure from the technical people who enjoy developing more and more ingenious weapons systems? I do not know the answer.

Nuclear weapons are a deterrent primarily against nuclear attack. They may not deter aggression by "conventional" (i.e. non-nuclear) forces, because in the event of, say, the Soviet Army invading Western Europe, it is not credible that the United States would pull the nuclear trigger, inviting a nuclear holocaust. Yet even here nuclear weapons may exercise a slight inhibiting force, because of the uncertainty as to when the opponent will get mad enough to strike, regardless of the consequences.

The situation here has become confused however, by the introduction of "tactical" nuclear weapons. These are small weapons, intended, not for terror attacks against cities or industries ("strategic weapons"), but for use on the battlefield. To say that they are smaller is a relative statement: they are still very powerful weapons. They come in various kinds and sizes, but most have a range of destruction which in populated areas would cover many non–combatants. Their primary purpose is to act as a deterrent against conventional attack in circumstances where the big strategic weapons would not be a credible deterrent. Whether the threat of tactical nuclear defense against non–nuclear attack is credible depends on the likelihood of a "limited" nuclear exchange, of the tactical nuclear war not escalating into a full exchange of strategic weapons.

There are great uncertainties here, particularly since the Soviet Union has never admitted that there is an important distinction between strategic and tactical weapons. In spite of this, it is NATO policy to rely on the use of tactical nuclear weapons in a situation where their conventional forces are in danger of being overrun. This decision was strongly criticized by the late Lord Mountbatten and by Lord Zuckerman. It is understandable because conventional forces need a great deal of manpower, and are therefore politically embarrassing. It seems unlikely that the Warsaw Pact countries will want to start an invasion of the West in the near future, but this is the main contingency for which NATO plans exist. What is worrying is that the Communist countries, if they did desire such an adventure, might not believe in NATO's readiness to employ tactical nuclear weapons. Suppose there were an invasion. NATO would then have no choice other than to concede defeat or to use these weapons; and even if using these weapons did not lead to escalation and a general holocaust, it would almost certainly involve the devastation of the battleground—presumably much of Western Europe.

Limiting nuclear war by restricting it to tactical weapons is a very speculative idea. There is another way of limiting it, which has more credibility: the superpowers might avoid attacking each other's territory, thus limiting the war to Europe—and such a limitation might endure. From a European point of view, it is a most unattractive thought.

Thinking about deterrence necessarily involves thinking about the use of nuclear weapons, and this exercise can easily lead military planners to think of such a use as something actually intended as part of a future war: one now hears talk of tactical weapons, in particular, being acceptable in themselves and not as a deterrent to be held in reserve for an ultimate catastrophe. Coupled with this there is much talk about survival in a nuclear war, and civil defense measures, either communal or individual, to improve our chances of survival. This is frightening, not because it is wrong to think of the fate of the survivors as long as we cannot entirely rule out the possibility of nuclear war, but because it presents a war of this kind as something other than an unbearable calamity.

What are the particularly British problems in this situation, and what is the role of the British nuclear arsenal? The latter question is analyzed in Lawrence Freedman's serious and thought–provoking book, *Britain and Nuclear Weapons*. To the first question, why Britain should have nuclear weapons, he cannot discover a very clear answer. The arguments have changed over the years, but the policy of keeping the weapons, and bringing them up to date, has continued unchanged. It is often stressed that they provide additional strength for the NATO alliance. But, as strategic weapons, they are, from the point of view of NATO, only a marginal addition to the enormous U.S. stockpile: as I have already said, counting weapons of this kind makes no sense, once there are enough to deter. Another argument one hears is that they constitute an independent deterrent, meaning that the decision whether and how to use them rests ultimately in British hands. This would be important if the United States were to turn isolationist again and lose interest in the security of Western Europe. Is the expenditure on the weapons, and on the submarines which carry them, intended for this contingency? And would Britain plan to stand up to Soviet pressure without support from the U.S.? Another aspect of this independence is that the UK would have the physical possibility of precipitating a nuclear exchange if the U.S. were hesitant. It is hardly an attractive thought for the Americans that the game of mutual annihilation could be triggered by their allies, but the same thought may, for the Russians, make the deterrent marginally more credible.

There remains the "top table" argument: the possession of these weapons is supposed to guarantee their owners a seat at the top table, i.e. some weight in international negotiations. This thought must be voiced with some discretion, because Britain supports the idea of Non–Proliferation, which involves dissuading non–nuclear countries from acquiring nuclear weapons. If they are essential for us to retain our influence in world affairs, should not others (without aggressive intentions) follow our example? Also it is by no means obvious that membership of the various negotiating committees is necessarily the same as carrying much weight in them.

Freedman reviews the arguments for and against remaining "nuclear," the latter being mainly the cost and effort, which could otherwise be devoted to conventional armaments, and finds them fairly evenly balanced. He also reviews the choice of particular equipment, and the present plan of replacing the submarines carrying Polaris missiles by a new generation carrying the more effective Trident missiles. Compared to alternative options, he thinks this a good choice.

This is strongly questioned by Air Vice-Marshal Stewart Menaul. To him, the decision to rely on submarine–carried missiles when the V–bombers became obsolete was misguided, and the choice of Tridents even worse. The only sensible option now, he maintains, would have been Cruise missiles. The views of a professional on such matters should be treated with respect. He tends to see problems in black and white, and he betrays many emotional prejudices. He is grieved by control of the nuclear deterrent having passed from the RAF to the Navy, and his comments on Mountbatten, who "was dedicated to ensuring the Navy's survival, if necessary at the expense of the RAF," make him sound only a little less evil than the "disreputable minority" of the Campaign for Nuclear Disarmament, started by "pro–Communist elements both in the government and the country" in the late 1940s. His historical research seems to have been less thorough than one might wish, to judge by a number of minor slips, such as that Robert Oppenheimer escaped from Germany, or that the uranium 235 bomb is started by firing two hemispherical pieces together. He blames Alexander, Minister of Defense in 1947, for basing military planning on the ill–conceived assumption that there would be no major war for at least ten years. It does not seem to matter that events have proved this assumption correct. The tone of the book makes me somewhat prejudiced against its conclusion: but it does not follow that his arguments on the relative merits of different weapons systems might not be sound.

If Menaul sees matters in black and white, James Avery Joyce has no doubts about seeing them in white and black. He uses strong language—in my view rightly—to describe and denounce the arms race with its senseless overkill capacity, and with the frightening possibility that some accident, error or wrong speculation might one day unleash the horrors of a full–scale nuclear war. Few will quarrel with his description or disagree that a halt to the arms race, and ultimately disarmament, are imperative. Many people are ready to use their voice, and any influence they may possess, toward these ends. If books like these contribute to the public awareness of the situation, and win more supporters to the cause of disarmament, this is all to the good. But if one wants to know what can be done about it, the book is rather disappointing. It quotes the marginal limitations already achieved, such as the partial Test Ban, and advocates the negotiations now proceeding toward a comprehensive test ban and other desirable measures. It does not attempt to analyse in depth the factors now blocking progress. Joyce advocates a World Disarmament Conference, and/or a European Disarmament Conference, but it is not clear how one could ensure that such conferences would be more productive than those of the past. He would like more use made of the peace keeping machinery of the United Nations, but we know that, under present rules, there are many situations in which these peace–keeping forces cannot be effective. He says: "We must work hard on that 'moral equivalent to war.'" Many of us are ready to work hard, but how is it done?

Protest and Survive (a title inspired by the Civil Defense Pamphlet *Protect and Survive*) is a collection of essays, edited by E. P. Thompson and Dan Smith, with emphasis on the effect of nuclear weapons on Europe, and particularly on Britain, and on current statements dealing with civil defense. An impressive article by Alva Myrdal discusses the function of tactical nuclear weapons, and argues that their real purpose would be to keep nuclear devastation out of the territory of the superpowers. An essay by Dan Smith along similar lines argues that NATO is not, as is claimed, inferior to the Warsaw Pact in conventional forces, and that tactical nuclear weapons are therefore not essential for the security of Western Europe. E. P. Thompson points to the danger which the presence of American cruise missiles will constitute for Britain, and attacks the revival of civil defense. I would emphatically agree with him that any realistic civil defense measures can at best have a marginal effect. Whether the presence of the cruise missiles makes a nuclear attack on this country more likely is debatable, given that they are not the first American missiles to be stationed here and that a pre–emptive Soviet strike against them would make no sense unless other means of retaliation could also be eliminated. The real argument against the cruise missiles, it seems to me, is that they are just a further addition to a strategic arsenal containing a large capacity for overkill, and therefore part of the futile practice of counting weapons. We should indeed argue against these weapons, and against NATO's reliance on tactical nuclear arms, but not as Thompson would like it, as an exercise in unilateral nuclear disarmament which, he hopes, would be infectious.

In "The American Arms Boom" Emma Rothschild describes, and castigates, the growing expenditure on arms research, development and production, and its motivation. Elsewhere, Mary Kaldor discusses the conversion problems which would arise as a result of substantial disarmament. Ken Coates, restating the arguments against tactical nuclear war, claims that at one time U.S. plans provided for the possibility of nuclear attacks against targets in allied countries "to deny their resources to Soviet troops." This might perhaps be taken with a grain of salt, since the information comes from a document leaked by the Soviet Union which is supposed to have reached them through a spy.

Forty Years into the Atomic Age

I n 1945 we heard a lot of talk about the "atomic age," which had burst into the open with the attacks on Hiroshima and Nagasaki. These events were the visible culmination of development which had taken place in secret on a gigantic scale. The first tangible proof of the possibility of releasing nuclear power—or "atomic energy," to use the contemporary jargon—was the successful operation of a nuclear reactor by Fermi's team in Chicago on December 2, 1942. The fortieth anniversary of this event was commemorated in a meeting at the University of Chicago, the proceedings of which have now been published by the University.

How different are public attitudes now from those at the birth of the Atomic Age! In Britain and the United States, at any rate, the dropping of the first atomic bombs was greeted with joy at the ending of the war and with pride at the achievements of the scientists and engineers. This joy was tempered by awe at the might of the new power, and horror at the suffering and devastation of the Japanese cities. The new weapon had made war so terrible that a future global war appeared unthinkable. At the same time the newspapers and radio carried stories about the bounty expected from the new energy source. The scientists were given credit for what had been achieved, and for what was to come. They were respected, given support, and even sometimes listened to.

Now, forty years on, the mood has changed completely. Fear of nuclear war is a persistent nightmare. Nuclear power is regarded with suspicion by some and passionately rejected by others. In the United States, this opposition has led to a practically complete abandonment of plans for further nuclear power stations. In Great Britain the opposition is less strong, but still very

This is a review of *The Nuclear Chain Reaction Forty Years Later*, edited by Robert G. Sachs. Chicago: The University of Chicago, 1984. For the complete reference, please see the Acknowledgments section.

vocal. In France, where the lack of coal and oil deposits has made nuclear power more attractive, its development is continuing.

Along with the different reactions to "atomic energy," the status of scientists has changed. They are seen by many as the irresponsible inventors of evil new agencies, not caring about the damaging or dangerous effects of their activities.

There are, of course. many other branches of science and technology that have developed spectacularly in recent years, and many of their products, from computers and air travel to detergents and antibiotics, have had a far greater impact on our everyday lives than atomic energy. Some of these produce side–effects which arouse hostility, and are blamed on the scientists. But we shall look at atomic energy in particular.

One of the developments arising from the work on atomic energy is undoubtedly beneficial, but is not mentioned much in public debates. This is the use of radio isotopes and neutrons for research and for medical treatment. The practical value of this work is perhaps best demonstrated by the fact that it has been found worthwhile to build a considerable number of sophisticated nuclear reactors dedicated to research, without any relation to the production of weapons or nuclear power. The privatization of the British government–owned firm marketing radioactive isotopes has recently been in the news. But while the commercial aspects of this transaction were controversial, there is no disagreement about the value of the activities to which it relates.

However, to most persons "nuclear" means either nuclear power or nuclear weapons, and around both of these feelings run high. Most of the passionate opposition to nuclear power comes from the possible hazards associated with it. These have three distinct aspects: radiation leaks during normal working, radioactive waste, and major disasters.

The first of these, the possible health hazards from nuclear power stations working normally, should be easy to control. One difficulty is to maintain in an expanding industry adequate standards of caution and discipline. It is not possible to make safety arrangements completely immune to human error. There have been some errors; for example, the recent discharge of excessive radioactivity into the sea at the Sellafield plant in Great Britain. There is as yet no clear evidence of any injury to the public. Even if there was, it has to be remembered that the alternative energy sources, coal and oil, involve necessary accidents and damage from pollution, and compared to these the health record of the nuclear power industry is probably superior.

Public opinion has some difficulty accepting this conclusion, for two reasons. One is the fear of radiation as an insidious, invisible agent. The story, is told that when the Atomic Energy Research Establishment at Harwell started, residents in the neighborhood were perturbed by the presence of a tall chimney from which there was never any visible emission. This suggested that something sinister was being discharged. The problem was solved by arranging to discharge some smoke or steam from the chimney. It then seemed normal, and local fears were allayed.

The other difficulty is that much of the information about radiation levels and their effects comes from experts connected with authorities responsible for the plants, who may be regarded as interested parties. Here matters are not helped by the past record of official statements, both in the United States and the United Kingdom, which rather too readily gave assurances on the basis of inadequate information. Present controversies in both countries concerning damage to the health of military personnel during nuclear weapons tests add to the suspicion of official information.

Radioactive waste is an unavoidable product of nuclear power stations, and, since it will remain active for many hundreds of years, its disposal raises serious questions. It is easy to find containers which will shield the radiation, but not easy to be sure that the containers will not corrode or otherwise deteriorate over the centuries. Burying the waste requires certainty that movement of earth or underground streams will not carry activity back to the surface or into wells or mines. Much research is being done on these problems, and it is fortunate that decisions can safely be postponed, since the amounts of waste produced so far can be kept in supervised holding sites for ultimate disposal. When decisions have to be taken, the question of trust in the competence and conscientiousness of the authorities will again arise.

The aspect of nuclear power arousing the most emotional reactions is no doubt the possibility of catastrophic accident. There is no possibility of a reactor causing a nuclear explosion similar to that of a nuclear bomb, but one can imagine a failure that would spread radioactive material over a wide area, with numerous casualties. The question therefore arises how to ensure that this does not happen. No serious expert would claim that safety measures can make such an accident quite impossible, only that it can be made very unlikely.

Public opinion is not accustomed to thinking in terms of probabilities, and to be told that such disasters are possible, even with an extremely small probability, is frightening. To put the matter in perspective, it has been pointed out that on this basis of a "worst–case" analysis, other parts of technology offer equally frightening perspectives. One example often cited is that of a jumbo jet crashing on a crowded football stadium, an event which is certainly not impossible, if very unlikely. Another is an explosion of a chemical plant. This has indeed happened, though perhaps not in the worst imaginable circumstances. These possibilities have not led to demands that aircraft no longer be allowed to fly or that chemical plants be closed, only to an insistence on precautions that make such disasters as unlikely as possible.

But how unlikely are such accidents? There exists a considerable literature on this subject. A major breakdown of safety would require the simultaneous failure of several components, and since most of the relevant components occur also elsewhere in industry, one can judge their failure–rate from industrial experience. Several authors have made calculations on this basis, and while their estimates differ, they all find extremely small risks. However, this analysis seems incomplete without taking into account the human factor. The two important near–disasters to nuclear power stations, at Browns Ferry and at Three Mile Island, were both caused by unforeseen human actions. In the first case, it was the thoughtless action of a technician, in the second a fairly general disregard of safety rules. I have heard it said that the fact that there was no real disaster on these two occasions, when several of the safety barriers had been breached, only proves how safe the nuclear industry is but, I do not find this convincing, as in both these cases only luck prevented worse troubles.

The problem of eliminating human error is complex. For persons whose normal occupation involves handling dangerous equipment, it is very easy to forget about remote possibilities of danger, and to become casual about safety rules meant to forestall them. More study is needed of ways to keep such workers alert and careful, and of ways of making equipment more foolproof.

A further objection sometimes raised against nuclear power is that encouraging its development in other countries might lead them to develop the technology for producing nuclear weapons. This point too, is controversial. Some say that a country with technology capable of producing nuclear weapons will be able to do so whether or not it has a nuclear power industry. The link between power and weapons would perhaps become closer if the "breeder reactor" became widespread. This reactor is capable of making the raw material, uranium, go much further, and is therefore economically attractive, but it is technically more complicated and raises more difficult safety problems, because it uses highly enriched material, like a nuclear bomb. It requires a reprocessing plant which extracts plutonium from the exposed fuel elements. Such a system therefore involves handling large quantities of plutonium, some of which is suitable for weapons, and the reprocessing plant is an essential and expensive part of a weapons program. At present France leads in the development of a breeder; Britain has a prototype plant, while the American breeder program is in abeyance.

The Chicago symposium included an excellent discussion of these problems by Dr. Alvin Weinberg, who has played an important part in the development of nuclear power, though for my liking he does not pay enough attention to the problem of the human factor. Other interesting talks were by Dr. A. Carnesale, who concentrated on the political and commercial aspects of the problem, and Dr. Walter Massey who described the contribution of the

Argonne Laboratory, the successor of the "Met. Lab." in which Fermi started the first chain reaction.

The subsequent panel discussion was mainly concerned with the prospects for nuclear power in the United States, which are at a very low ebb, though there was some rather wistful mention of the situation in France and Japan, where the urgency of nuclear power development is much greater, and where popular opposition does not seem to impede development.

The aspect of nuclear energy that gives rise to the most serious worry is undoubtedly nuclear weapons. The decision to use the first available bombs on Japanese cities caused unending controversy. It was defended because it cost fewer lives, even fewer Japanese live, than an invasion of Japan, and it produced fewer casualties than an earlier fire–aid on Tokyo. On the other hand, many have argued—and I agree—an attempt should have been made to use the first bomb on a much more sparsely inhabited target, to demonstrate the power of the bomb in the hope that this would be enough to bring Japan to surrender. In any event, the argument goes, the second bomb on Nagasaki was surely unnecessary.

Many of the scientists working on the atomic bomb project argued against the use of the bomb on cities, at least initially. It is interesting that the center of this agitation was in Chicago, not in Los Alamos, where the bomb was made. Perhaps the reason is that the Metallurgical Laboratory in Chicago had completed their urgent wartime task, and its members had more time for reflection, while those in Los Alamos were still working under great pressure. In any case, the Chicago paper, written on the initiative of James Franck and Leo Szilard, never reached the decision–makers.

After the war there was great enthusiasm for the idea of international control of atomic energy. The government of the United States was to give up voluntarily its temporary monopoly of nuclear weapons, in return for an agreement by all other countries to develop nuclear energy only through an international agency. This plan, worked out by the Acheson–Lilienthal committee, with Robert Oppenheimer playing an important part in the drafting, was presented to the United Nations by Bernard Baruch, and became known as the "Baruch Plan." It had meanwhile undergone a subtle change of style: Baruch was aware of the need for the proposal to be accepted by the United States Senate and therefore emphasized the points in America's favor, but this made it less attractive to the Soviet Union. This did not really matter, because there never was a chance of it being accepted. For the Soviet Union to accept a plan proposed by the United States at a time when the one state had nuclear weapons and the other did not, would have been humiliating; the Soviet Union suspected that it would somehow be put at a serious disadvantage.

Next the talk was of disarmament, and many liked to see this as complete disarmament, at least complete nuclear disarmament. Numerous international meetings, mostly under the auspices of the United Nations, discussed numerous proposed disarmament treaties without any real progress, and even today complete nuclear disarmament is claimed to be the ultimate aim. But the difficulty is that neither of the great powers would risk dismantling its nuclear arsenals without reliable verification that the other side was complying with the treaty, since a few clandestinely preserved nuclear weapons would give an overwhelming military advantage if the other side had completely destroyed its own stockpile of such weapons. Therefore, any such treaty requires an effective international inspection procedure and this has been unacceptable to the Soviet Union, for whom the restriction on foreigners roaming around the country is an essential part of security. It is also by no means obvious that the required pervasive inspection would have been acceptable to the United States Congress or industry.

At the present time, the most that international negotiations aim at is arms control, i.e., treaties which would reduce nuclear arsenals, or at least stop their expansion. There have been some marginal successes. The partial test ban treaty has at least stopped the two major powers from testing weapons in open air and has therefore halted the radioactive pollution—a very desirable result, even if the treaty has not banned underground tests and thus has not stopped the development of new weapons. The SALT I treaty put some very slight restrictions on the numbers and types of weapons to be deployed, and stopped the development of anti–ballistic missile systems, which would have led to further escalation of the arms race. All these negotiations are in abeyance in mid–1984.

Meanwhile the arms race continues, and the stockpiles of both sides have reached grotesque dimensions. Everyone is convinced that nuclear war cannot be won against an adversary with nuclear weapons. Such a war would result in devastating damage to both sides. This is true, notwithstanding occasional assurances by governments that their country could fight a nuclear war and survive. These are statements to keep up morale.

The only purpose served by nuclear weapons in the present situation is as a deterrent against nuclear attack. But then why accumulate the present excessive stockpiles and race to get more and better weapons? One may argue about just how much is needed for a deterrent, but it is beyond question that the present amounts are many times more than is needed. If one side has enough bombs to wipe out the major cities of the adversary five times over, what would it matter if the adversary has enough to do so ten times?

There is one important proviso. In public debates the idea of a "first strike" is often bandied about. This term denotes an attack which would wipe out the other's nuclear missiles and thus forestall retaliation. If such an operation were feasible, it would indeed put a premium on pre–emptive attack. It would also make it vital to have large numbers of highly accurate weapons. In that situation, no country would feel safe unless it was guaranteed superiority or at least equivalence in nuclear arms, and the arms race would have to continue to the limit of each country's technical and economic possibilities.

Fortunately, the whole concept of a nuclear "first strike" belongs in the realm of science fiction. It ignores the existence of nuclear weapons carried by submarines, and at present there is no effective way of dealing with these. Even if there were spectacular progress in the technology of submarine detection, it would be unlikely to lead to the possibility of disposing of all of them in a short time, which would be essential for an effective first strike. In addition, there are bomber planes carrying nuclear weapons, which, even during the short available warning time of an attack, could take off, and survive.

Even ignoring submarines and bombers, a first strike would always remain a most risky undertaking. It would rely on the functioning of thousands of missiles, all in a short time, and allowing for only a small failure rate. Such an operation is easily carried out on a computer, and the performance of the missiles studied on the test range, but reliability in the heat of battle and in a gigantic mass operation is something very different. The American space shots, which are serviced by the cream of the technical manpower of the United States, have frequently to be postponed because of the malfunction of some component. No government would stake the survival of its country on the success of such an uncertain operation.

In fact, none of the military experts believe in the feasibility of a first strike. How can one explain the continuing race for more and better weapons? It is easy to understand the motivation of the scientists and engineers in the weapons laboratories who, often initiate new technical ideas, which satisfy their professional pride. It is easy to understand the support from industrialists, for whom the arms race means more business. But why do so many politicians, and why does public opinion, support this wasteful and dangerous activity?

Is this a failure to appreciate the new situation created by nuclear weapons? In assessing the prospects of naval battles or tank warfare, numbers are indeed vital, but one has to understand that nuclear weapons are not battleships. Or is it a naive belief that the threat of retaliation will deter more if it is backed by greater numbers of—quite unnecessary—weapons of retaliation?

Whatever the reason, it seems hard to get across the simple truth that each side could safely afford to freeze its stock of nuclear weapons at the present level or reduce it by a considerable factor, although not below the minimum for an effective deterrent, without depending on equivalent reductions in the opponent's stockpile.

There has indeed been support, particularly in the United States, for a "freeze" movement, but usually it is not made clear whether this is meant to be a unilateral freeze or a negotiated one. A negotiated, and verified, freeze would be an excellent thing, but the negotiations to achieve it are not likely to be easier, or more promising, than other recent negotiations for arms control.

Recently, all these questions have been overshadowed by novel speculations about defense against missiles, based on lasers, or particle beams, or other techniques using the latest scientific devices. Few, if any, scientists outside the weapons laboratories regard such plans as credible, and those with relevant expertise who have looked at the ideas claim that they would involve prohibitive cost, and be vulnerable to relatively cheap and easy countermeasures by the attacker. Even the government of the United States has now lowered its target from the defense of the country to that of missile launching sites, relevant only in the event of an attempted first strike.

The problems of nuclear weapons formed the third part of the symposium at the University of Chicago, and while their discussion occupies only 32 of the 283 pages of the proceedings, they are the most fascinating to read. Talks by John A. Simpson, Marvin Goldberger, and M. H. May gave impressive and impassioned presentations of the arguments. A panel discussion with these speakers and Richard Garwin and Hans Bethe followed this up and showed the audience, as well as all the speakers, to be in agreement on the futility of the arms race, the urgency of arms control negotiations, and scepticism about the "star wars" talk. There were differences of opinion only on details. Professor Eugene Wigner maintained a belief in the effectiveness of civil defense, and expressed the fear that by instituting an effective civil defense system the Soviet Union could make itself immune from nuclear retaliation, and could therefore blackmail the United States. He found no support from any of the speakers.

If we count the blessings of the "atomic age" we must remember the many constructive applications of radio–isotopes and neutrons in research and in medicine; we may argue about the advantages of nuclear power, but at least we have the choice between using or not using the new source of power. Above all, we must be aware of the possibility of nuclear war. As was stressed at the meeting at Chicago, we are fortunate that we are still here, and that for nearly 40 years there has been no global war, a fact for which the credit must probably be given to nuclear weapons in making everybody frightened of war. This gives our world a certain stability, but we must work for a better system which does not also contain the possibility of utter disaster.

The Case for the Defence

E dward Teller has often been called the father of the hydrogen bomb. Today he deserves the additional title of grandfather of "Star Wars," since he is believed to have originated the idea which was publicized by President Reagan. Readers who expect, from the title of his new book, to find a full account of Teller's views on "Star Wars," will be disappointed. It is a collection of 33 essays, of which only five are concerned with that topic, including the one that gives the title to the book.

Thirty-nine pages is a short space in which to discuss a controversial subject in depth, and this is aggravated by the fact that the five essays were evidently written for different purposes (we are not told whether and where they were previously published), and there is much overlap and repetition. This section of the book therefore does not add much in the way of new arguments.

Teller makes much of the claim that the Soviet Union is well advanced in research on anti–missile defence, but does not give sources for this information. Presumably it comes from intelligence sources, but American intelligence has been known to exaggerate estimates of Soviet strength, presumably in order to support American defence projects.

The author is too good a physicist to claim that success in developing an effective defense is certain, and this is shown by his use of cautious phrases, for example: "Attempts to deploy a defense will be made only if there is a chance that defense will be less expensive than the compensatory deployment of aggressive weapons or the means to destroy the defenses"; or, "If we find real cost–effective ways for defense...". Yet in other places he sounds much more positive: "The genie that produced the sword of modern times

This is a review of *Better a Shield Than a Sword: Perspectives on Defense and Technology*, by Edward Teller. New York: Free Press, 1987. For the complete reference, please see the Acknowledgments section.

can also produce the shield"; or "Today there is increasing evidence that modern technology can produce successful defense measures." Perhaps the words "increasing evidence" are meant to convey a reservation, but this is not how the average reader will understand them. Here is even more definite: "[the] development [of lasers] has produced the unexpected result of making defense feasible."

Many critics of the project have pointed out that the demands that the anti–missile defense make on computers far exceed in complexity anything so far achieved. Teller disposes of this argument in rather a cavalier manner by claiming that the American telephone system is much more complex than the required defence management computer. It may be true that the phone system connects more subscribers than there are inputs and outputs in the defence system, but the computing functions in the latter are enormously more sophisticated. Moreover, the phone companies have to maintain an army of engineers to correct faults, whereas the defence computer would have to function reliably first time. All its parts could be tested separately, but the functioning of the whole can be rehearsed only in the real battle situation, when it is too late for corrections. There is no discussion of counter–measures, except for the use of decoys; none of the articles mentions the obvious response to laser beams against missiles of making the missiles spin.

I do not believe that these essays will convert any skeptic who doubts seriously whether the vast effort devoted to the Strategic Defense Initiative can make a contribution to anyone's security.

Several essays deal with the history of the atomic and hydrogen bombs, and it is interesting to have the recollections and reflections of someone so closely involved with both. During much of the wartime period at Los Alamos, Teller continued thinking about the hydrogen bomb, although this was not given any priority as it could make no contribution to the war. He explains the reason at one point by saying that Bethe asked him to take charge of the calculations on the implosion but, "I refused. I knew there were others better at it. I also wanted to pursue more novel plans, including fusion." This account agrees with my recollection; I took over that work instead, though I would not claim that I was better at it. However, in another essay we are told that Oppenheimer urged Teller to continue exploring the thermonuclear field, and he says, "That was not easy advice for him to give or for me to take." In looking at recollections of events of some forty years ago one has to allow for uncertainties in human memory. Some little absent–mindedness is illustrated by his referring twice to his stay at "City College" in London, when he means University College.

The essay "Seven Hours of Reminicences" is a criticism of the BBC series on Oppenheimer (presumably written at about the time of the screening of that programme) which discusses Teller's view of Oppenheimer, whom he admires but finds hard to understand. It also outlines Teller's testimony at the Oppenheimer security hearings, where, in reply to the question whether he believed Oppenheimer to be a security risk, he explained that he often disagreed with him, and could not understand the basis of his actions: "In this very limited sense I would like to express a feeling that I would feel personally more secure if public matters rested in other hands."

In the essay he summarizes his "reluctant testimony": "I considered Oppenheimer loyal but because his actions appeared confused and complicated I would personally feel more secure if public matters would rest in other hands." At the time, his testimony caused indignation among many of Teller's colleagues, who felt that his reply to the very clearly understood question whether there was a security risk was confused and complicated, and hard to understand.

The essay about the hydrogen bomb stresses in its title "the work of many people" and makes polite references to the contributions of Stanislaw Ulam and others, but leaves no doubt that the important invention was Teller's. "A Ray of Hope" comments enthusiastically on a plan for international control of atomic energy proposed by "a board of consultants to the State Department," although Teller suggests some modifications. This essay must puzzle the reader, until it is discovered, looking at the list of copyright credits, that it was written in 1946!

Other essays, too diverse to summarize, always present original and unorthodox views, which are interesting to read, whether one agrees with them, as with Teller's view that excessive secrecy does more harm than good, or disagrees, as with his ideas on the possibilities of Civil Defence in a nuclear war.

The Making of the Atomic Bomb

M any books have been written about the history of nuclear weapons, and they have become more detailed as more information, previously inaccessible, is made available. But Richard Rhodes's book seems unique, not only for its length of 886 pages (788 without the notes), but for his unusually broad interpretation of what is relevant background material. No person is mentioned without a paragraph or so about physical appearance and the essentials of biography. These characterizations, though terse, give a lively picture of the person, and for all those whom I have known, an accurate one, with very few lapses (Heisenberg did not have red hair). For the leading actors in the story there are more detailed profiles. The one of Niels Bohr goes back to his grandfather.

An introductory chapter is devoted mainly to Leo Szilard's realization in 1933 that the existence of the neutron, just discovered, might make a chain reaction possible. He expected that there might be nuclei which, on being hit by a neutron, would give up further neutrons, and that energy would be released in this process. These neutrons would hit other nuclei, and the result would be a chain reaction, which would continue until the number of suitable nuclei was reduced below a critical amount, or the material was dispersed. Szilard expected that the nucleus of beryllium might serve for this purpose, but this was a misguided idea. Such a result, Szilard felt, would justify the speculations by H. G. Wells and others about the liberation of atomic energy on a large scale for industrial and military purposes, although Lord Rutherford—who had discovered the atomic nucleus—had called such speculations "moonshine." Szilard was, in fact, not the only one to see this point; others

This is a review of *The Making of the Atomic Bomb*, by Richard Rhodes. New York: Simon and Schuster, 1987. The book review is reprinted with permission from *The New York Review of Books*. Copyright © 1987 Nyrev, Inc. For the complete reference, please see the Acknowledgments section.

did so, including the Soviet theoretical physicist Lev Davydovich Landau. But in the early 1930s nobody knew by what nuclear process such a chain reaction could be implemented.

The full story involves the elements of nuclear physics, and Rhodes takes us through this subject, starting with the discovery of radioactivity by Henri Bequerel in 1896, to the discovery of fission and of the secondary neutrons from fission. The story necessarily involves technicalities not easily accessible to the non–scientist. Rhodes manages to explain the relevant points briefly in simple language, so as to give the reader a feeling for the nature and relevance of the argument. He often explains the nature of an experiment or a device by a happily chosen analogy.

So we are taken through the development of nuclear physics. Some of the explanations contain minor errors, which show that the writer is not a physicist, but they will not prejudice the understanding of the nonphysicist reader. Since the narrative is more or less in chronological order, the history of nuclear physics is interspersed with the relevant political history—including the air raids and gas warfare of the First World War, and the Gallipoli adventure in which Harry Moseley, a brilliant physicist who provided experimental confirmation of Rutherford and Bohr's model of the atom, was killed.

Later comes an account of the rise of the Nazis in Germany, introduced by a brief history of the Jews in Europe and of anti–Semitism in Germany. The refugees included many scientists who played a large part in the atomic–weapons program. It has been suggested that the reason for the large number of refugees in this work was the fact that they had special reason to hate Hitler's Germany and to be afraid of Hitler getting the bomb first; but the more plausible reason was, at least in Britain, that the locals were already engaged in urgent war work, and refugees were the main source of available manpower, until the project gained sufficient urgency for people to transfer from other war research.

The academic physics story ends with fission. Enrico Fermi, who had shown the way to the use of neutrons to study nuclei, believed his uranium experiments led to "transuranics"—elements heavier than uranium—and did not accept the explanation based on fission when this was suggested by a German chemist, Ida Noddack. The discovery of fission had to wait for the work of Otto Hahn and Fritz Strassmann in Berlin. Simple experiments on fission were immediately started in most nuclear physics laboratories of the world, and soon the presence of secondary neutrons was found in Paris and New York.

The great excitement came not only because one was dealing with a completely unexpected phenomenon in physics, but also because now a chain reaction had become a reasonable possibility. From now on the narrative is interrupted by glimpses of world events, as well as by shifts of the scene between many centers, including Germany, the Soviet Union, and Japan. It feels almost as if one is watching a three–ring circus, but Rhodes's racy presentation rivets the attention.

Quite early on, Bohr had argued that the observed fission was caused predominantly in the rare isotope uranium 235, present in natural uranium only in a proportion less than one percent. Only very fast neurons, faster than most of those produced in fission, would cause fission in the dominant isotope, 238. Fermi did not believe Bohr until about a year later Bohr was proved right by direct experiment.

From this it followed that no explosion could be caused in natural uranium. This left only two possibilities open: one was to slow down all the neutrons until they were thermal, by using a "moderator," a light substance with whose nuclei the neutrons would collide and be slowed down. In that case the chain reaction would proceed very slowly, and the developing heat would blow the uranium apart before the chain reaction had got very far. This method was therefore useful only for a controlled chain reaction, which could produce power.

The other was to work with nearly pure 235. This would yield a new weapon of fantastic power, but required separating the isotopes of uranium, a procedure that was "evidently" prohibitive in cost and effort.

A number of groups decided to try for the slow chain reaction, although they all thought of the eventual development of a weapon. The most active group was that of Fermi and Szilard at Columbia University. Szilard had been dreaming of a chain reaction all the time, and now it had become more real. He was also particularly afraid of Hitler getting the bomb first. Fermi probably was first attracted by the challenge to start the first chain reaction (which he did) after missing by an inch the discovery of fission.

Other groups with similar aims were in Germany (Kurt Diebner and Heisenberg) and in Paris (Fréderic Joliot and coworkers). In England a group under George Thomson concluded after preliminary work that a chain reaction in natural uranium was not practical, even with the best available moderators to slow the neutrons; so the group adjourned. Physicists in the Soviet Union and Japan were interested and did some experiments with fission, but did not then attempt to develop a chain reaction.

These are some of the circus rings that Rhodes presents to us in chronological order, but most of the space is rightly taken up by the developments in the United States. Fermi and Szilard decided that the most promising "moderator" to slow down the neutrons, so as to make them most effective, was graphite, but tests showed that the graphite swallowed up so many neutrons that the chain reaction would not take place. Fermi had the hunch that the graphite they were using still contained impurities in amounts too small to detect by chemical analysis, but large enough to contribute strongly to the neutron loss. Efforts to purify the graphite further proved him right.

In Germany the nuclear physicist Walther Bothe was accused of wrong measurements on graphite. Probably his measurements were correct; but he did not have Fermi's intuition that the "pure" graphite was still too impure. The German physicists therefore decided on heavy water (water in which the hydrogen is replaced by its heavy isotope, deuterium). Heavy water is very difficult to extract from natural water; supplies for experiments were obtainable only from a plant in Norway, which was sabotaged by Norwegian patriots. The German scientists, led by Heisenberg, never succeeded in producing a chain reaction. They did not attempt isotope separation on an industrial scale, which they (probably rightly) regarded as too big an effort for Germany in wartime.

Fermi and Szilard, by now with several collaborators, needed larger quantities of graphite to extend their experiments toward an actual chain reaction. They appealed to various government agencies for support, but the bureaucracy moved very slowly. Even after Szilard and Edward Teller had persuaded Einstein to write a letter to Roosevelt the only result was that a committee under Lyman Briggs was appointed, which did not proceed with any sense of urgency.

In England it was pointed out by Otto Robert Frisch and the present reviewer that a bomb of uranium 235 would not be as large as had been assumed intuitively, and that its explosion would release enormous energy. Their reasoning was theoretical, since experiments on the nuclear behavior of uranium 235 were not made until much later, but they relied on the very convincing theory of Niels Bohr. The British authorities showed growing interest. Further research was commissioned by the MAUD committee, whose report indicated that the separation of the uranium isotopes in the necessary quantities was not impossibly difficult, and that it would lead to a new powerful weapon.

Information on what this report contained reached the United States in various ways but does not seem to have had much effect until Ernest O. Lawrence, whose enthusiasm was aroused by a personal report from Mark Oliphant, the Australian physicist, started to agitate for urgency. His pressure, added to all the others', caused a change in pace. The work was put in the overall charge of Colonel Leslie Groves, of the Army Corps of Engineers, who was promoted to brigadier general in the process. The problem of isotope separation was now tackled seriously by three different methods, and the entire project was firmly directed toward a bomb.

However, it had been found in the meantime that, in a chain reaction, some neutrons hit uranium 238 without causing fission, but cause it to change into a new element, neptunium, which in turn decays to the rather long–lived plutonium. According to Bohr's theory, plutonium should also be capable of a fast chain reaction, so it was an alternate weapons material to uranium 235. The slow reactors, larger versions of the assembly of uranium and graphite by which Fermi produced the first chain reaction in December 1942, were now seen as suppliers of bomb material.

There were still many hurdles to be overcome. Rhodes runs through the troubles of the gaseous diffusion plant at Oak Ridge. Gaseous diffusion was one of the ways of separating isotopes, which depended on membranes or "barriers" with very fine pores, yet strong enough to stand a pressure difference and inert enough to withstand the very corrosive uranium hexafluoride, the only known gaseous compound of uranium. He tells us about the electromagnetic separation method pioneered by Lawrence, which needed for its magnet coils more copper than was available in wartime, so that the coils were instead made of silver borrowed from the U.S. Treasury. The third method of isotope separation, liquid thermal diffusion, was proved by Philip Abelson rather late. It involves a large but very simple plant, which consumes a prodigious amount of power. All three methods contributed to yielding the separated uranium for the Hiroshima bomb.

We are told of the reactor design problems, of arguments over the most effective method of cooling, and of the surprise when one of the fission products, an isotope of xenon, swallowed too many neutrons, which would have made the reactors unworkable if they had not been designed with spare space for additional uranium.

Rhodes describes how Oppenheimer build up the Los Alamos Laboratory on a mesa in New Mexico for studying the fast-neutron physics and bomb design—and later for bomb manufacture. We are told how he inspired a large number of first–rate scientists with a sense of urgency and common purpose, and how he managed to keep down the natural friction between the civilian scientists and the army officers who were running the place. Here there was a surprise, too: it turned out that besides the isotope plutonium 239, which was the weapons material, the Hanford Engineer Works in Washington state, which consisted of reactors for its production, yielded also some plutonium 240, which undergoes fission spontaneously very fast. This meant that, to start the explosion, the parts of a bomb would have to be put together very quickly, otherwise the fission of a plutonium 240 atom would initiate the reaction when the parts had approached only close enough to make conditions just critical, and a weak explosion would result.

So a new technique for assembling the bomb had to be adopted: the "implosion" method, in which a sphere of plutonium is compressed by the detonation of an explosive shell around it, instead of firing the two halves of the bomb at each other in a gun barrel, as was sufficient for uranium 235. The implosion method is much more subtle than the other, and much harder to test without an actual bomb explosion. It was decided to test such a weapon at Alamagordo (the "Trinity" test). There were doubts up until the last minute whether it would work. The test confirmed that the design was right, and showed the enormous explosive power of the new weapon.

Meanwhile there had been many discussions about what to do with this weapon. There was little doubt that it would be used on Japan as soon as it was available. One argument for this was undoubtedly that the expenditure incurred by the project, without the knowledge of Congress, by now added up to $2 billion, and the military would be blamed if this had made no contribution to the war. More substantially, the argument was that the use of this weapon would surely lead to the surrender of Japan, thus obviating the need for an invasion, which, it was said, was going to cost more lives (even more Japanese lives) than the bomb.

The idea of a demonstration in the presence of invited observers was considered, but rejected. A scientific panel made up of Lawrence, Fermi, and Oppenheimer, among others, concluded in July 1945 that "we can propose no technical demonstration likely to bring an end to the war; we see no acceptable alternative to direct military use." There was also as yet no complete confidence in the reliability of the design. The Trinity test, it was said, was not conclusive proof that the bomb would explode when dropped from an airplane, and if the Japanese were told of the weapon and it then failed to explode, this would be exploited by Japanese militarists. The obvious alternative of dropping a bomb unannounced on a thinly populated area of Japan, to demonstrate its effect, was not, as far as I know, ever considered. It would have involved killing some people and destroying some buildings; the effects of an explosion in the desert, as at Alamagordo, were stunning to the experts, but would not mean much to the observer.

In reporting the discussions that led to the decision to drop the bomb on Hiroshima and Nagasaki, Rhodes reminds us of the history of the terror bombing of cities. From the early widespread indignation at the bombing of Guérnica and Rotterdam, attitudes changed to such an extent that the deliberate fire raids on Hamburg, Dresden, and Tokyo were considered an acceptable form of warfare. Without this background the atomic bomb raids on Japan might not have taken place.

A group of scientists in Chicago tried to prevent the bomb being used and wrote a memorandum (known as the "Franck Report," after the physicist James Franck) which never reached the decision–makers in Washington. At the time it was taken to Washington, nine days after Roosevelt's death, the top people were too busy and passed the matter on to men of no influence. Rhodes does not mention this report. It is interesting that the Chicago scientists, who were concerned with reactors, had stronger feelings against the use of the bomb than those in Los Alamos, who were making the bomb. The reason may lie in the presence in Chicago of a few people like Franck and Szilard, or it may be that the Chicago group, whose work was essentially complete, were free to discuss fundamental issues, while the Los Alamos people were working all–out to finish their job and had little time or inclination to consider them.

Niels Bohr, who had escaped from German–occupied Denmark in 1943, when he was about to be arrested, and joined the atomic energy project, was thinking of the years ahead. He saw the danger of a nuclear arms race developing after the war, and sought ways of trying to counteract this by creating more confidence, and more openness between the West and the Soviet Union. Roosevelt listened to him with seeming interest, but did not act on his suggestions; Churchill completely failed to understand him and became very hostile. In 1950 Bohr addressed an "Open Letter" to the United Nations, again without much response.

Most of the book is written objectively, leaving judgment to the reader, but in a number of places the author's personal opinion is apparent. Leo Szilard is his hero, although he fairly describes his many eccentricities, and his occasional arrogance, which must have been trying for the administrators. But Rhodes rightly admires his imaginative grasp, and his indefatigable insistence that the threat of nuclear weapons must be taken seriously.

It is also clear that Rhodes disapproves of the trend to regard the bombing of cities as a legitimate form of warfare. He is critical of the British and American air force officers and statesmen who took pride in the efficiency of their operations against cities, and justified them on the dubious grounds of "breaking the will to resist."

He evidently does not think highly of the process by which the use of the bombs was decided. The ancient city of Kyoto was removed from the list of targets, he notes, only by the personal insistence of Secretary of War Stimson, who happened to be familiar with its cultural significance. The conduct of the war, he says, had been "stupid and barbarous," and "the barbarism was not confined to the combatants or the general staffs. It came to permeate civilian life in every country" and, "was perhaps the ultimate reason Jimmy Byrnes, the politician's politician, and Harry Truman, the man of the people, felt free to use and compelled to use a new weapon of mass destruction on civilians in undefended cities." I happen to share the author's opinions in all these respects.

Nuclear Weapons:
How Did We Get There,
and Where Are We Going?

A s everyone knows, nuclear weapons were invented during World War II. I was involved in the early stages, when O. R. Frisch and I pointed out to the British authorities that an "atom bomb" was feasible and not prohibitively large, and what the consequences would be. We knew about fallout and about the numerous civilian casualties that would result from almost any kind of use of this weapon. This, we said, "might make it unsuitable as a weapon for use by this country." Yet it had to be developed because of the fear that Nazi Germany might get there first. The basic scientific discoveries had been made before the war and were common knowledge. There was no defense other than deterrence, the threat of nuclear retaliation.

The fear of a German nuclear weapon proved groundless, but the Anglo–American development continued. It seemed wrong for the Allies to forego a weapon that could end the war quickly. It did, though the decision to use it without warning on two cities is still argued about. It was clear that the monopoly of this weapon would not last long, and that this would give rise to problems in the post–war world. In this respect the most far–sighted person was the great Danish physicist Niels Bohr, who foresaw the arms race, and tried to persuade government leaders to inform the Russians that a nuclear bomb was being developed. He was able to talk to Roosevelt, who seemed sympathetic, but Bohr's interview with Churchill proved a complete disaster. Churchill did not understand what Bohr was saying and became suspicious he would on his own give the "secret" of the atom bomb to the Russians.

Later Bohr wrote an open letter to the United Nations, pointing to the inevitability of a nuclear arms race unless nations made a common effort to prevent this, and this was not compatible with great military secrecy—he urged there should be an Open World. This letter, too, was largely ignored, but it is gratifying that in the last few years the world has come much closer to the openness which he advocated.

As soon as the Soviet Union had succeeded in making nuclear weapons, the nuclear arms race started, and it has reached utterly crazy proportions, both in the number of weapons stockpiled by the "superpowers" and in their destructive power, increased a hundredfold by the thermonuclear, or "hydrogen" bomb.

What are these stockpiles for? Everybody knows that a war in which both sides possessed and used nuclear weapons would be disastrous for both, and would leave only losers, no winners. Such a war would completely annihilate civilized life in both countries, and, if recent speculations about the "nuclear winter" are correct, in other countries as well. The idea of a disarming "First Strike," which eliminates the victim's power to retaliate, is seen by nearly all experts as impossible, or at least so uncertain that only a madman would stake the survival of his country on such a venture. Reagan's proposed Strategic Defense Initiative ('Star Wars') to create a defensive umbrella protecting a whole country has practically been abandoned as unrealistic, and replaced by a limited scheme, which might somewhat reduce the effects of a nuclear attack, but would not make it acceptable.

So the only purpose of the possession of nuclear weapons remains deterrence, the threat of retaliation against possible attack. This is particularly important against a nuclear attack, but there is also the idea of "extended deterrence," which means nuclear retaliation against an attack by non–nuclear means. This arises particularly in Europe, where NATO fears that its armed forces are not strong enough to withstand a "conventional" attack from the Warsaw Treaty powers. At the present time the likelihood of such an attack appears extremely small, but it is the business of military leaders to be prepared for all eventualities. Originally the idea was to counter such a hypothetical attack by using the intercontinental nuclear missiles, but it was realized that such a threat had little credibility. Retaliation by such strategic bombardment would of course result in similar action in return, with intolerable damage and casualties to both sides.

This led to the idea of tactical, or "battlefield," nuclear weapons which could be used in Europe without escalation into a full nuclear holocaust. The trouble is that there exists no clear definition of tactical, as against strategic, weapons. They are not all small; some have powers many times that of the bombs used on Japan. A definition is evidently essential if both sides are to refrain from going beyond the tactical weapons. Besides, while they would be aimed at military targets, there would be heavy civilian casualties, and this thought is not popular in Europe, particularly in West Germany.

The main object of deterrence remains retaliation against nuclear attack, and the question therefore arises how many weapons are necessary for this purpose. Estimates of this "minimum deterrent" have been quoted as a few hundred to a thousand weapons on each side. Remembering that the majority of the weapons in the present stockpiles are twenty to fifty times more powerful than the "small" nuclear bombs dropped on Japan, this seems a generous estimate, even allowing for some failures or missed targets.

The arsenals of the superpowers hold at present around 25,000 such weapons a side. Even the UK's "Independent Deterrent" is well in excess of a minimum deterrent. These excessive numbers are of no conceivable military advantage to their possessors. They have resulted from a race, in which both sides insist they must at least have equal, if not superior numbers to the other. This is a misunderstanding of the nature of deterrence. If you have enough weapons to kill every person on the other side, it does not make you safer to have enough to kill them five times over.

Why then do military experts and statesmen on both sides insist on matching the weapons of the other? An interesting study of this problem has recently been published by a psychologist, who interviewed many of the experts and decision makers in the U.S. (Stephen Kull, *Minds at War: Nuclear Reality and the Inner Conflicts of Defense Policymakers*, Basic Books). One factor is evidently the intuitive belief that bigger is better. Numbers were of course vital in the days of naval battles, when the numbers and sizes of the opposing battleships almost determined the outcome. It seems that some of the experts have not yet understood that nuclear weapons are not battleships.

Another factor may be the idea that a show of strength in the form of masses of unnecessary weapons, will raise prestige and frighten off potential aggressors, and establish confidence amongst allies. The accumulation of weapons may also serve domestic politics by showing the public how much the government is doing for their security.

The last few years have seen important improvements in attitudes, and the INF treaty to reduce the number of medium–range missiles in Europe is a welcome step, not so much because of the actual reduction in numbers, which amounts only to a few percent of the total stockpiles, but because the verification and inspection provisions of the treaty have been accepted and their implementation proceeds in a most amicable way.

However the controversy over the "modernization" of the NATO short–range nuclear weapons may threaten the progress resulting from the INF treaty. Such modernization cannot raise the destructive power of the weapons—many are already too powerful for battlefield use. It would increase their range, right up to the limit laid down in the treaty, or conceivably improve their accuracy, which with their great area of destruction, is not important except for "hardened targets," of which there are not many in Europe. The pressure for modernization seems another example of the "battleship mentality," or perhaps reflecting an illusion that fighting a nuclear war might in some circumstances actually make sense.

The "START" (strategic arms reduction talks) negotiations, aiming at a fifty percent reduction in strategic nuclear weapons, are also a step in the right direction, though a small step, since even that reduction will leave excessive numbers compared to a minimum deterrent. Both the INF and the START negotiations have been and are carried out with the aim of reaching an assured balance; each side to be assured by verification that the other side does not have a superiority in number or capacity. This may be a political necessity, but in fact a balance is of no importance for security as long as each side has at least the minimum deterrent.

While the large numbers, as I have argued, do not increase security, it may equally be argued that they do not diminish it, and that they therefore constitute an expensive, but harmless, luxury. But this leaves out of account the risk of accidental or irresponsible firing of such a weapon, which is increased by their numbers.

Meanwhile the number of countries having, or capable of acquiring, nuclear weapons is growing, and with it the risk of their irresponsible use. The only remedy for such proliferation would seem to be persuasion or political pressure by the superpowers, but while they remain excessively armed this cannot be very effective.

At this time the arms race does not take the form of increases in numbers as much as in the development of new, more sophisticated weapons. To stop this a treaty banning the testing of nuclear weapons has been proposed, (which might also help with the proliferation problem) but so far the U.S. and the UK are strongly opposed to this. There have been claims that underground tests would be difficult to verify, but the technique of seismographical detection has made substantial progress. Experiments have been conducted both by U.S. scientists near the Soviet test site, and by Soviet scientists near the test site in Nevada. One cannot expect that the verification might not occasionally miss a single illicit test, but a significant series of tests would certainly be picked up. The objections to a test ban are in fact based on a desire to go on testing. It is sometimes thought that testing must continue to check the quality and reliability of the weapons in the stockpile, but according to the experts there is no need for that. The continued testing serves only the continuation of the arms race.

Public opinion, which tended to feel strongly about nuclear weapons, is apparently reassured by the fact that there are now promising disarmament negotiations. These, and other aspects of East–West detente, are welcome, but the world is still full of unnecessary and dangerous weapons; large numbers of scientists and technologists spend time and effort developing "better" ones. Dare one hope that some day governments will understand the simple basic facts? So far Mr. Gorbachev seems to have come closer to this understanding than any other world leader.

Atomic History

B ritish scientists contributed to the work of the Manhattan Project (the code–name given to the secret wartime effort in the United States to make atomic weapons) in three ways. First, they persuaded the British government, and thus indirectly the U.S. government, that an atom bomb could be built, and that one was urgently needed as a deterrent against a possible German equivalent. Second, British teams took part, after 1943, in the bomb research at Los Alamos and in the work on electromagnetic isotope separation at Berkeley and at Oak Ridge, Tennessee, as well as acting as consultants to the Kellex team in New York, which designed the gaseous diffusion separation plant at Oak Ridge. Finally, some groups in Britain worked on special problems for the Manhattan Project. In *British Scientists and the Manhattan Project*, Szasz mentions and commends all these contributions. But, as the subtitle of the book indicates, his real interest lies in the British mission to Los Alamos.

The secret laboratory established in Los Alamos in 1943 was without doubt an exciting place. There, in a remote corner of the New Mexico mountains, was gathered a unique group of talented researchers (including at least eight actual or future Nobel laureates) faced with the challenge of building bombs. It was especially exciting for the British participants, many of whom were new to North America, all to the New Mexican desert. By collecting a tremendous amount of source material, Szasz has succeeded in painting a vivid, albeit kaleidoscopic, picture of those years. One of his aims is to assess the value of the British contribution, but as he is not himself an expert, he has to

This is a review of the books B*ritish Scientists and the Manhattan Project: The Los Alamos Years*, by Ferenc Morton Szasz. New York: Macmillan/St Martin's Press, 1992; *The Los Alamos Primer: The First Lectures on How to Build an Atom Bomb,* by Robert Serber. Berkeley: University of California Press, 1992. For the complete references, please see the Acknowledgments section.

quote various people's opinions on this question. And naturally there are various opinions.

Readers familiar with the story will notice a few errors in dates and other facts. For example, the author tells us that the Los Alamos laboratory was set up in an isolated place so that scientists could talk to one another freely without having to be compartmentalized. In fact, the purpose of isolation was to afford greater security and lessen the risk of leaks. The absence of compartmentalization was a concession to Oppenheimer after he had agreed to the isolation. Some of the other inaccuracies relate to the way of life at Los Alamos; here Szasz fills in gaps in his sources by using common sense. But security and military administration do not always operate by common sense. These slips, however, are not serious enough to distort the overall picture.

Szasz concludes with a sketch of the postwar history of U.S.–UK atomic relations, brief résumés of the later careers of members of the British mission to Los Alamos, and an appendix containing the text of two memoranda written by Otto Robert Frisch and myself to the British government in 1940 to draw attention to the possibility of nuclear weapons. Altogether, he tells the remarkable story of the Los Alamos adventure in clear, if not always elegant prose.

An interesting document from those times is *The Los Alamos Primer*. This contains the notes of a series of introductory lectures that Robert Serber gave to members of the laboratory when it started and to later arrivals. It is a clear and concise exposition of what was known at the time and what problems there were to be solved, testifying to the vision of the team of physicists who under Oppenheimer had prepared the ground for work on an atomic bomb. Although some details have of course changed since the primer was written most of it could still serve as a first introduction today.

The primer was declassified in 1965 and now makes its appearance as a published book. The text of the original document, actually notes by Edward Condon of Serber's lectures, occupies less than 30 pages. It is interspersed with explanatory notes, aimed presumably at readers with only a rudimentary knowledge of physics, if any at all. At the beginning Serber provides an estimate of neutron multiplication using the diffusion equation (which assumes that the mean free path is small compared to the dimensions of the system). This is in fact a very poor approximation for the purpose. He then goes on to develop somewhat laboriously a better method. This is surprising, because at the time there was an exact calculation published in the open scientific literature.

Serber has now added an autobiographical foreword in which he tells how he came to work with Oppenheimer and about the early research into atomic energy before the days of Los Alamos. A historical introduction is provided by Richard Rhodes (the author of *The Making of the Atomic Bomb*, Simon and Schuster, 1986). About the only similarity between the primer and Szasz's book is that the Frisch–Peierls memoranda, which were not known at Los Alamos when the lectures were given, are reproduced in an appendix. Nearly 50 years on, the primer still makes interesting reading, and will be particularly appreciated by historians and others studying the development of nuclear physics.

"In the Matter of J. Robert Oppenheimer"

T he official documents about the Oppenheimer "case," which were published last summer by the U.S. Atomic Energy Commission,[1] have already gone through several printings and are likely to become bestsellers. They concern the position of Oppenheimer as a consultant to the A.E.C., and his "clearance"; this is a domestic issue for the United States and no business of outsiders. But the problem is far wider; in the first place Oppenheimer is well known and respected amongst the scientists of the world, including those who enjoyed the privilege of working under him in the wartime atomic energy project. In addition, the issues of "security" and "loyalty" in the relations between a scientist and the official machine raise matters of principle far transcending national boundaries. Lastly, the testimony before the Personnel Security Board (publication of which was rather surprising) represents a human document of absorbing interest. Many readers will wonder how, in the place of Oppenheimer, they would have stood up to the experience of hearing, for weeks, every small incident of a busy and varied career dissected and analyzed, and being questioned about details of past correspondence and conversations, with the original letters, or recordings of the conversations, produced afterwards for comparison. They will try to see themselves in the place of the scientific witnesses and wonder how well they would have succeeded in presenting their views about the important points of principle, or whether they would have resisted any better than some of the witnesses, the temptation to allow their personal emotions to color their testimony.

---

[1] *In the matter of J. Robert Oppenheimer*, Transcript of Hearing before Personnel Security Board. U.S. Government Printing Office, Washington, D.C., 1954.

The sequence of official action is briefly as follows: In December, 1953, Oppenheimer was notified in a letter from Major–General Nichols, the General Manager of the A.E.C., of a number of "items of derogatory information" (there were strong objections when in the hearings these were called "charges") which cast doubt on his eligibility for clearance. His clearance, *i.e.*, his access to secret information, was at that time suspended. Oppenheimer replied in a statement in March, 1954, that these matters could not be "fairly understood except in the context of my life and my work." The statement outlines his career and the development of his views in relation to the information in General Nichols' letter. He exercised his right to hearings before a "Personnel Security Board" which met from April 12 to May 6, 1954. The board consisted of Dr. Gray (chairman), Dr. W. V. Evans and Mr. T. A. Morgan. Oppenheimer had the assistance of counsel, and there were counsel for the board, who played the part of prosecuting counsel, except that the hearings began with witnesses for the defense, the adverse witnesses coming mostly toward the end of the proceedings.

Two members of the board (Gray and Morgan) wrote a majority opinion recommending against renewal of Oppenheimer's clearance, whereas Dr. Evans added a minority opinion for clearance. Oppenheimer did not appeal from the majority finding to a "Review Board," and the papers then went, with an argument by his lawyers, to the Atomic Energy Commission. It thus became the duty of the General Manager to recommend to the commissioners a course of action, and he recommended against the reinstatement of clearance. The commissioners were divided four to one, the majority accepting the General Manager's recommendation partly in joint and partly in separately written opinions. The fifth, Dr. Smyth, wrote a minority opinion supporting his vote in favor of Oppenheimer.

Before we consider details of the "derogatory information" it is as well to remember the purpose of security tests. They exist to assure that "individuals are employed only when such employment will not endanger the common defense and security." What are these potential dangers? A person may willingly give information to agents of potential enemies. That this is a realistic possibility has been demonstrated by the trials of May, Fuchs and Greenglass, and by other cases. Since it is not easy to find evidence of the transmission of information and impossible to prove the intention, it has become usual to investigate whether a man's political views and activities show him to be the kind of person who is likely to form such an intention. It is for this purpose that the government agencies are instructed to take account of the "character, associations and loyalty" of the individual. These tests exist to find whether a man is likely to be, or to become, a spy for a foreign country. Another danger arises when a man with access to important secrets is indiscreet and liable to disclose information unintentionally through carelessness, under the influence of drinks or drugs, or as a result of blackmail.

Apart from the question of leakage of information, disloyal men in responsible positions might endanger security if they misused their position to sabotage the efficiency of government agencies. There do not seem to be any proved instances of this on record (though charges of such sabotage are a familiar feature of trials in totalitarian countries) and it would not be easy to imagine this to happen in practice. It is rare that an individual can take important decisions individually without having to convince his colleagues of their merits. A consultant, who acts in an advisory capacity, would find it hard to succeed in a course of action which is calculated to damage his country's interests unless both his fellow-consultants and the officials whom they advise were equally disloyal or unusually incompetent.

It might also be thought that a disloyal person could be a danger because he would not obey instructions in security matters. This is a course of action which would scarcely appeal to a sympathizer with a potential enemy. (The record of the Fuchs case shows that he was most punctilious and enthusiastic about security regulations.) In any event, the proper way to enforce security regulations in detail would seem to be by direct disciplinary action rather than through a "loyalty" investigation.

The list of "derogatory information" before the board consists of 24 items (our numbering corresponds with that in the board findings) most of which are found proved by the board. Indeed the facts about most of them are not in dispute; the question is as to their significance and whether they are indeed derogatory. The items differ in weight. The most strikingly irrelevant is No. 11, which states that "... in 1945 Frank and Jackie Oppenheimer" (his brother and the brother's wife) "were invited to an informal reception in the Russian Consulate... (held)... for the purpose of introducing famous American scientists to Russian scientists who were delegates to the United Nations conference... and that Frank Oppenheimer accepted... ." The board solemnly "concludes that this allegation is true."[2]

Of the more substantial items, the last is concerned with Oppenheimer's attitude to the thermonuclear or "hydrogen" bomb. Briefly, it claims that he opposed the decision to develop it, that after it was decided to go ahead, he continued to oppose it, that he refused to cooperate fully in the project, and that he persuaded others not to work on this. "... the opposition... of which you are the most experienced, most powerful and most effective member, has definitely slowed down its development."

---

[2] *In the matter of J. Robert Oppenheimer*, Text of Principle Documents and Letters of Personnel Security Board, General Manager, Commissioners. U.S. Government Printing Office, Washington, D.C., 1954.

For many pages the testimony goes into the recommendations of the General Advisory Committee under Oppenheimer's chairmanship, the procedure leading to the recommendations, and the merits of a "crash" program, *i.e.*, the rapid expansion of production facilities in the absence of an established design, as opposed to continued research on basic problems. It is stated by many experts that the plan, opposed strongly by the committee, envisaged equipment much too bulky for use as a weapon. One particular allegation, that Oppenheimer had copies of the committee report circulated to staff of the Los Alamos Laboratory in order to influence them against the work, is disproved.

The board finds that he was opposed to the development of the thermonuclear bomb, but did not oppose it once the decision had been taken to proceed with it. But, according to the majority report "had Dr. Oppenheimer given his enthusiastic support to the program, a concentrated effort would have been initiated at an earlier date."

This particular allegation caused indignation among scientists and other sections of the public, best expressed by the cartoonist Herblock in a drawing which shows a poster with the slogan "Think" in the wastepaper basket, and its place on the wall taken by one reading "Enthuse."

Another matter, not contained in the original list of 24 items, occupied a good deal of the time of the hearings. This is the part taken by Oppenheimer in a number of study groups concerned with national strategy. The reports of these groups are, of course, not made public but they seem to have been concerned with the necessity of tactical atomic bombs (*i.e.*, bombs suitable for use against military formations in the field as against those used for "strategic" bombardment), with keeping the arrangement for the allocation of fissile material to different purposes flexible, and with the balance between offensive and defensive weapons.

According to the majority opinion of the board: (Oppenheimer) "may have departed his rôle of scientific adviser to exercise influence in matters in which his convictions were not necessarily related to the protection of the strongest *offensive*" (my italics) "military interest of the country." In another context it is stressed that, if people give advice in other than a technical capacity it is important that "underlying any advice is a genuine conviction that this country cannot... have less than the strongest offensive capabilities in a time of... danger."

The reply by Oppenheimer's lawyers enquires whether it is wrong for a scientist to believe "in the wisdom of maintaining a proper balance between offensive and defensive weapons."

The General Manager's memorandum is somewhat on the defensive. "... at no time has there been any intention on my part or the board's to draw in question any honest opinion expressed by Dr. Oppenheimer. Technical opinions have no security implications unless they are reflections of sinister motives. The evidence establishes no sinister motives... in his attitude to the hydrogen bomb...". He adds, however, that the evidence calls in question Oppenheimer's veracity. This relates to the controversy whether the General Advisory Committee's report was opposed only to a "crash program" or to any work of the hydrogen bomb. It also concerns the unanimity of the committee. The report was unanimously agreed to by those present, but one absent member (Dr. Seaborg) wrote a dissenting opinion in advance, in which, however, he stressed that he could not be definite without hearing the discussion. Had Oppenheimer omitted to mention this preliminary dissenting opinion?

The majority opinion of the A.E.C. also stresses that Oppenheimer's opinions on the thermonuclear bomb are not in question. The added statement by Commissioner Zuckert says his decision was not influenced by the hydrogen bomb question or by the advice given by Oppenheimer on national security affairs. There are similar statements by Commissioners Campbell and Murray.

This disposes of this particular item, which raised stronger feelings and more controversy than any other. It is also the only important item relating to Oppenheimer's recent actions.

Of the remaining allegations probably the most important single item is No. 23, "that prior to March 1943,... Peter Ivanov, secretary of the Soviet Consulate... approached George Charles Eltenton for the purpose of obtaining information, that Eltenton subsequently requested Haakon Chevalier to approach you... Chevalier thereupon approached you, either directly or through your brother... Chevalier finally advised Eltenton that there was no chance whatsoever of obtaining the information." Oppenheimer did not report this to the authorities for several months, and did not state that he himself had been approached or disclose Chevalier's identity. He did this only when ordered to do so by a superior. "... upon your return to Berkeley you were visited by the Chevaliers on several occasions;... your wife was in contact with Haakon and Barbara Chevalier in 1946 and 1947."

In his written reply Oppenheimer describes this incident: (Chevalier) "told me that George Eltenton had spoken to him of the possibility of transmitting technical information to Soviet scientists. I made some strong remark that this sounded terribly wrong to me. The discussion ended there. Nothing in our long standing friendship would have led me to believe that he was actually seeking information; and I was certain that he had no idea of the work on which I was engaged." He admits he should have reported the incident at once.

The board hearings show that in order to avoid mention of Chevalier and of himself, Oppenheimer invented details, e.g., that the unspecified intermediary approached three members of the Radiation Laboratory. These he now describes as "a cock–and–bull story." Asked for the reasons he says, "Because I was an idiot." He was concerned with drawing the attention of the security officers to Eltenton, who, he felt, might be acting as an agent, and he saw at the time no need to have Chevalier involved.

This incident is probed extensively at the hearings. Transcripts of his various talks with security officers are reviewed and tape recordings of these conversations are gone over several times to establish precise phrases.

This matter clearly plays a large part in the final, adverse decision and, in the opinion of the A.E.C. majority, is aggravated by the fact that Oppenheimer visited Chevalier as recently as 1953. That he was wrong in not reporting the incident at once, and in fact distorting details, he admits fully. In evaluating this, one must, remember than in 1943 few people visualized Russia as a potential enemy, and that the mere mention of transmitting information to Russia did not, at that time, have the significance it would have today. To most scientists the need for secrecy on such projects as atomic energy was then obvious (though their first concern would have been to keep the information from reaching the Germans) and they took great pains to protect the secrets. In this the activities of security officers were often felt to be a nuisance. They appeared to the scientists as amateurs, who could not appreciate the nature of the work, who often tried to insist on unnecessary formalities, which delayed important work, and at other times missed obvious risks of leakage. Perhaps the Greenglass case, which Colonel Lansdale, the chief Security officer of the wartime atomic energy organizations describes as "the outstanding blunder of the century" provides a measure of justification for this impression. Few people, scientists or otherwise, would in 1943 have had that reverence for security procedures which seems to be implied by much of the discussion.

The remaining allegations tend to establish that Oppenheimer had close associations with left–wing people and activities, mainly in the years before he worked on the atomic energy project. No. 19 alleges, in addition, that he was responsible for the employment of Communists by the project. All this evidence has been on the record for a long time, except perhaps for details. Some questioning at the hearings aims to find out to what extent this information was available to the A.E.C. when on many previous occasions, they considered Oppenheimer's clearance. The position remains somewhat vague, but little doubt is left of the existence of a sizeable file on Oppenheimer, examined by many people on many occasions, which must have contained substantially this kind of information.

Oppenheimer's written answer describes, and his testimony elaborates, the development of his views on these matters, the period of his "short but intense" association with left–wing activities and subsequent disillusionment.

The board finds unanimously that he is loyal and discreet, but neverthe-less, in the majority view, he is a security risk. The members of the board seem to have been rather concerned with their definition of a security risk, and enquired of several witnesses whether a man can be loyal to his country and yet a security risk. In the unanimous part of their report they discuss this question and raise as the only issue relevant to the Oppenheimer case as an aspect of the security system "the protection and support of the entire system itself." In other words a loyal citizen who was not sufficiently cooperative with security officers might thereby become a security risk. They are willing to overlook lack of cooperation in the days when the security problem was new.

The majority report then goes on to stress Oppenheimer's more recent personal contacts with Chevalier and others. "It is not important to determine that Dr. Oppenheimer discussed with Chevalier matters of concern to the security of the United States... his association is not the kind of thing that our security system permits... one who... has access to information of the highest classification." General Nichols' memorandum similarly refers to "... the obligations necessarily imposed by an adequate security system" and the A.E.C. majority says "Oppenheimer has defaulted... upon the obligations that should and must be willingly borne by citizens in the national service." Thus associations cease to be a factor which may throw light on a person's opinions and loyalty and become a matter of right and wrong in themselves. This is reminiscent of the primitive view that heresy (in this case Commu-nism) is a communicable disease with which contact must be avoided, per-haps even a disease of which there are "carriers" who are not themselves affected. The minority report of Smyth does not accept this principle. "The security system has... neither the right nor the responsibility to dictate every detail of a man's life... If Dr. Oppenheimer has misunderstood his obligation to security the error is occasion for reproof but not for a finding that he should be debarred from serving this country. Such a finding extends the concept of security risk beyond its legitimate justification."

A further reason for the board majority's adverse recommendation is "sus-ceptibility to influence." This is based in part on the "Peters letter." As a witness before a Congressional Committee Oppenheimer repeated a state-ment he made in wartime to a security officer, describing Dr. Peters as a "dangerous Red and former Communist." This statement got into the papers at Rochester, where Peters was working, and led to attacks on Dr. Peters. It was pointed out to Oppenheimer that this was unfortunate, in that it endan-gered Peters' job at the university. He therefore wrote a letter to the paper trying to correct the impression which had been left. This letter is read by the board majority as repudiating his testimony, and this interpretation is shared by Nichols and the A.E.C. majority. Smyth points out that the letter "of which a copy was sent to the Congressional Committee... was a manifestation of a belief that political views should not disqualify a scientist from a teaching job."

The last reason given by the board majority is that "Oppenheimer has been less than candid in several instances in his testimony." No details are given, but the General Manager's memorandum lists six instances, which are repeated by the A.E.C. majority. Smyth analyses some of these in detail and says of the others "these contradictions have been given undue significance." "Unless one confuses a manner of expression with candour, or errors in recollection with lack of veracity, Dr. Oppenheimer's testimony... has a ring of honesty."

The dissenting member of the board, Dr. W. V. Evans, stresses above all that the charges of left–wing associations were old and known when the commission cleared Oppenheimer in 1947. "They took a chance on him...and he continued to do a good job. Now the job is done we are asked to investigate him for practically the same derogatory information."

This argument is answered by the General Manager. "Consideration of the complete record plus a cross–examination of Dr. Oppenheimer under oath were not accomplished by anyone prior to the...hearing in 1954." The importance of this distinction depends on the weight one attaches to the various "contradictions" which are listed by Nichols.

Throughout the mass of papers there is nowhere any evidence given, nor any opinion expressed that Oppenheimer is not loyal to his country except for a letter by a Mr. W. L. Borden, who, after private study of information available to him was moved to write a letter to the director of the Federal Bureau of Investigation stating as his "exhaustively considered opinion...that more probably than not J. Robert Oppenheimer is an agent of the Soviet Union." This letter appears to have been, at least in form, responsible for setting the whole process of investigation in motion. But when Mr. Borden testifies, the board takes pain to stress that there is no evidence whatever known to them to substantiate his conclusion. The impression is such that Oppenheimer's lawyers do not even trouble to cross–examine Mr. Borden.

Mr. Borden's position is logically simple, but the reader who has gone through the 992 pages of testimony and the subsequent findings does not see any clear indication what particular danger to security the others would expect if Oppenheimer's clearance were restored.

One suggestion comes from the testimony by Dr. Teller, the chief opponent of Oppenheimer on the hydrogen bomb issue. He is asked if he would see any danger to national security if Oppenheimer had access to restricted data, but were to give no advice. If he really were to refrain from giving advice, Teller believes there could be no danger. It is an interesting idea that advice given to an organization which is not obliged to follow it, may be a danger to security. This is another facet of the principle that ideas can contaminate.

The board majority go almost as far. Anybody is entitled to his opinion, but a scientist must be careful when he gives advice on matters outside his technical province, he may still give such advice provided it is based on approved principles. So one of the dangers is non–technical advice given on the basis of wrong opinions. This reasoning is not supported by the A.E.C., and their main arguments are based on the criteria of "character and associations." It is laid down in the clearance rules that character and associations must be taken into account in determining whether a man's clearance would endanger security, but they seem to have become ends in themselves.

In this context the board raises the question "Can a different test...be justified in the case of the brilliant technical consultant than in the case of the stenographer or clerk?" Surely this question is answered by the length and intensity of the hearings, the eminence of the witnesses and others concerned with the decisions, and the many earlier investigations. From the beginning of his work at Los Alamos Oppenheimer was accompanied by men who were said to act as bodyguards, but whose functions really were to report on him to the security officers. Such attention cannot be given to every clerk. In the case of a clerk whose associations make a prima facie case that he may be committed to extremist views and actions it may be necessary to play safe and remove him to less sensitive work. Such a transfer will make much less difference to him, and to the organization employing him, than for a high–level expert. Even the clerk may suffer unjustly if the reason for his transfer becomes widely known; this is one of the distasteful consequences of a security system which has to act on probabilities rather than on facts established under the normal safeguards of legal procedure. But if it is unavoidable to do injustice to some, is justice served by doing an equal injustice to others when it would be practicable to avoid it?

Most of the space in the documents, and therefore in this review, is taken up with the logic of the allegations and their relation to the rules. But even a superficial reading of the testimony shows that behind this there lie many other factors, tendencies and emotions. This is what gives the record of the hearings its fascinating human interest. There are witnesses who sound as if they disapprove of Oppenheimer because he opposed their particular pet schemes, or because he has too much influence by being persuasive in discussion. (Could envy at his success be a factor here?) There are witnesses who seem to identify, in all sincerity, the interests of Strategic Air Command with the interests of the United States, and hence think Oppenheimer (and others) dangerous because he advanced arguments which might lead to atomic weapons being placed at the disposal of other branches of the armed services, or because he discusses the balance between defensive and offensive weapons.

There are also parts of the record which make refreshing reading. One of them is the testimony of Dr. Vannevar Bush before the board. He tells its members that they ought not to conduct the investigation on the basis of allegations which may be read as accusing a man for the opinions he has expressed.

There is, above all, the minority opinion of Commissioner Smyth, who, in four pages, gives an impressively clear and complete analysis of the facts and replies to all the main arguments of the board majority, the General Manager and the other Commissioners. Those readers who are distressed by the study of all this material may take a little comfort from the procedural accident which let Smyth have the last word.

As far as can be seen, the matter is now closed. In the words of the board, "this case puts the security system of the United States on trial." The verdict in this trial will always remain unsettled; the jury, which is public opinion, is not likely to agree. The results include a cruel ordeal as a reward for one of the most valuable servants of the Atomic Energy Commission and over a thousand pages of, at least, doubtful linen washed in public. What discouragement to the recruiting of scientists will result from this investigation remains to be seen. Smyth left the A.E.C. shortly after the end of the case and its seems a reasonable inference that his resignation had some connection with it. Perhaps the danger of losing many scientists for the government service as a result of this case has been overrated. In the words of General Groves' testimony, referring to the scientists who left Los Alamos at the end of the war, "They all started to get itchy feet...almost every one of them has come back into government research because it was just too exciting, and I think still is exciting." In any event the work of the scientists on atomic and thermonuclear weapons has developed so far that perhaps the loss of a few of them would not too greatly endanger national security. But if this case were to shake, even slightly, the enthusiasm of the educated people of a country for its way of life, if it were to sow any doubt as to the survival of freedom, fairness and reason and the importance of bringing sacrifices for these principles, then the loss of moral strength which it would cause would be the equivalent of many superbombs or ships or planes.

"Security" Troubles

On a number of occasions I experienced difficulties with the problem of security clearance. Basically, the need for clearance comes from the desire of the government to keep communist sympathizers and other potentially disloyal people out of sensitive jobs. In occupations concerned with military secrets one appreciates the need for such measures and sympathizes with the people charged with the "vetting." If we could fail to recognize the views and activities of Klaus Fuchs, a close personal friend, the task of the outside investigator must be hard indeed. But the problem of ensuring the loyalty of people entrusted with sensitive, confidential information is sometimes confused with a desire to keep "undesirables" from entering the country or any particular profession. This shows a failure to appreciate that the strength of a stable democracy rests on its citizens' understanding of the basic issues and their ability to reject simplistic extreme ideologies. Marxist or fascist ideas are not like an infectious disease that can be contracted by exposure to it. On the contrary, familiarity with them helps to give a firmer basis to one's own convictions, and makes one's arguments against such ideas stronger.

My first encounter with such problems was in 1951, when I applied for a U.S. visa to attend the Nuclear Physics Conference in Chicago. My application met with a long delay, and I heard informally that it was going to be refused. This was of course during the McCarthy era, when such refusals were very common. Sir James Chadwick, who had official contacts, told me that the reason for the delay was probably due to some statement in the files that I had been seen at a certain date in a place where I was not supposed to be (and where I certainly was not). The problem was solved by a coincidence: at approximately the same time, there was to be a meeting in Washington on the declassification of papers about atomic energy, to which I was to go as one of the British representatives. I thus had the status of "government official" and was entitled to a visa. Once in the United States, I was able to go to the Chicago conference.

Early in the following year I was to go to the Institute for Advanced Study in Princeton, and I thought naively that after having been granted a visa once, it would be easy to get another. However, there again was an inordinate delay, and I was getting ready to cancel my visit to Princeton when the visa came through at the last possible moment. I do not know what went on behind the scenes to get the visa for me.

Another disturbing episode occurred about 1953, when Mark Oliphant had resigned from his chair in Birmingham to go to the new Australian National University in Canberra. A suggestion was made that the faculty board should invite Cecil Powell, the discoverer of the pi–meson and one of the outstanding physicists of the country, to succeed Oliphant. The dean of the faculty and I were encouraged to call on Powell and sound him out about the prospect. He was quite interested, and made only one major condition—that he should be able to bring his research group with him, which was exploiting the photographic emulsion technique he had developed. Their rather modest equipment was provided for by a government grant, and their need for space could be met, so it looked as if the proposal would be accepted. But then someone pointed out that Powell had rather left–wing views and was, possibly, a member of the Communist Party. This turned many of my colleagues against him, some on general principle, others because they feared that his presence as head of the department would result in the loss of government research contracts. I was shocked, because the work of the department had nothing to do with military secrets or with politically sensitive problems. I came close to resigning from the university, and I would have done so if I thought that political prejudice was mainly responsible for blocking the appointment. However, some members of the faculty board were opposed because they feared he would concentrate on the work of his own research group and not take the responsibility for the big machines in the department seriously enough. I disagreed (and indeed, when some years later Powell was one of the U.K. representatives on various committees at CERN, the European accelerator laboratory, he pulled his weight very well), but this was a substantive argument, on which reasonable people could hold different views. Powell did not get the appointment, and Philip Moon was appointed to the chair; he proved to be an excellent head of the department, but this did not diminish my concern with the behavior of some of my colleagues.

The Oppenheimer case of 1954 was by far the most serious of these security troubles. I was only a distant spectator without direct involvement, but the affair was a great shock to me, as it was to many other scientists. Briefly, Robert Oppenheimer had made many enemies, including some in the Strategic Air Command, when he argued for more effort on defensive measures and for tactical nuclear weapons. He had upset Lewis Strauss, who was now the chairman of the Atomic Energy Commission, by making him look foolish in hearings before a congressional committee investigating the possibility of exporting radioactive substances. The General Advisory Committee of the AEC, of which Oppenheimer was chairman, had opposed a crash program to build a hydrogen bomb. He was accused of disloyalty and his security clearance was withdrawn. When he appealed, a quasi–judicial board was set up to examine him. In three weeks of grueling hearings, every facet of his activities was discussed, including some early indiscretions, which had long been known to the authorities.

One of many witnesses was Edward Teller, who was asked whether he thought Oppenheimer was a security risk. The question clearly meant whether he might be disloyal to the country, or might sympathize with a foreign government. Teller evidently could not assert anything of the kind, and said so in his testimony. However, he added, "In a great number of cases I have seen Dr. Oppenheimer act ... in a way which for me was exceedingly hard to understand. I thoroughly disagreed with him in numerous issues, and his actions frankly appeared to me confused and complicated. To this extent I feel that I would like to see the vital interests of this country in hands which I understand better, and therefore trust more. In this very limited sense I would like to express a feeling that I would feel personally more secure if public matters would rest in other hands." (From *In the Matter of J. Robert Oppenheimer*, M.I.T. Press, 1971.)

The three-man board resolved by two votes to one to condemn Oppenheimer for lack of enthusiasm for the hydrogen bomb program, but found no evidence of disloyalty. The AEC knew that any ruling against a scientist for views on technical questions would cause concern among scientists. They therefore did not accept the findings of the board, but decided to bar Oppenheimer from access to classified information because of "grave defects of character." One member of the commission, the physicist Henry de Wolf Smyth, disagreed and wrote a minority report. In 1963, when the McCarthy era was an embarrassing memory, Oppenheimer was awarded, as a gesture of reconciliation, the Enrico Fermi Prize, a high–level award then conferred by the President of the United States on the advice of the AEC.

As a footnote to this story I might recall the remark of George Placzek, then a member of the Institute for Advanced Study. At the height of the crisis, a representative of the AEC came to offer Placzek a research contract on some problem within his field of interest. Placzek terminated the conversation by asking, "Who do you think I am?"

Then, in 1957, another irritating episode occurred. I was still a consultant to Harwell. I was not concerned with very weighty or very secret problems, but I still had clearance. Then, when my appointment as consultant came up for renewal, I received a letter saying that for reasons of administrative convenience I would no longer be able to see classified documents. There was something odd about the letter, and I inquired whether this was really a routine measure, or whether there was more behind it. In reply I was told that the same action was taken in all cases of senior consultants whose contracts were coming up for renewal. This was technically correct, because mine was the only contract due for renewal at the time. In fact, there had been a message from American security asking that I be given no more access to American secret documents. Eventually I was told about this, because there was to be a conference at Harwell with American scientists, and if I had turned up to attend, I would have been refused admission. Actually, I had no intention of going there.

I was not upset by the American ruling. One knew they were inclined to act on unsubstantiated reports, and after the Oppenheimer affair I felt I was in good company. I also had no regrets about seeing no more secret documents. But I did resent that the Harwell authorities had tried to deceive me on this matter. Presumably the message from America specified that I should not be told, but they could have insisted that they must tell me, as in the end they did. The incident showed that I did not have the full confidence of Harwell, and I resigned my consultancy. After my move to Oxford, when the proximity to Harwell led to frequent contacts, I reconsidered my position. By now all the people involved in the offending episode had left, and there seemed no point in keeping up resentment against an abstract body. So I became a consultant again, but no important secrets have come my way.

The inclination to see "Reds under the bed" can reach grotesque proportions, as I discovered in 1979, when a book, *The British Connection*, written by the journalist Richard Deacon, was about to be published by Hamish Hamilton, a very reputable publisher. The book dealt with the alleged activities of Russian revolutionary agents in Britain since 1889. It contained many unsubstantiated allegations against well–known people, including, for example, a completely unfounded slur on Lise Meitner, the well–known nuclear physicist. But nearly all the individuals mentioned were no longer alive, so it was safe to make careless remarks about them, since in English law there is no libel against dead people. But for some reason the author thought I was dead, too, and made some extremely damning and quite unjustified statements about me. These statements were mentioned in a prepublication review in a student magazine, and some of my shocked colleagues brought them to my notice. The review had been published before the book, because the book had been delayed. Another person about whom unpleasant things were said in the book complained to the publisher and the pages concerning him were replaced. Because of this I was able to take legal action very early, and a writ was served on the publishers and the author a few days after publication. The matter was settled out of court very promptly; the distribution of the book was stopped at once, so that the few copies that were sold are now collector's items. I received a "substantial sum" by way of damages. The speed of action was impressive: the settlement was announced in the High Court just thirteen days after I first consulted my solicitors. The publishers could have reissued the book in amended form, but they decided to abandon it.

News of my legal action leaked out, and I was approached by many reporters. Usually one does not comment in public when legal action is pending, but when a journalist asked for my reaction to the statement that I was dead, I could not resist the temptation to say, "It is about as accurate as the rest of the book." My favorite comment on the episode was made by Paul Foot in the *New Statesman* (vol. 98 [1979], p. 129): "Something really must be done to protect ordinary working journalists who want to write books about espionage, but who can't possibly have access to obscure reference works such as *Who's Who, Burke's Knightage*, or the telephone directory."

III. PHYSICS, POLITICS, AND PLEASURES

The Concept of the Positron

I n undertaking to read Professor Hanson's book about the positron I was under a serious misapprehension. The discovery of the positron was one of the important and exciting developments of modern physics. It therefore seemed natural to suppose that a physicist familiar with the discovery, its background and its consequences, should possess the necessary qualifications to understand and appreciate the book. But the study of the book makes it very clear that its field is the philosophy of science, and that this is a game played with rules and based on motives which a scientist does not necessarily understand or appreciate. This review is therefore necessarily amateurish in character like a layman's review of a technical monograph. It is not necessarily informative about the book but only about a physicist's reactions to it. Naturally the physicist's eye is caught by statements on which the scientific philosopher seems to disagree with the physicist. Some of these are errors of fact, or errors concerning the meaning of scientific arguments; they give the impression of being rather fundamental and of implying that the author has failed to grasp what the work of the physicist is about. This impression, of course, may be misleading; the philosopher of science may be perfectly justified in analyzing physics from his own point of view regardless of whether the physicist can follow this analysis as long as his logic and his motivation make sense. On such questions there is room for more than one opinion.

Only the last chapter of the book deals directly with the positron. The rest is a discussion of the interpretation and logical structure of modern physical theory, in particular of quantum theory, leading up to the question of the

This is a review of *The Concept of the Positron. A philosophical analysis,* by Norwood Russell Hanson. Cambridge, Mass.: Cambridge University Press, 1963. For the complete reference, please see the Acknowledgments section.

positron. This structure of the exposition might give the impression that the discovery of the positron is presented in some sense as the culmination of the whole development, but I do not believe this is the author's intention. In describing the discovery it is necessary to build up the background and to explain both the situation in which the discovery occurred and the ideas that were used in the theory associated with the discovery. For this purpose a physicist would have explained the state of knowledge, the limitation of experimental techniques, and the nature of the theoretical difficulties to which the existence of the positron provided a solution. Some of these things are mentioned, but the emphasis is throughout on the conceptual structure of quantum theory and the wave–particle duality.

In discussing the history of the understanding of light the book goes back to Newton, who favored a corpuscular picture of light, and the author criticizes physicists for forgetting this view and wrongly accepting the wave theory on the basis of Maxwell's equations. This is a complete misunderstanding. Until the beginning of the twentieth century there was no evidence known which supported the particle view of light against the wave picture. It must be remembered that a wide range of phenomena is described by the approximation of geometrical optics, in which waves can form sharp beams so that their propagation appears indistinguishable from the motion of particles. It is only when one tries to locate the line of propagation to an accuracy comparable to a wavelength that the properties of waves differ from those of particles.

At the time of Newton these results of geometrical optics were not yet clearly understood, and therefore Newton believed that the ordinary laws of optics required the presence of particles. In the light of the clear evidence for diffraction and interference phenomena, of which Newton had found the first example himself, and the mathematical results about the derivation of geometrical optics from wave optics, Newton clearly would not have maintained his particle view without physical evidence calling for it. There is no reason for assuming, as the author does, that in the light of the Young–Fresnel experiments Newton "might have said" that light must be both wave–like and corpuscular. To reconcile both requires a readjustment of customary ways of thinking which it is wrong to attempt in the absence of a compelling reason. It is true that Newton was troubled by similar dilemma because he could not visualize the connection between wave and geometrical optics. Had he been shown the mathematical argument he would undoubtedly have accepted the wave picture as fully as it was by later physicists.

The first realization that the wave theory did not contain the whole truth came from Planck's analysis of the radiation law. The author refers to the infinite energy content predicted by Rayleigh's law, but this was not conclusive because it only stated that there could be no equilibrium between hot bodies and radiation with any finite amount of radiant energy. This leaves it open whether in practice such equilibrium is ever reached, and also whether the arguments of statistical mechanics underlying the conclusion were applicable. It was only when experiment had shown clearly that there was a genuine equilibrium for black–body radiation and when Planck had concluded that the arguments of statistical mechanics were inescapable, that the difficulty was clearly established.

Plank is also given credit for inventing a special furnace and light detectors and doing experiments proving the emission of radiation in distinct pulses. In fact, Planck was a theoretician who made deductions from other people's experiments, and the discrete nature of light was then only an inference not demonstrated directly by observation.

Space does not permit me to list all the small inaccuracies in the discussion of the wave–particle duality. For example, the fact that on the corpuscular theory light should move faster in refracting media arises only because at the time of Newton one could imagine only the law $p=mv$ relating momentum and velocity for a particle. Experience with relativity has shown that this is not the only possible law and this would, in fact, get one out of the difficulty. The deflection of electron beams in a magnetic field does not prove the corpuscular nature of electrons, but, for example, Millikan's experiment does.

Chapter 2 deals with explanation and prediction, a difficult and interesting subject on which it is hard to generalize because, with the development of physics, physicists have had to change their views about what is meant by explanation and what kind of predictions are possible. At the head of the chapter there is a quotation from Hempel relating explanation and prediction. As a layman I am not familiar with Hempel's writings, but I would assume that in the quotation the word prediction is not necessarily meant in the deterministic sense. Modern physics cannot predict when and in what direction an alpha particle will emerge from a uranium nucleus and in that sense it has not "explained" the behavior of this particular alpha–particle. It can, however, predict the distribution in time and space of the alpha-particles from a large sample of uranium and this prediction can be tested by observation. In that sense we have explained the phenomenon. In an attempt to clarify the difference between explanation and prediction the author uses the concept of "post–diction" which involved calculating something about the earlier state of a physical system from observations at a later time. From the physicist's point of view there is, in this sense, no difference in principle between prediction and post–diction except in practical convenience, and it is not clear why one should have a different relation to explanation than the other.

Here and elsewhere reference is made to the attempts at finding a deterministic interpretation of quantum theory in terms of "hidden variables," as contrasted with the "Copenhagen view" of quantum theory. There is no specifically "Copenhagen view"; this is the structure of modern quantum mechanics; it relates to Copenhagen only in so far as Niels Bohr made vitally important contributions to the clarification and exposition of the situation. One can question its validity in different ways. One way would be to expect that all results of quantum mechanics may remain valid but could be described in a different language which left room for deterministic concepts. In this view a modified theory would never lead to any predictions different from those of quantum mechanics, but the substitution of different words in talking about the problem would then be quite an empty change; in particular, it would then not be possible to predict individual atomic events with certainty since this would mean making predictions different from those of quantum theory, and the new interpretation would be no more deterministic than the current one.

Alternatively, one can take the view that quantum theory will undergo modifications in the future. It certainly needs further development to deal with the problems of elementary particles and it is likely that in the course of this development some of our fundamental ideas will change. Some of the authors who want to see a different interpretation of quantum mechanics think it worth while to speculate on an alternative theory whose practical predictions could duplicate those of quantum mechanics over a wide range of phenomena but would have a basically deterministic structure, and whose new conclusions would be an improvement on quantum theory in the fields in which quantum theory is not, as yet, a complete scheme. One can never prove that such attempts might not succeed. One cannot prove that this approach is right until it has been successful. Whether the attempt to move in that direction is sufficiently promising to justify the investment of effort is the kind of personal judgment about which it is impossible to argue, although in general it is of the greatest importance for the development of the subject.

Chapter 3 on "picturing" seems to miss the point that one can visualize atomic events only in terms of familiar everyday experience. The discovery of quantum effects means that the concepts of space, time, motion and observation, which our intuition has developed from everyday experience, are not applicable on the atomic scale, and it is therefore impossible for us to visualize the contents of quantum mechanics fully though it is often useful to make pictures which will help us visualize some part of it. To ask whether an electron "is" a particle or a wave is using words in a situation to which they do not apply. The recognition that our power of visualization is limited by our macroscopic experience is absent or not sufficiently stressed in this chapter.

Chapter 4 on the correspondence principle is based on a complete misunderstanding. The author quotes the statement that quantum physics contains classical physics as a limiting case, but relates this only to the fact that in an abstract mathematical sense the equations of classical physics also appear as equations between quantum mechanical operators. If this was all, his criticism of the statement would be fully justified, but the real point is that where classical physics yields the equation for the motion of a planet as mathematically precise equations, quantum theory makes in this case approximate statements which agree with the classical ones to within an accuracy that far transcends the possible accuracy of any practical observation. Even if we could improve the accuracy with which we can detect the speed and position of a billiard ball both by a million times, the resulting accuracy would not yet conflict with the uncertainty principle. This does not alter the fact that in principle the uncertainty exists so that quantum mechanics never becomes identical in content with classical mechanics. What is important to the physicist is that the results are reproduced to a sufficient accuracy so that quantum theory does not conflict with any of the tests which served to prove the validity to the necessary accuracy, and in the proper domain of classical physics.

Chapter 8 deals with the equivalence between wave mechanics and quantum mechanics first established by Schrödinger in 1926. The author also queries this statement for three reasons. First, wave mechanics is not necessarily, and was not in 1926, based on the probability interpretation of the wave function. This is correct, but it has since become clear that the original interpretation suggested by Schrödinger in terms of real waves like radio waves or sound waves cannot be reconciled with physics. While in 1926 one could have attempted to build a theory which might lead to different results differing from present–day quantum mechanics, today there is only one wave mechanics, and to this the old proof applies. Secondly, matrix mechanics was not complete in 1926 and not applicable to certain situations, *e.g.*, the case of a continuous spectrum to which wave mechanics could be applied. This also is true, but it has since been completed (the author's remark that the matrix method cannot be applied to continuous spectrum is out of date; the way to do this was shown by Dirac) and it is now completely evident that the two systems can be translated into each other. The essence of the argument is still that of 1926, though it had then of course not been spelled out in respect of the parts of the formalism not then developed. Thirdly, there is the question whether two descriptions can be called equivalent if they are, in fact, identical. The substance of this argument escapes me. As a physicist I am happy enough if two methods apparently different in form turn out to be identical in their results in all respects; if I am told I must not call them equivalent, I shall be glad to comply.

The final chapter about the positron gives a somewhat distorted picture of this discovery. In asking why physicists were so reluctant to admit the existence of a positive electron, the author overlooks the fact that it is the normal rule for scientific research that new objects in physics should be accepted only if every attempt to interpret observations in terms of the known objects has clearly failed. The literature is full of spurious "discoveries" by physicists who disregarded that rule. It was true that, long before its discovery, the positron must have been photographed many times, but in circumstances where it could be interpreted as an electron travelling in the opposite direction, and it was perfectly sound to accept this more conservative interpretation until forced to abandon it. One might blame the experimentalists of that time for lack of intuition in not following up these odd events to make sure. It is the mark of a great (or perhaps a lucky) experimentalist to suspect the odd little thing which could well be due to all sorts of secondary effects and follow it up. Anyone who wanted to follow up any odd observation in complicated circumstances would get hopelessly lost.

Once Anderson had shown good evidence for the positron, physicists showed no special reluctance to accept it. About the same time the theory had led to the idea that there was reason to suspect the existence of a positron with predictable properties. It seemed natural that the two might be the same though, obviously, this had to wait for definite proof. Many years later great confusion was caused when the prediction of Yukawa of the existence of a meson was linked to the discovery of the μ-meson, which turned out to have quite different properties from the π-meson found later to correspond with Yukawa's ideas. This situation is somewhat overdramatized in the book by referring to three distinct particles as observed by Anderson and Blackett and theoretically derived by Dirac, which only later were found to be the same.

The summary of the theoretical position is rather obscure. "Dirac ... tried to 'cook' the negative energy solutions so that they might be construed as protons." This statement obscures the fact that it was the *absence* of particles in negative energy states which was to be understood as the protons. A little later a reference occurs to the name "donkey electrons" which refers not to holes in negative energy states but actual electrons in such states. After referring again to the idea of "holes" the author quotes an extract from a paper discussing the difficulties encountered without thinking of holes. This is all most confusing. The logic of positron theory is not thought out clearly, and this is in line with such misleading statements as "the 'hole theory' as such never served to *predict* the existence of any positrons." It is much easier to understand physics than the philosophy of physics.

The Scientist in Public Affairs: Between the Ivory Tower and the Arena

I n recent years scientists have taken an active part in discussions and in
agitation on problems of public concern, and sometimes they have
even been listened to. This, no doubt, originated from the great importance, to problems of public interest, of new developments, particularly the
existence of nuclear weapons and other novel forms of warfare, where scientists had special knowledge. Also, information about these possibilities was
available to some scientists before it was disclosed to the general public, so
that they had had more time to consider the consequences and formulate their
reactions to them. As time passes, these factors are diminishing in importance, at least on a number of subjects. For example, the power and effects of
nuclear and thermonuclear weapons have been described in many nontechnical publications adequately to allow any intelligent person to think about the
international problems and problems of policy which arise. Information on
quantitative details is often not generally available, but this is usually kept
secret and also not accessible to many scientists.

In other areas, for example problems concerning chemical or biological
warfare, this stage has probably not yet been reached, and the technical knowledge of appropriate scientists makes their participation in such discussions
very desirable.

But independently, another factor has emerged, and that is that scientists
seem to be well qualified to make contributions to debates in which they are
not in exclusive possession of relevant facts. (This has been brought out particularly at Pugwash Conferences.) It also seems true that the views of scientists on such problems are widely, though by no means universally, respected
and have some influence.

It is interesting to reflect on the reasons for this. I believe that the main factor here is that scientists are capable of somewhat greater objectivity and detachment in discussing such problems. This is not to say that scientists are necessarily wiser, more honest, or less prejudiced than other people. The important characteristic of scientists, in this context, is that by their training and tradition they have acquired certain attitudes.

These include first of all a fundamental respect for facts. We all know from experience that a scientist who is known to be careless about facts is likely to lose in reputation, and usually will not be listened to by colleagues. A scientist found to have falsified or misrepresented facts will usually cease to be a scientist. Similarly, scientists are trained to exercise care in the use of language, so as to state their facts unambiguously, to state their arguments clearly, and to distinguish knowledge from conjecture, reasoning from intuition, hypothesis from proof.

With this training one tends to acquire the capacity for respecting points of view differing from one's own. A competent scientist who feels sure of his conclusions will gladly listen to an argument that appears to lead to a contrary conclusion, because he expects that the clarification resulting from that discussion will strengthen his own case. He also knows that he might be mistaken, and if so, he has nothing to gain by refusing to listen. He undoubtedly has experience of situations when he was proved wrong and when he gained from being openminded and recognizing his error readily.

When it is not a matter of deduction but of hypothesis or conjecture, the scientist's position is not weakened but strengthened by considering the alternative hypothesis constructively. In many cases this will help to reach a decision on which hypothesis is more reasonable, and therefore either to strengthen his original position or to help him discard an untenable hypothesis.

These attitudes can be, as we know, valuable outside the range of scientific problems and can make the participation of scientists useful in debates on general affairs and on such problems as international relations or disarmament.

We must not, of course, believe that merely by being a scientist, however eminent in one's profession, one is thereby qualified to judge such public issues. While the detached attitude of the scientist can be helpful, he also needs a sound knowledge of the relevant facts, of history, of the existing organizations and constitutions, etc. Above all one requires an ability to understand, and to form intuitive judgments of, political and human issues not capable of purely rational analysis. There is nothing to prevent scientists from possessing or acquiring such knowledge and such understanding and intuition, but they don't have it merely by being scientists. Nor should world affairs be settled by scientists exclusively.

If scientists are to make the most useful contribution to these problems, it is important that they do not lose the objectivity and the detachment which characterize them as scientists. They are also human beings, and their emotions, their prejudices and their political convictions are as strong as those of others. They are therefore under the temptation, when confronted with these issues, to discard their scientific tradition and allow their emotions (and who does not have strong emotions on problems of peace or war, of justice or want?) to become confused with their command of facts or with their power of reasoning. To do so may, on a particular occasion, help them to be more eloquent and perhaps more effective, but it will inevitably result in their losing one of their principal assets, and ultimately losing then the special influence they may have as scientists.

It is equally important that they should not only behave as scientists in forming their views and their arguments, but that they should be seen to be doing so. In other words, they must be as careful in expressing themselves in public as they would be in a learned journal, though of course the nature of the subject and the language appropriate to it will be very different. This raises a special difficulty when one speaks through news media—newspapers, radio, television—since the media tend to be impatient with the scientists' preoccupation over a careful choice of words. To a reporter the distinction between, say, "I know this to be true" and "I believe this to be true" seems academic and pedantic. In the interests of brevity and dramatic effect he is likely to render either as "scientist asserts... ." It takes great patience and some understanding of the point of view of the reporter or of the reader to guard against serious embarrassment. There are few scientists who tend to speak in public and who have not been caught out in this way.

There is therefore a dilemma for the scientist who is anxious to participate in the discussion of public issues. To preserve his integrity as a scientist, and his detachment, the safest way would be to stick to the topics on which he has professional expertise, but in that way we would give up the option of being of any influence on wider questions. This extreme course would guarantee that we retained the characteristics which can make us useful in public debate, but without applying it. The other extreme, which I have described above, means maximum participation and consequently the loss of our special position. Our problem is therefore to find the right balance between the ivory tower and the public arena.

One particular way in which scientists could exert a valuable influence on public affairs arises from their traditional attitude to facts and to the objective truth. Political or international controversy often is not primarily concerned with facts, but there are cases when disputed facts are of vital importance in a controversy. Sometimes the facts are generally in doubt and would be hard to ascertain. But when different parties to a conflict of views make mutually contradictory statements about facts which must be perfectly well known to both sides, one must inevitably conclude that at least one side is deliberately distorting the truth. To scientists, it is always surprising how indifferent public opinion is, particularly after some lapse of time when the problem has lost its news value. To distort the truth when it is expedient seems a behavior that is perfectly normal for governments or for politicians, one that is expected of them. I have mentioned sanctions that affect the standing of a scientist who has either been careless in making sure of his facts or falsified them deliberately. For public leaders, such sanctions, if they exist at all, are incomparably weaker, and as a result it is a perfectly normal course of action for them to hide or misstate ascertainable facts.

We can all quote such cases from recent history. In order to avoid the impression of prejudice I choose deliberately an example against my own country. In the Suez crisis of 1956 the government of Great Britain at first denied firmly that it had been involved in any advance planning or consultation about these events. A few years later a member of this government published a statement that there had in fact been such consultations. He was of course criticized, but it is interesting to note that some of this criticism took the line, not that he was not right, but that it was disloyal for him to publish the facts.

In singling out this one example I do not especially want to attack the British government of the time or its supporters, since one could easily add many similar examples from elsewhere, and the problem is precisely that this is accepted in practice and therefore presents itself to any government in a suitable situation as a perfectly normal course of action.

It may be that scientists could play a useful role by working toward a greater respect for truth in public and international affairs, by making it their job to collect and publicize cases where such respect had obviously been lacking. To be effective this would have to be completely unbiased, that is, it would have to confront public statements with the facts, where known, irrespective of the identity of the person or government making the statements and irrespective of whose case benefits by them. It would also, of course, have to restrict itself to cases in which the facts were established by clear and noncontroversial evidence.

In this discussion I have used the word science primarily in the English sense of the word denoting natural science. As a physicist I have perhaps looked primarily at the attitude of someone working in the physical sciences, where perhaps the detachment I have described develops most easily because it is easiest to reach agreement about the validity of the interpretation of an experimental result and controversies are usually short–lived. But I believe the difference between this and, for example, the biological sciences, where the relation between an observation and its interpretation can sometimes still be more controversial, is not substantial.

Born–Einstein Correspondence

The years from 1916 to 1955, over which *The Born-Einstein Letters* extends, contained dramatic developments in physics. Max Born and Albert Einstein were, in different ways, leading actors in this drama. The same period also saw dramatic developments in world events, and the two men, as well as Mrs. Born, who participated actively in the correspondence, shared their thoughts, their hopes and fears, for the future of humanity and of their friends and colleagues. Evidently the book contains much of interest to historians of the period and to the historian of science. However, its greatest fascination lies in the picture that emerges of the personalities of the writers.

Not all the letters have been preserved—it is perhaps surprising that so many of them could still be collected, as the writers had, at the early stages, certainly no thought for the interest that posterity might have in their letters. There are some gaps, there are some letters with answers to lost questions and vice versa. But the difficulty which this might cause to the reader is met by Born's commentaries, which summarize the background to each letter, indicate the main points of missing letters where they are essential for the context, and explain the significance of the scientific parts of the letters to the layman. The last purpose is greatly aided by Born's skill in the popular exposition of science.

Physics, of course, occupies some of the space in the letters. There are, as one would expect, comments on new developments, on new puzzles and new ideas. Some comments are based on misunderstandings and later corrected. Readers who do not believe in excessive hero–worship will be pleased to see examples in which even Einstein, with his enormous breadth and depth of

This is a review of *The Born–Einstein Letters*, commentaries by Max Born. New York: Macmillan Publishers. For the complete reference, please see the Acknowledgments section.

Max Born in 1918. (Photograph published courtesy of AIP Emilio Segrè Visual Archives, Landé Collection.)

understanding in physics, could slip up and develop a completely spurious argument.

Running right through all this is the fundamental disagreement about the statistical interpretation of quantum mechanics. The apparent contradiction between the particle and the wave description of matter had been resolved, to the satisfaction of most physicists, by Born's postulate that the waves represented the probability of finding particles in different places. This idea had first been tentatively expressed in a remark by Einstein about "guiding fields." The contradiction disappears if one accepts the fact that it may not make sense to ask questions to which, in principle, no experiment can give an answer, and this is again based on a lesson learned from Einstein's insight concerning the question of simultaneity of distant events. In quantum mechanics the question of the shape of the orbit followed by an electron inside an atom is meaningless. Einstein never accepted this—to him it seemed essential that all these questions had to have answers. While he accepted that quantum mechanics was a great step forward, he wanted it to be superseded

Albert Einstein standing by bookcases in his study, Spring 1920. (Photograph published by permission from and courtesy of Bilderdienst Süddeutscher Verlag, Munich.)

by a more deterministic theory. He realized that this was not easy and wrote off the solutions proposed by de Broglie and Bohm as "too simple." Born struggled to convince Einstein, but to no avail.

Many of the letters concern questions on how to help friends and colleagues in difficulty. In the years after the First World War there was not much money for science or for the scientist in Germany. One has to read between the lines of the letters to understand the selflessness with which Born was giving public lectures on relativity, in which there was then great public interest, and using the proceeds to buy equipment for his department or to assist penniless collaborators. Later, with Born in England in not too easy financial circumstances, much of his time was spent on helping or advising other refugees. Einstein was then more famous and had more influence, so frequently it is a question of Born asking whether Einstein could find help for someone in need. Often, of course, this was not possible, and the letters bring out clearly the pain and helplessness which both felt then.

An interesting episode concerns a book about Einstein written by a journalist on the basis of interviews. Einstein was not enthusiastic but did not regard the matter as important and was not going to try to stop publication. To the Borns and to other scientists it seemed that the publication of the book would ruin Einstein's reputation, and maybe that of scientists in general, not so much because of its contents, but because Einstein's permission for the publication of a book about him would show him as publicity–seeking—a disgraceful attitude for a serious scientist. They persuaded Einstein to withhold his permission, but eventually the book was published anyway without any of the predicted disasters. (I read the book long before I knew anything of relativity; it did not teach me to understand relativity, but it did give an impression of Einstein as a person which was not far wrong.) I doubt whether many people today would understand the emotion felt at the time.

Apart from arguments about such specific issues, Einstein and the Borns understood each other intimately, and shared common attitudes on most matters, world affairs, human relations, literature, music and others. But there were a few very profound differences. Born was inflexible in his opposition to any work on nuclear weapons, and distressed to discover after the war that Einstein, by his letter to Roosevelt, had played an important part in getting the American government's support for such work. Einstein remained inflexible in his dislike of Germany, and could not understand Born's decision to live in Germany after his retirement.

The translation by Born's daughter, Mrs. Irene Newton–John, is sensitive and reads well. There is a short note by Max Born which shows that he lived to see at least most of the English text. There is a preface by Heisenberg, and a short foreword by Bertrand Russell, who stresses the impression the book gives of Born's and Einstein's personalities. "Both men were brilliant, humble and completely without fear in their public utterances. In an age of mediocrity and moral pygmies, their lives shine with an inner beauty. Something of this is reflected in their correspondence, and the world is the richer for its publication." It would be futile to try to improve on this appreciation.

Is There a Crisis in Science?

There is with us today much talk about a crisis in science. What is that about? Imagine that one of the great scientists of the past, whether it be Newton, or Pasteur, or even from more recent times, Einstein, were with us today and could look at the way scientists work, and how they are situated. He would see, very differently from his own experience, the enormous number of scientists at work, and the great range of expensive equipment available to them. He would see the great size of our scientific teaching institutions, and the great number of scientists trained in these institutions who have positions in scientific research and are well looked after. At the same time he would see scientists depressed about the position of science, talking in tones almost of impending disaster.

To understand this, we have to look at history. I am not a historian, and shall not take you back very far. We must just remember how things were at the beginning of the century, and in many respects even up to the Second World War.

Scientists were not then in general provided generously with expensive equipment. Their numbers were limited. You did not expect to become a full–time scientist unless you were exceptionally able and lucky. There was a static situation in the universities. They, and other research institutions, were not expanding, and vacancies arose only from retirement or death. In public life scientists did not carry much weight; they were expected to live in their ivory tower. There were exceptions, of course, when the conclusions of some, like Darwin or Einstein, were upsetting accepted ideas. Then their names became known, and their sayings were reported in the popular press. But this was very exceptional and did not reflect the position of scientists as a group.

This is not in contradiction with the fact that at the time there was an unquestioning belief in the benefits of progress. Progress was not seen as very directly connected with science and the work of scientists. This was largely because of the long time lag between the basic scientific discovery and the practical application, a time lag long enough to hide, in the public mind, the close connection. The work of Clark Maxwell, and of Heinrich Hertz prepared the way for radio, but in the public mind the invention is associated with the names of men like Marconi, who helped to make it a technological and commercial reality.

In the inter–war period things changed a little, perhaps. For example, the close connection between basic research in chemistry and the progress of the chemical industry, and also between the chemical industry and the economic well–being of a country became clear and obvious. In this respect chemists were probably the first scientists whose academic work became recognized as practically important.

But, by and large, progress was not seen as anything scientific. There was no doubt that it was valuable. There was no doubt that, as a result of "progress," life was getting better from generation to generation. This concept involved not only a rising standard of living through technological progress, but also rising standards of education, and through better education, a rising standard of moral behavior.

Then came the Second World War, and this accelerated developments which had a profound influence on the position of science and of scientists. Things like radar, atomic energy, antibiotics, and later rockets and space travel, came along to impress the public mind and gave scientists an image as people who were terribly important. They seemed a modern kind of magicians, who could create power where there had been none before. One of the reasons why the part played by science was more easily recognized than in the past was the shortening of the gap in time between discovery and technological realization. Very often the same scientists who had been involved in the basic discovery were able to help in developing the practical utilization. Very often this in turn helped to shorten the time scale.

It was now realized that society needed scientists. They were evidently useful people who could make vitally important contributions to the technology, and hence to the standard of living and trade balance of their country. This change in attitudes affected the position of science, and of scientists, profoundly. They found it much easier to obtain financial support and elaborate equipment. They could easily undertake research projects which would have been unthinkable a generation earlier, because the cost would have seemed prohibitive. There was a great demand for training in sciences; there were queues of prospective science students outside our universities. In most countries the universities expanded to accommodate more students. The need to staff the universities and other new scientific institutions created an increased demand for trained scientists, so any trained scientist skilled at research could generally choose between many attractive posts.

This did not mean that scientists were paid higher salaries than people in other professions. This was probably just as well. It has often been stressed that scientists should be looked after, but that it would be a disaster if their salaries were substantially above those of other people, because then some would turn to a scientific career as a means of making money. I am told by Russian colleagues that there was a period in the Soviet Union when you could earn more money as a scientist than in most other professions; the scientists in their laboratories were greatly relieved when this was readjusted. I do not, of course, want to imply that scientists should be seriously under-paid. It is right that scientists should bring some small sacrifice for the enjoy-ment which their profession can bring them, but one should not ask them (as some of our authorities might be inclined to do) to bring unreasonably severe sacrifices.

Scientists were now listened to with greater attention. They could acquire a voice in public affairs. They were respected. But, and here are the seeds of later trouble, there were very inflated expectations of what scientists would be able to produce. The phrase "atomic age," very common in the late 1940s, is a good example of this oversimplification. Once the possibility of atomic power was assured, this, some of the popular press and some of its readers seemed to think, would revolutionize our lives. We would have electricity at practically no cost and would drive our cars on little pellets of uranium. These stories were not invented or spread by scientists. Perhaps in some cases the scientists did not do enough to counteract such wishful thinking. There were in some countries cases of nuclear physicists, for example, who were quite pleased that their government believed building a nuclear physics laboratory would result before long in achieving new possibilities of nuclear power or nuclear weapons. In most cases scientists were sensible enough to discourage this kind of confusion, but perhaps not firmly enough.

This was the general picture until recently, when a tendency developed for society to become somewhat suspicious of the scientists. This was due to a combination of factors. One of them came from the exaggerated expecta-tions of the miracles to be performed by scientists. If one had these expecta-tions, the scientists were not delivering the goods. Better–informed people, who where not expecting miracles, pointed out that one needed a proper balance between basic and applied research. Short–term benefits could not in general be expected from the support of academic pure research. In the United Kingdom, for example, one started looking at the number of scientists work-ing directly on useful applications. There seemed to be a shortage of people willing, and qualified, to go into work that would benefit technology directly and that needed scientific thinking, if not necessarily scientific research.

This coincided with a period of economic recession in a large part of the world. Money was becoming shorter, and there was a natural tendency to ask what returns could be expected from the expenditure on science.

A third factor appearing at the same time was a tendency in the minds not so much of governments, as in the minds of the public, particularly of the young, to blame science for the troubles of our society, particularly those which had a vague scientific or technological connection. Physicists became the people who make atom bombs and create radiation hazards. Medical doctors are the people responsible for the lowered mortality, and therefore for the population explosion whose consequences we must fear. The chemists produce synthetic foods (or foods grown on synthetic fertilizers) which may damage our health in ways we don't yet understand, and they develop pesticides which, we are now being told so insistently, are ruining our environment. Biologists are dangerous people because they can produce means for germ warfare; they may also do genetic experiments or other work leading to a dangerous interference with natural life processes. Automation causes unemployment—and so it goes on. Science became identified with technology, and technology with pollution and the general damage to our environment.

Of course, the problems of pollution, the problems caused by a short–sighted expansion of technological usage without thinking of the consequences are real enough. But they are not the fault of the scientists. Scientists have often been the first to warn of dangerous consequences of various courses of action. They are also often responsible for inventing anti–pollution measures, often to a greater extent than their direct involvement in practices that create pollution. Pollution is a subject which today seems to engender more strong emotions than rational thought. I live in England, and it is very noticeable that the air over London is now most of the time much cleaner than it was a few generations ago. This is not an accident, but the result of deliberate measures based on scientific studies and rational thought. Similarly, London's river, the Thames, is cleaner now than it has been, and work is still going on toward further improvement. I cannot claim, of course, that such measures are being taken by the scientists, but those who are working to ease pollution use the help of the scientists more than did those whose actions caused the pollution.

It is not reasonable to blame scientists for these troubles, but it may be true that scientists could take more initiative, and raise their voices more loudly in warning of dangers which they may be in a position to appreciate before they have become obvious to the general public.

The problems involved in the relations of the scientist to public authority are complicated, and the difficulties find their extreme in matters which have a terrific impact on life. I was involved during the Second World War in work on atomic weapons. My attitude at the time was to believe that the scientist had the duty to show the possibilities of such weapons, but with it also to point out their effects, and to make sure the people in authority fully understood the possible consequences of their use. One then had to leave the actual decisions to the statesmen and military leaders, who in a democracy had come to their positions by a democratic process, and could be expected to be of good will. They, and not the scientists, had the knowledge about the state of the war, and about the military factors which were relevant to such decisions. Now, some twenty–five years later, I still hold the same view basically. However, looking back, I think we overestimated the efficiency of communications, and the power of imagination of the leaders. Given the same situation again, we would insist much more strongly on debating with higher authorities all the arguments, to make absolutely sure they were aware of all possibilities and all implications.

Today, rightly or wrongly, the attitude of the public to science has changed in the way I have indicated. The consequences for the position of the scientist are real and have to be faced. At the same time science faces some serious internal problems. One of these is the so-called "information explosion," the unprecedented growth of the scientific literature to a point where a scientist finds it extremely hard to keep up to date with the material published in his or her own particular line of work, let alone beyond it. Partly but not entirely as a result of the information explosion, we face the other problem of overspecialization. Scientists today often do not know, and sometimes even do not want to know, what goes on in a field right next to theirs. There is a tendency to forget that the most important progress, even on a narrow front, is usually due to people who can look over the fences, who thereby form a better understanding of basic principles and can see their problems in a wide context. This trouble of overspecialization is accompanied by a tendency to overcomplication, so that people may become engrossed in detail to such an extent that they do not see the wood for the trees.

These are the two ingredients to our "crisis." What should be done about them? The answer does not lie in a single step, of course, but in many. Basically, we have to adapt ourselves to the present situation. We should have learned from the past experience of scientists being blamed for the misuse of the results of science to raise our voices whenever we can see dangers to society from such misuse and raise our voices high. We should use what influence we have, and what access to public opinion we can achieve. We must try to get society to understand what scientists can do. This must, of course, be related to an understanding of what society wants from us.

By this I certainly do not mean that all scientists should suddenly turn to practical problems, to work on how to make industry more efficient or how to cure diseases. These are important problems, but we must not neglect basic science. The point is that in seeking support from the community for basic science it should be sought for the right reasons. The motivation for this is a mixture of two ingredients: on the one hand efforts in basic science may one day lead, directly through discoveries, or indirectly through its educational effects, to applications of great practical value. I have already mentioned radio and nuclear energy as arising from studies undertaken with no thought of any practical consequences present, or even possible. X–rays and penicillin are other familiar examples of this kind, and one could list many more such cases. In addition our academic tradition rightly stresses the value of contact with scientists engaged in the search for new knowledge for the education of the engineer, the doctor and other practitioners of applied science.

On the other hand many members of the public appreciate the extension of knowledge provided by basic science; they can share to some degree the curiosity which drives the scientist to seek to extend this knowledge. Many people want us to tell them what is now the best picture of the structure of matter or the nature of the universe, or the basic nature of living structures. These two motives are inseparable, because the scientist never knows when some pure research may lead to a useful application. To satisfy the second need we must be ready to communicate our findings in such a form that someone not trained in science but with a serious will to understand, can follow. This is a great art, and it is becoming more difficult with the present trend to specialization and complication. Scientists who have not learned to communicate with their colleagues down the corridor who work in a slightly different field are not in a strong position when they try to communicate with the general public. Here is a place where bridge-building is evidently needed, both within science and from science to the outside world.

In addition to this, scientists, and people trained in science, can be useful with problems that are not scientific in the strict sense. This is because of the way the scientist is trained to look at things from first principles, to distrust personal prejudices (it is not true that scientists have fewer prejudices than others; we have plenty, but we are trained to recognize them and allow for them). The scientist has experience in accepting other people's arguments and in being proved wrong. This is something we all learn very early—it is a painful, but instructive experience. For these reasons scientists can often make valuable contributions on non–scientific questions.

This can be exploited in two ways. One is to educate young scientists in the expectation that they will not necessarily spend their lives in the laboratory, but may be willing to go out and do work for which they have not been specifically trained, but for which the kind of experience that one gains from doing research, from pursuing a scientific project, can be particularly valuable.

In this respect we science teachers have an important responsibility. In the last 25 years, the period I have described, the impression was formed amongst students that, once you had started on scientific training, certainly once you had started on research training, you were set on a career in which you would continue doing research, preferably in the area in which you had done your Ph.D. work. During that period such an impression was not so far wrong; with the expansion of science departments in universities, and other institutions, there was such a demand for scientists that a large proportion of them could remain within the system that had produced them. It was always clear this could not go on indefinitely—but it stopped more suddenly than we expected. I have heard graduate students complain often about the tragic loss of an old tradition, and I had to remind them that this old tradition was barely older than they were themselves. This is a change we have to live with, and our task is to persuade young people that there are many interesting and challenging things to do outside of academic research, and get them to keep an open mind so that they are ready to consider many possible careers.

The other way in which we can respond to the present situation is that some of the academic scientists should be alive to the existence of practical and urgent problems in the outside world, to which they can make occasional contributions, both because of the value of such contributions to the community, and because this keeps them in touch with the kind of thing they would like to encourage their students to think about. I believe that in this respect, in having academic scientists participate in non–academic problems of urgent public importance, Israel can show an example to many other countries.

At the same time we have to fight the effects of the information explosion and the overspecification. This is not the place to go into a detailed discussion of this very difficult problem. I would, however, mention one relevant point: We should encourage more able scientists to give time to the task of writing, of presenting areas of science in accessible form. There is a tradition that a good scientist should produce new results, and that writing is a less constructive activity, to be left to those who are past their active research period, or who are not quite able enough to do good research. This may have been true at some time; it is not true today. To write a clear-minded review of a part of a scientific field, which will allow people outside that area, or sometimes even those within it, to form a clear picture of what is known without spending all their working hours in the library looking at the literature, is doing a positive service to science that may contribute far more to its progress than would tackling a specific problem or building a new piece of apparatus. I do not say, of course, that we should all start writing review articles and monographs, but more of us should do so and do it well.

This should go together with an effort to make society understand what we are doing, what we are capable of doing, so as to get society to make realistic demands of science. This, in a way, is selling science. We have, in that respect been living until recently in a seller's market. Then anyone can sell. We are now, for a while at least, in a buyer's market. That is much harder work, but we have to learn to be our own promoters.

Together with that we have to attend to the other problem in communications, that is to set out not our methods, our capabilities, but our findings, our ideas, our understanding of nature, in a way accessible to those of the general public who want to understand it, and that has to be at many levels, from the broadly popular to the serious exposition for the intellectual non–scientist.

So for the scientist of today there is a struggle ahead. Science has always been involved in struggles. Struggles with problems, with techniques, struggles for funds, except in the relatively short recent period. There have to be some changes in our ways of doing things, but science has always been changing. Over the centuries even the questions which the scientist is trying to answer have been changing and if change were to cease, if we settled down to doing the same things in the same way, it would be the end of science.

Changes are healthy, struggling is healthy for the state of science even if it means hard work.

There are a great many points between which communications need improving, and rather than one bridge, what we need is more in the nature of a telephone exchange, establishing the many connections that are required.

But one of our most important problems, which has always been important, but is so especially at this time, and particularly in connection with our "crisis" in science, is this problem in communication, and it is essential that we all take it seriously, and tackle it with all our power.

The Jew in Twentieth–Century Physics

In any recent list of distinguished scientists, Jewish names are far more prominent than would correspond to the number of Jews in the general population. However, this trend shows striking variations with the subject, the period, and the country. In this article, which is concerned with physics, of which I have some knowledge, I shall not attempt to analyze this phenomenon, which poses an interesting challenge to the social scientist, and even less attempt to find the causes, but confine myself to describing some typical cases and to commenting on the variations.

The outstanding example is, of course, Albert Einstein (1879–1955), the most famous scientist of our century. His name is widely known among more or less educated people, who usually have not the least idea of the nature of his work. This has even proved very convenient to physicists like myself when asked in social contact to explain one's profession. The occasion usually does not warrant a dissertation on the nature of physics; the answer "Einstein was a theoretical physicist" generally satisfies, even if it does not enlighten, the questioner.

What is, then, in simple language, the basis for Einstein's fame?

The theory of relativity, which bears his name, was the first of two major revolutions in the basic concepts of our understanding of nature that helped the twentieth–century scientist to deal with questions that predecessors were unable to answer and sometimes were even unable to ask.

The difficulties that required the first revolution had arisen from the problem of the relation of light to moving bodies. The laws of motion had been settled since the work of Galileo and Newton in the seventeenth century, and in the nineteenth century the studies of Michael Faraday, James Clerk Maxwell, and others had led to a full description of electricity and magnetism and had shown that light consisted of electromagnetic waves. These were at first thought to propagate in a hypothetical medium, the "ether," and this suggested the possibility of finding out whether we were standing still in space or moving. If the earth was moving through the ether, so that the ether appeared to be streaming past, light would appear to travel a little faster "downstream," with the ether, than "upstream," against the ether flow. This possibility aroused much interest, because until then there was no way of telling a state of rest from that of steady motion. (In a fast plane, for example, in steady air, we are not aware of moving, and there is nothing to show us the high speed with which we are going around the axis of the earth and the earth around the sun.) This was a general feature of the laws of nature, except, it seemed, for light.

When the test of this possibility showed that the expected difference in the speed of light did not exist, and that nature seemed to have conspired here, too, to prevent our telling motion from rest, one faced a contradiction. Other great thinkers, such as H. A. Lorentz in Holland and Henri Poincaré in France, came close to the answer. But it was Einstein who came to the conclusion that we had to revise our ideas about space and time, and that, in particular, it was not possible to say which of two events taking place a large distance from each other was earlier or later—the answer might depend on how fast the observer who was timing the events was moving. The possible ambiguity in time is very small, less than the time it takes light to get from the place of the one event to that of the other, and so for practical purposes this ambiguity can be ignored. No wonder, therefore, that our intuitive sense of time has not developed in a way that leaves room for this.

The new laws of motion that Einstein developed from this basic idea found immediate confirmation in new tests. For example, it was predicted that light should be deflected by the gravitational attraction of the sun, and that during an eclipse of the sun stars appearing very close to the darkened disk of the sun should appear to have shifted slightly from their normal positions, because the light from them had passed very close to the sun—and this prediction was confirmed.

Today these ideas of "relativity" firmly underlie all the physicist does. They matter particularly when motion at speeds close to that of light is involved. We can today accelerate electrons to speeds short of that of light by much less than one part in a million, and their behavior is then completely different from what it would be according to pre–Einstein ideas. There is, then, no room for doubt that Einstein was right.

What I have sketched is called the "special" theory of relativity as distinct from the "general" theory, which Einstein developed later and which deals with gravitation and with the possible structure of the universe. This is another great intellectual achievement, but its immediate consequences for the practical scientist are less important. Einstein also made important contributions to other areas of physics, including the second great revolution, to which I shall return.

It is not surprising that Albert Einstein is universally admired by scientists, but it is less easy to understand the extent of his fame among the general public. This may be due to a misunderstanding of the term "relativity," which may suggest an attitude of general skepticism or nihilism—"everything is relative." In fact, his contribution to science is the very opposite, in showing that the laws of nature have an absolute validity for observers in very different states of motion, provided the influence of their state of motion on their observation is duly taken into account.

Some philosophers objected to Einstein's theories because of this misunderstanding, and others because they felt they were the guardians of the basic ideas of space and time and could not allow a physicist to modify them. Nonetheless his fame was tremendous, and he was aware of this. He did not seek the publicity, but given that it existed, he felt it his duty to use the attention his words could command to support causes that seemed to him deserving.

Einstein was born in the small southern German town of Ulm in 1879 and grew up in Munich. His parents were not practicing Jews and did not observe any of the dietary or other restrictions. His interest in science was not stimulated by any parental influence or by school, where he resented the rigid discipline and was not an outstanding pupil. His first contact with mathematics and physics appears to have been through books chosen for him by a student who was a friend of the family. Even later he preferred his own reading to lectures and other instruction. His advice to others was often to study on their own. He never had any pupils in the usual sense, and few collaborators.

He was conscious of his dual background, as is evident in the famous answer he gave to a reporter who asked about his nationality. This was at the time of the first test of relativity during an eclipse. "If the test confirms my theory," he said, "the Germans will regard me as German, and the French as a Jew. If my theory is disproved, I shall be a Jew to the Germans, and a German to the French."

He was a confirmed pacifist until the rise of Hitler and then changed his mind—fighting the evil of Nazism justified war. For this reason he agreed in 1939 to help persuade President Roosevelt to authorize research into the possibility of an atom bomb. It is sometimes said that the possibility of an atom bomb, of obtaining energy from nuclei, is a consequence of relativity, because of Einstein's conclusion that mass and energy are one and the same thing. This is misleading, since the loss of mass of the nuclei in the reaction is incidental.

If we ask how Einstein's work affects not only our intellectual heritage but our practical life, the answer must be found not in relativity but in his contributions to the quantum theory, which, by giving us a full understanding of the behavior of atoms, has led directly to many tangible products of technology, such as the transistor and the laser.

This brings us to the second revolution in physics. Relativity showed that our conventional concepts, based on our everyday experience, failed when we were dealing with very great speeds, up to the speed of light. The quantum theory, originally advanced by Max Planck, corrected a similar failure when dealing with very small objects, the size of atoms or less. This German theoretical physicist, in thinking about the behavior of radiation from hot bodies, which disagreed with what the laws of physics then said, was led to the unexpected idea that light, or other radiation, was not infinitely divisible but consisted of definite quanta. As the implications were studied, it became clear how profound a change this meant in our basic concepts, and Planck was reluctant to see such a drastic upheaval. He tried hard to minimize the consequences of his own idea. Einstein realized that the major upheaval could not be avoided and took the new ideas seriously. He drew conclusions that helped to show that the experiments of Arthur H. Compton in America on the detection of gamma rays in passing through matter gave further confirmation of the idea of light quanta.

It was the Danish physicist Niels Bohr (1885–1962) who took the next step in this development. Bohr was working in 1912 with Ernest (later Lord) Rutherford in Manchester. Rutherford had just discovered that the atom consisted of a positively charged nucleus surrounded by negatively charged electrons. Bohr succeeded in applying the new idea of quanta to the motion of the electrons, and this explained the spectral lines of atoms, in particular of hydrogen, which up to then had been mysteries.

In its original form Bohr's theory of the atom was, in spite of many successes, unsatisfactory in that it mixed old and new laws in a seemingly inconsistent and arbitrary manner. To some questions it gave manifestly wrong answers, and others it was unable to answer. Bohr realized all these difficulties clearly and worked hard to overcome them. The final step was taken in 1926 to 27, following two very different approaches: one started by Werner Heisenberg in Göttingen, the other by Erwin Schrödinger in Zurich, who used a brilliant but incomplete idea of Louis de Broglie, who was working in Paris. There was now a consistent set of laws, but their interpretation was not too clear.

Over the next few years Bohr took a strong lead in the discussion of the significance and interpretation of the new quantum mechanics. His Institute for Theoretical Physics in Copenhagen had become a center where the world's physicists came to learn and to expose their ideas to comments, particularly to those of Bohr. Unlike Einstein, Bohr tended to work with others, and his contribution to physics consisted in part of his influence on the thoughts of others. He was fluent in English and German as well as in Danish, but he used all these languages in a slightly personal way; one could always recognize that a colleague had spent a few months in Copenhagen by his having picked up some of Bohr's ways of speaking. Bohr's arguments and his papers were not easy to follow; as he himself explained, truth and clarity are complementary. He meant that if you want to speak clearly and simply you must oversimplify and what you say will not be quite correct. The closer you stick to the truth, the more complicated the statement that will emerge.

In this period he had many arguments with Einstein, who did not believe in the quantum theory, which he himself had helped to start, and always tried to disprove it. He believed that the right answer would ultimately be a "unified field theory," which for the rest of his life he attempted to construct, but in vain. Bohr enjoyed the challenge of Einstein's objections, and his refutations remain very instructive reading.

When in the 1930s it became possible to carry out experiments on the atomic nucleus, Bohr took a great interest in these problems and was responsible for understanding certain nuclear reactions whose explanation had been missed by everyone else. After the discovery of nuclear fission, he saw, working jointly with John Wheeler of Princeton, how to interpret this new phenomenon, and in a sense this paper paved the way for the work on the atom bomb.

After the Nazi occupation of Denmark, Bohr regarded it as his duty to remain and look after the members of his institute. But in 1943 the German command decided to arrest all Jews (this included Bohr, who was certainly a Jew in Hitler's definition, for his mother was Jewish; he was also known to be anti–Nazi). Fortunately the news leaked out, and all those to be arrested (except for a few bedridden people) fled to neutral Sweden. Bohr was flown to England and then joined the atomic–energy team in the United States. By then his main concern was the effect of atomic weapons on the post–war world, and he worked hard for an imaginative move toward an "Open World." He was able to talk to Roosevelt and Churchill; he did not have much effect on Roosevelt, and the interview with Churchill was an utter failure, because Bohr's way of expressing himself was quite incomprehensible to Churchill.

The distinction of Robert Oppenheimer (1904–1967) was of a rather different kind from that of Einstein and Bohr. He was born in New York City, where his father, who had immigrated from Germany, was a prosperous businessman. His parents were cultured people with a taste for art but no special interest in science. Though conscious of their Jewish background, they were not religious. Robert was a brilliant student and completed the Harvard four-year course in three years, learning many subjects, including Sanskrit, besides his professional subject, theoretical physics. He spent his years of post-graduate study in Europe, where he took part in the work of consolidating the new quantum theory on the foundations laid by Heisenberg, Schrödinger, and others. On returning to America he settled in California. He divided his time between the University of California at Berkeley and the California Institute of Technology at Pasadena and continued actively in the front line of the development of theoretical physics. He attracted a large number of students and became one of the great teachers of the subject.

One of his great qualities was his quick understanding; in talking with him one hardly ever needed to complete a sentence, because halfway through he knew what you were saying and was ready with his comment. Another was his ability to find the right word, to sum up a scientific or a human situation perceptively and clearly. He was perhaps the most articulate of physicists.

At the beginning of the war, when many started speculating whether the newly discovered phenomenon of fission could be used to make an atom bomb, he was one of the first to appreciate clearly the essence of the problem. He was put in charge of the theoretical work on the way the bomb would function, and he later built up and directed the Los Alamos Laboratory, which was to design, make, and test the first bombs from the materials produced in the large plants built elsewhere. It was an unusual idea to put a theoretician—and a rather highbrow and seemingly impractical one at that—in charge of a laboratory that had to get results in a hurry, but he was outstanding as director. His quick mind and clear perception helped in that, and so did his capacity for finding the right word.

After the war he was much in demand as adviser on many military matters. In 1947 he became the director of the Institute for Advanced Study at Princeton, where he did no more active research himself but still had a profound influence on physics. He remained conscious of his responsibility for the atom bomb and for the death and destruction it had brought—he had been a member of a committee that in 1945 agreed to, or at least acquiesced in, the decision to drop the bomb on Japan.

When, in 1949, it was proposed to start a crash program to develop a hydrogen bomb Oppenheimer opposed this, partly for technical and partly for moral reasons. Soon he was accused of disloyalty—for having been sympathetic to communism in the 1930s. This became a *cause célèbre*, and he was judged unreliable and barred from access to secret information. This was in the witch-hunting days of Senator Joseph McCarthy in the 1950s. When by the early 1960s some sanity had returned, some amends were made. Oppenheimer was awarded one of the nation's highest honors, the Enrico Fermi Award.

Lev Davidovich Landau (1908–1968) was the greatest Russian theoretical physicist. He grew up in Baku where his father was an engineer in the oil industry. His high ability in mathematics and science showed very early. He completed the undergraduate course in physics at the University of Leningrad in 1927 at the age of eighteen. This was the time of the rapid growth of the new quantum mechanics and Landau was at once at home with it. His way of studying new material was to look at an important paper briefly, to get a general idea of the author's aim and method, and then work out the results for himself. He rapidly developed a very deep understanding of all aspects of physics.

After finishing his research training under Yakov Illich Frenkel, a man of great versatility and originality. Landau spent two years abroad, working particularly with Wolfgang Pauli in Zurich and Bohr in Copenhagen. By this time he had already gained attention as a result of a number of papers, of which the most important was his explanation of the diamagnetism of metals, where he resolved very simply an old problem that had puzzled many. After his return to the Soviet Union he continued to tackle problems of fundamental importance, ranging from the internal structure of stars to colli-

From left to right, (unknown), Issak Y. Pomeranchuk, Arkadii B. Migdal, Sir Rudolf E. Peierls, and Lev Davidovich Landau. (Photograph published courtesy of AIP Emilio Segrè Visual Archives, Rudolf Peierls Collection.)

sions of atoms in gases, from cosmic radiation to the strange behavior of liquid helium at very low temperatures. His theory of the "superfluidity" of liquid helium was cited when he was awarded the Nobel Prize in 1962, but it must have been hard to choose from among his many important ideas the one to single out for this purpose.

He was a brilliant if demanding teacher, and most of the best theoretical physicists of the Soviet Union are his pupils. He was extremely critical of unsound or careless arguments, but he was guided more by physical intuition than by mathematical rigor. He could be very scathing to unfortunate colleagues who did not accept his intuitive insight and demanded proof. He had equally strong views, equally hotly defended, on many subjects outside physics, including poetry, ethics, and human relations. As a young man he had theories—developed quite as seriously as his physical theories—about what a satisfactory relation between man and woman should be, and he patiently but firmly explained to many acquaintances nearly twice his age that their particular marriage did not measure up to this standard and should they not get a divorce?

Politically he was for the revolution and regarded the capitalist society of the West as rather stupid, particularly the British monarchy. But he did get incensed at many features of the late Stalin era—he himself had to spend a year or so in jail. He was violently against religion, Jewish or any other.

In 1962 Landau was severely injured in a car accident, and in hospital he was several times pronounced clinically dead. He made a partial recovery as a result of intense efforts by medical teams supported and urged on by numerous pupils and colleagues who volunteered their services to run errands, solicit special drugs from friends abroad, and help in any other possible way. However, he never recovered his intellectual powers, and he remained an invalid until his death in 1968.

The four men sketched so far were all theoreticians, and it is probably true that the weight of Jewish contributions has been greater to theoretical than to experimental physics. But there were also great Jews on the experimental side. Heinrich Hertz, who was born in Hamburg in 1857 and died in 1894, was impressed by the ideas of Maxwell, who had succeeded in combining all known facts about electricity and magnetism into one consistent set of laws (the "Maxwell equations"). This predicted the existence of electromagnetic waves—today called radio waves.

While many physicists still had doubts about Maxwell's ideas, Hertz was convinced of their truth and set out to produce the predicted waves and to demonstrate their existence. He succeeded; he was able to transmit signals over a distance of some sixty feet. He used this discovery to study the nature of these waves, but it was left to others to develop this phenomenon into a means of communication. Thus the name of Maxwell, who predicted, and that of Hertz, who discovered, the waves, tend to be forgotten by the general public; one chiefly remembers Marconi, who pioneered the practical application.

Hertz would undoubtedly have received the Nobel Prize if it had existed in his time. The list of Nobel Laureates in physics includes James Franck (1882–1964), who was a professor at Göttingen until Hitler came to power. His greatest piece of work, done jointly with Gustav Hertz was the discovery that, in collisions of electrons with atoms, the electrons lose energy only in definite amounts, equal to the amounts of energy the atom in question can receive according to the Bohr theory. At a time when the Bohr theory was still in an unsatisfactory state and physicists were groping for a more consistent description, this discovery was of great importance in demonstrating that some detailed parts of the theory were indeed right, and therefore helped to channel further discussion along more definite paths.

In 1933 Franck left his chair in Göttingen, before he was dismissed (which undoubtedly he would have been), and settled at Johns Hopkins University. He later joined the atomic–energy project, of which he was a valuable member. He is remembered particularly for the "Franck Report," a document written by scientists under his chairmanship that urged strongly that the atom bomb not be used. The effort failed, but it is characteristic of the spirit of Franck, who was not only a great physicist but a person who had won the affection of all who knew him.

We should remember Otto Stern (1888–1969), who won the Nobel Prize in 1943 for his invention of the atomic–beam technique, by which he and Walther Gerlach were able to follow the behavior of single atoms in magnetic fields. This again confirmed some predictions of the Bohr theory, and at the same time raised, as a challenge to the theoreticians, questions that the Bohr theory was not able to answer consistently.

This is a small sample of a long list of physicists of Jewish extraction. I have tried to detect common characteristics among them. Not of course in the substance of their physics—the idea of "Jewish physics" or "German physics" is safely buried in the ruins of the Nazi ideology—but there is a personal style in one's work, the kind of problems a scientist selects, the boldness with which he tries new ideas, the way he works with his colleagues or his pupils. In this the Jewish scientists seem to be all different; the range of styles seems to me to be as wide as among all physicists.

The interesting general point is already mentioned at the beginning: the surprisingly large number of Jews among the leading physicists. A closer look shows that this strength is not uniform. There is little of it in the nineteenth century, where few examples come to mind apart from Heinrich Hertz and H. Rubens (1865–1922), who, in Berlin, did important work in optics. The other surprising observation is the enormous variation from country to country. The abundance of Jews among the physicists of Germany (before 1933), of Russia, and of the United States, particularly since the 1930s, is as clear from a casual glance at any list, as the relative scarcity in Britain, France, or Italy. Any such comparison should of course exclude immigrants, at least those who moved as mature and established scientists, since the exodus of refugees from Germany and Austria added numerous Jews to the ranks of scientists, particularly in Britain and the United States. To be sure, there are eminent Jewish physicists in England, including the latest Nobel Laureate, Brian Josephson, but their proportion among physicists is hardly greater than that of Jews in the general population.

What could be the reasons for the prominence of Jews in science, and in particular in physics? The suggestion has been made that the intricate arguments of modern physics have some resemblance to Talmudic studies, and that people who grow up in an environment in which such studies are traditional find this mode of thought natural. However, quite apart from the question of how real or deep is such a resemblance, one notices that a large proportion of twentieth–century Jewish physicists came from families in which religious customs have not been observed for generations. Perhaps this attitude was bred by a kind of natural selection, since it was for so many generations a cause of pride for a family if a daughter married a learned scholar. One could judge this hypothesis better if more were known about the genetic character of intellectual traits. It sounds just possible, but a little farfetched.

What could be reasons for the great differences between countries? It is not very likely that some subtle selection factor might be at work which would result in a different incidence of the ability needed to be good at science. It is more likely that the other requirements, motivation and opportunity, particularly educational opportunity, which do depend on the environment, might be responsible.

Getting to the right school or the right university can be an important condition for getting the right start in science (much less important for an Einstein than for lesser men), and in England and France the "pecking order" of schools and universities is strong and was even more pronounced one or two generations ago. The selection process for getting into the right channel could conceivably act against Jewish students, even in the absence of any actual discrimination.

But that is speculation without evidence. As a scientist I prefer to register the interesting phenomenon and to admit that I am unable to offer an explanation.

From Winchester to Orion

The most striking thing about this book is how well it is written. Each word is right for its place, the images are apt, and the quotations expressive. In explaining that his style is not that of the social scientists, the author says: "The methodology of this book is literary rather than analytical. This is the result of my upbringing and background." The outcome is a work which it is a pleasure to read, even in places where one cannot agree. For believers in the "Two Cultures," this writing by a scientist would be hard to classify.

Freeman Dyson is a scientist of great distinction. He is best–known for his share in laying the foundations of quantum electrodynamics, the discipline which describes the interplay of charged particles, such as electrons, with each other and with radiation. In the pre–war years, when other parts of our understanding of atoms had reached an impressive fruition, the extension to electrodynamics appeared to be beset by insurmountable difficulties. In the late 1940s there was a breakthrough. This development was set in motion partly by new and sensitive experiments, notably that by Willis Lamb at Columbia University, which posed specific questions for the theory to answer. Responding to this challenge, Julian Schwinger and Richard Feynman invented novel ways of tackling the problem. Some of the steps had already been taken earlier by Sin–Itiro Tomonaga in Japan.

Dyson was the first to understand the approaches both of Schwinger and of Feynman, and to demonstrate that, while different in form, they were identical in content. This made it possible to accept the mathematical rigor of Schwinger's work, while using the intuitive and much more manageable techniques of Feynman. Today these methods are the standard tools of every theo-

This is a review of *Disturbing the Universe,* by Freeman Dyson. New York: Harper and Row, 1979. For the complete reference, please see the Acknowledgments section.

retical physicist, and they are mostly used in the form set out and explained by Dyson.

He tells in the book about this success, achieved when he was at the very beginning of his career, barely a year after he had started learning modern physics. He talks about it without false modesty, but without exaggeration. Since then his research has borne fruit in many other ways, in the areas of nuclear and solid–state physics, astrophysics and cosmology. He has made his contributions to the struggle with the deeper problems of subatomic physics which are not as yet solved. But among his colleagues his name is associated predominantly with his great work of 1948–49. The book does not tell us much about the rest of his work in pure physics. It is in part autobiographical, but it is selective, and the emphasis is on his thoughts and ideas; the story stresses those experiences which were relevant to the development of these ideas.

His parents come out as remarkable people: Sir George Dyson, composer, conductor, and director of the Royal College of Music, had been a music teacher at Winchester, and Freeman breathed the Winchester atmosphere even before he became a scholar there himself. Childhood memories are scanty, but he tells us about a book by Edith Nesbitt, whose fantastic world becomes a symbol for some later ideas. We hear about his early interest in mathematics and relativity, teaching himself the art of differential equations in order to understand Einstein. At the same time, in the atmosphere of impending war, there starts a concern with moral problems. He was inclined to be a pacifist, but decided it was his duty to participate in the fight against Hitler, and chapter three finds him in Operational Research for the air force. He comes out of this experience an angry young man, shocked by the waste of lives in what he sees as senseless bombing of cities, and equally by official obtuseness in failing to accept measures that could have reduced the casualty rate among air crews.

This chapter is followed by one about Frank Thompson, whom he knew at Winchester, and who was executed in Bulgaria for fighting with the anti–Nazi resistance. The portrait of Thompson belongs here because it is part of the war, but perhaps also as a contrast with Dyson's own wartime record, with which he is evidently not very satisfied. He is enough of a realist to know he could not have altered the bombing policy, and probably not even the tactics of bombers, but, he seems to say, he could have put up more of a fight. He paints a vivid picture of Frank Thompson's faith (in the cause of Communism), his enthusiasm and courage. We are not told how and where Dyson managed to spend time at Cambridge and to make his mark in pure mathematics, but in the next chapter he has gone to Cornell University for an apprenticeship in theoretical physics. There is a legend around, probably apocryphal, that about that time a fellow student at Cambridge, who was working in theoretical physics, remarked that modern theoretical physics was getting too complicated, and he had decided to switch to pure mathematics. According to the legend, Dyson replied: "It is true that theoretical physics is getting too complicated, therefore I have decided to switch to theoretical physics."

He went to work with Hans Bethe at Cornell, and the group there also included Feynman. The personalities of Bethe and Feynman and the spirit of the group stand out very vividly. The friendship with Feynman is consolidated in a joint drive west to Albuquerque. Later on that trip Dyson joined a summer school at Ann Arbor, and had the chance of learning directly from Schwinger about his new methods. In the bus back east comes the insight that allows him to unify the Schwinger and Feynman approaches.

This new understanding was not accepted at once. Dyson was now joining the Institute for Advanced Study at Princeton, and Robert Oppenheimer, its director, at first refused to believe in the new theories, or in Dyson's exposition. Oppenheimer usually had a very quick and deep perception, but when he did not accept an argument he could be very cutting in his comment and make it hard for the opponent to present his case. Eventually he was won over and became infected by Dyson's enthusiasm. The portrait of Oppenheimer is one of the strongest parts of the book. Dyson is critical of many of Oppenheimer's qualities, but he writes with affection and respect, and he describes how to him real greatness became evident in the way Oppenheimer approached death. One cannot write about Oppenheimer without referring to his "trial" and to the part played in it by Edward Teller. Dyson does not condone Teller's actions, but shows some understanding and sympathy, and remains a friend of his. The portrait of Teller does not come off, however, compared with those of many other people in the book, who are made to seem solidly alive.

Dyson became involved with some practical design projects, first on the design of a nuclear reactor (his first experience of working with Teller) and later, with much more enthusiasm, on Project Orion, the plan for a space–ship propelled by a series of small nuclear explosions. This kind of team work on applied–science projects was a new and important experience for him. His main research had been of a very abstract kind, and it was lone work, although requiring a close understanding of what other people were doing. It related to an empirical subject, and aimed both at understanding experimental results and at suggesting new ones, but the relation with what went on in the laboratory was always tenuous and indirect. So the more practical collaborative work was for him a little like the work of Goethe's Faust in the Dutch villagers' fight with the tides, to which he refers in the course of the book.

All this time Dyson was thinking deeply and seriously about the ethical problems of war and peace, and in particular about nuclear weapons. He is somewhat suspicious of the people who worked on the atom bomb in Los Alamos during the war. He interprets Oppenheimer's famous phrase about the physicists—they "have known sin"—as referring to the fact that they actually enjoyed their work on the weapon. I do not believe this is a fair interpretation of Oppenheimer or a fair charge against the Los Alamos scientists. Any group of people working intensely on a difficult task against time will find it hard to suppress feelings of comradeship and of pride in their progress, as long, of course, as they do not believe the objective to be fundamentally evil. These feelings are not inconsistent with an awareness of the seriousness of the enterprise, and responsibility for the consequences. Those of us who allowed this awareness to fade would merit Dyson's censure.

Dyson became involved in discussions about disarmament, partly while spending periods with the Arms Control and Disarmament Agency. His first major public expression of views was in an article in *Foreign Affairs* in 1960, in which he opposed the Test Ban Treaty then under discussion. He later changed his mind and became a supporter of the Treaty. Everybody, except for unusually narrow–minded people, changes his views occasionally: but not everybody discusses his change of heart, and comments on the previous, discarded views, as candidly as Freeman Dyson, who sees, and says, clearly that his article was influenced in part by his enthusiasm for Project Orion, which was made illegal by the Treaty, and in part by his friendship for colleagues who were working hard on "clean" weapons.

He is aghast at the present balance of terror, and at the idea of deterrence based on Mutually Assured Annihilation. In this he is not alone, though he presents his objections more powerfully than many. This leads him to advocate a policy based on defense: but this would only be feasible if a highly efficient anti–ballistic missile system were technically possible—and this still isn't clear. A great number of scientists (including myself) are extremely skeptical on this point, and believe that any alternative to the present frightening situation must be looked for elsewhere.

While recounting his experiences in test–ban and other negotiations, he gives his views, which are always responsibly thought out and vividly expressed, on such matters as nuclear terrorism, the dangers of biological warfare and the controversy about "genetic engineering." The last eight chapters contain speculations about the future. They include a vision of self–reproducing automata, which may alter the appearance of deserts, and ultimately perhaps of the solar system. He points out that the nature of such developments is critically dependent on time–scale, on whether a generation of these automata is comparable to a human generation, or as short as, say, one year.

The latter possibility seems to me to belong to extreme science fiction. There is also no mention of how one is to provide for dealing with faults, break-downs and design errors. Another extreme speculation is about a scheme for colonizing extraterrestrial lands, perhaps some of the asteroids. It is probably this pet scheme of his that gave rise to the title of the book, which derives from T. S. Eliot.

This miscellany of thoughts about the future also contains many items which are much more down to earth, both literally and otherwise. He stresses convincingly the contrast between the clumsiness and inflexibility of projects run by large bureaucratic organizations (including the present nuclear–power technology) and schemes pursued by teams of devoted and imaginative individuals, of which there are many examples to be found in the early develop-ment of new technologies.

There is a plea for diversity, and for preserving national, ethnic and lin-guistic differences; and a report on a visit to his son in the wild areas of North–West Canada, and the lessons to be learned from his kind of life. This again reflects the longing for a way of life based on individual activity rather than on the impersonal institutions of 'civilized' society. There are remarks about the relation between science and religion. And that is not the complete list.

It is a remarkable book but it does not achieve one of its stated objectives. On page 5, after reminding the reader that policy decisions about the use of science have to be taken by non–scientists, he says: "If you, unscientific people, are to succeed in this task, you must understand the nature of the beast you are trying to control. This book is intended to help you under-stand." The book is clearly not attempting to explain the content or method of science, although all scientific points mentioned are explained with great clarity and simplicity. Will it help the layman to understand scientists? One certainly learns much about Freeman Dyson, but he is hardly a run of–the–mill specimen of the genus, even if one only looks at the top range of distinc-tion. In fact, it is his uniqueness which gives the book much of its attraction. If the reader insists on wanting to learn about scientists in general he may be disappointed. If he enjoys getting to know one particular individual among the leading scientists and his thoughts, he will find the book rewarding.

The Physicists

C. P. Snow had an unusually wide range of interests and many different talents, but he probably was at his best in crisply summing up people and situations. I recall from my own experience the occasions when he used to visit the universities during the war, with his friend and assistant William Cooper, to decide the fate of students about to graduate in science. Each student had to be assigned to some branch of the armed services, or to a research project of military importance, or perhaps to further research training. On the strength of the paper record and a two–minute interview, he would give a thumbnail sketch of the personality and ability of the student, which usually could not be improved upon by the teachers who had known the student for years.

The same power of characterizing real people, a talent which is quite distinct from the ability to make fictional characters come alive, showed up in his biographical essays, for example about Einstein, or about the mathematician G. H. Hardy. But he did not, as far as I am aware, produce a book of such portraits.

When he died in 1980 he had been working on a book that would describe a half–century of physics and some of the leading characters in it. He was able to complete only a first draft, which has now been published. We are not given the name of an editor, but from remarks in the introduction by William Cooper it appears that he was responsible for arranging the text for publication.

The drama of twentieth-century physics, which Snow presents, comes in several acts. Up to the end of the nineteenth century physicists had not regarded the structure of matter as their problem. One knew there were atoms, but their nature was the chemist's problem, if anybody's. The beginning of

This is a review of *The Physicists*, by C. P. Snow. New York: Little, Brown, 1981. The book review is reprinted with permission from *The New York Review of Books*. Copyright © 1981 Nyrev, Inc. For the complete reference, please see the Acknowledgments section.

the new attitude came with the discovery of the electron, and the realization that the laws of mechanics could be applied to its behavior within the atom. From then on atoms and, what went on in them, became the physicists' main preoccupation. With improved tools experimental discoveries led to a rapidly growing body of information about the atom, one of the crucial steps, which Snow describes at the start of the book, being Rutherford's discovery of the atomic nucleus as the center of the atom, with the electrons orbiting around it.

But the new–found knowledge did not fit in with the laws of physics as they had developed since the time of Isaac Newton, and the resolution of these contradictions required the acceptance of revolutionary new concepts. These came through Einstein's theory of relativity, and through the quantum theory of Max Planck (whom Snow barely mentions), Einstein, and Niels Bohr. At first the quantum theory was a makeshift affair, grafting ad hoc rules on to the old mechanics, and it was only in the late 1920s that a drastic revision of the most fundamental concepts about space and motion led to a deeper understanding, the foundations of which were laid by de Broglie, Heisenberg, and Schrödinger. On these foundations, as Snow emphasizes, a complete edifice was built with breathtaking speed.

Rutherford took no part in this study of the mechanics of the atom, but continued to work on the nucleus, and succeeded in demonstrating nuclear transmutations with the aid of particles from radioactive substances. Later it became possible to do so with artificially accelerated particles. Thus started the use of large and ever–growing accelerators, initiating the era of "big physics." New chapters in the story were opened by the discovery of the neutron and of artificial radioactivity, and by Enrico Fermi's use of neutrons for producing nuclear reactions.

All this seemed a pursuit devoted entirely to the search for knowledge. Nearly all nuclear physicists were convinced that there could never be any practical consequences from their work. But this changed dramatically with the discovery of the fission of uranium, and the possibility of a chain reaction. Soon physicists in several countries were at work on the problem whether and how a nuclear bomb could be made. The decision to drop such a bomb on Japan will be debated by historians and others for many generations.

This work was done in secret, but when the news was released physicists found themselves in the limelight. Some blamed them for the power of destruction they had released, some admired them for having helped to win the war, but either way they had become important. Atomic bombs were the most spectacular, but by no means the only, important contribution by physicists to the war effort. Other work, including particularly radar, was also of vital importance, probably more decisive than atom bombs.

After the war physicists returned to their search for the basic laws of nature. Their work on nuclear physics and their new study of subatomic particles required the use of bigger and bigger accelerators. The new respect for physicists resulting from their wartime work helped to obtain the necessary financial support. Snow repeatedly quotes a comment by Mark Oliphant about the new position of physics: "We have killed a beautiful subject"; he disagrees because physics did after all return to the old spirit of inquiry. But Oliphant might well counter that the nature of "big physics" with its large machines and gigantic measuring equipment, requiring long–term planning and large teams, makes it a very different kind of activity.

But the rapid advance of knowledge resumed, and a whole world of new particles and their transformations was discovered. The physicists are making progress with discovering the underlying laws, but much remains to be done. There was also impressive progress in other branches of physics. New discoveries in solid–state physics have led to the transistor and other electronic devices, and these have led to the modern computer and other tools for automation. Astronomy has made spectacular advances, assisted by the use of radio telescopes, which benefit from the lessons learned from wartime radar work. The new science of molecular biology has arisen from the use by physicists of X–ray methods for puzzling out the structure of biological substances, and thus the genetic code was deciphered.

Snow's presentation of this story is brief, and he cannot be expected to do justice to details in an account of some eighty pages of text. But with his particular skill he makes the story lively and brings out the main features, perhaps like a charcoal sketch which may present the character of the subject even though details may be missing and some of the lines are not very accurate.

There are indeed many inaccuracies in the book, some amounting to very surprising errors of fact. One would assume that it was Snow's intention to check on some of these facts in later drafts, and William Cooper (or whoever had editorial responsibility) has not done Snow's memory a good turn by allowing these errors to stand. Cooper mentions in the introduction that Snow wrote the book from memory, and comments on how remarkable his memory was, if selective and with an occasional odd emphasis. But there are limits to the reliability of even the best human memory.

It is surprising that Snow would make some of the errors even in a rough first draft, as when he refers to the "British Nobel Prize winner A. H. Compton" (Compton was one of the most American of great physicists), or when he reports that the "ALSOS" mission to Germany at the end of the war to look into German atomic–energy work was led by Samuel Goudsmit and George Uhlenbeck (it was only Goudsmit; the other name came from an association of these two in their earlier discovery in physics). It is surprising to see the reference to Heisenberg as "an active spokesman for the Nazi faith."

Heisenberg was, as Snow remarks later, a German nationalist, and could be accused of making his peace too readily with the Nazi regime. But to call him an active spokesman is a gross distortion. General Groves is called "a singularly bad choice for his job," apparently on the strength of his error of judgment in claiming that the United States had a twenty–year lead over the Soviet Union in making atom bombs. This deplorable, but not uncommon, arrogance should not overshadow the merits of an administrator of unusual energy and courage.

Other errors, too numerous to list, are more understandable, but might well have been eliminated in a revised draft. Atomic physics was not becoming known as "particle physics" in the 1920s—this name for subnuclear physics came into use only after the Second World War. While the Soviet physicist Peter Kapitsa was out of favor with Stalin he did not stay in his house on the grounds of his institute in Moscow, but in his country cottage or "dacha." Murray Gell–Mann did not introduce group theory into physics (this was done by Wigner a generation earlier), though he made very effective use of it.

Some of the portraits of people known personally to Snow display much insight. The study of Rutherford penetrates beyond the superficial appearance of robust self–confidence, and Snow succinctly characterizes the personalities of Einstein and Bohr. Of Kapitsa only some aspects emerge (his contribution to physics is not mentioned, unless I have been misled by my inadequate memory, and the somewhat slipshod index), but what there is bears a true likeness. Others, such as Chadwick or Fermi, are recognizable; many appear only as rather shadowy figures. The single sentence about Wigner contains a curious juxtaposition: "Wigner was calm, judicious, ironic, temperate, mildly conservative: his sister was Dirac's wife."

In telling the story of the physicists one must of course refer to the problems with which they were struggling and to their solutions. This is by no means easy to do in layman's language in so brief a space. As a physicist I am badly qualified to judge how far Snow succeeds in making these explanations intelligible, but it seems to me that, apart from some rough spots, and some inaccuracies, his brief account does remarkably well.

No such difficulty in communication arises in discussing the physicists' role in the development of nuclear weapons. Snow does not go deeply into the technical aspect of this. I cannot refrain from noticing that he overrates the direct influence of the "Frisch–Peierls–memorandum" on the atomic–energy–work in the United States. This paper served to persuade the British authorities to take the possibility of a nuclear weapon seriously, and this may in turn have accelerated the American decision to back the physicists who were anxious to go ahead.

He is more interested in the moral issues. He does not blame the physicists for their part in creating these terrible new weapons. He regards this as an inevitable development. As for the German physicists, he criticizes the myth that they deliberately refrained from making bombs for ethical reasons. Instead he claims that they failed to have a clear idea of what was needed for the bomb. This view is not supported by recent historical studies: they did know, but considered—probably rightly—that there was no chance of Germany completing such a project before the end of the war. In fact they did not believe that the United States could muster the tremendous effort required. Snow does not think they would have refused to make the weapons if doing so had appeared feasible. "In comparable circumstances, American, English, Russian scientists would have felt that the evils of the regime counted for nothing against the evils of absolute defeat." Perhaps so, but there were German scientists who felt otherwise, and who, at tremendous personal risk, worked for Allied intelligence in Germany during the war.

About the consequences of the scientists' work, he points to the many beneficial results besides the destructive ones. Even the weapons have their positive side, since the fear of them has been a restraining influence in areas of potential conflict. While there cannot be a complete assurance that the world is safe from global nuclear war, he thinks the risk is small. However, with an increasing number of smaller nations acquiring nuclear weapons, he believes it is very likely that sooner or later there will be some local war with nuclear bombs. This and their possible use by terrorists are the risks he thinks we should worry about.

He ends without attributing to the physicists any special role in safeguarding the future: "Their own intellectual structure waits there to be added to, but is unshakeable. The application which has come out of that structure has left us with some threats and more promises. It is for the general intelligence of us all to make the best of both."

Yet one of three appendices included in the book, presumably not by Snow's choice, is an address he gave in 1960 with the title "The Moral Un–neutrality of Science." In this he assigns to the scientists a much greater share in the responsibility than they bear just as citizens, because of their close association with developments that can vitally affect the future, and because of their deeper insight. They "have a moral imperative to say what they know." He implies that the scientists can, by explaining the consequences of alternative policies (e.g., continuing the arms race or accepting some restrictions) help argue for the right course of action. Did he change his mind in the twenty years since the address was given or is this just a change in emphasis?

According to William Cooper's introduction, the book was written at great speed. This may account for some of its weaknesses, but it gives it what Cooper calls an unimpeded narrative impulse, which makes it a pleasure to read.

Fact and Fancy in Physics

According to their authors, both books under review set out to portray an epoch in physics through the medium of a biography. The period dealt with is almost the same, the late nineteenth and early twentieth centuries, though one story extends into the 1950s, whereas the other ends effectively in 1918. One is about American physics, the other about German physics and physicists. But the main difference is that one author has chosen a real physicist, in fact one who contributed to the development of modern physics, whereas the other book is about an imaginary person.

Robert Kargon's *The Rise of Robert Millikan* is a workmanlike effort. Millikan, born in 1868, was old enough to be educated during the reign of "classical" physics and influenced strongly by the belief of his teachers, including the famous Albert Michelson, that the aim of physics was to measure known quantities with greater and greater precision. He lived through the great revolutions in physics started by the discoveries of the electron, of X–rays, and of radioactivity, and leading to the theory of relativity and the quantum theory.

At the age of twenty-seven, when he had started on research with some minor, but sound, papers, he made a trip to Europe, and here met the beginnings of the "new" physics. This was not an accident, because then the center of physics was in Europe, and discoveries were mostly made there. He also had the opportunity of working with Walther Nernst, and published a paper based on it. But it was not until 1907 that he turned to the problem which would bring him his main success. The reason was in part that he had, in his

This is a review of the books *The Rise of Robert Millikan: Portrait of a Life in American Science,* by Robert H. Kargon. Ithica, New York: Cornell University Press, 1982. *Night Thoughts of a Classical Physicist,* by Russell McCormmach. Cambridge, Mass.: Harvard University Press, 1982. The review is reprinted with permission from *The New York Review of Books*. Copyright © 1982 Nyrev, Inc. For the complete references, please see the Acknowledgments section.

position at the University of Chicago, very heavy teaching duties, which he took seriously and which led him to write some successful textbooks. But more important, he could not decide on a topic on which major progress was possible. Some of his experiments were successful, but not exciting, others led nowhere. The result was a considerable lack of confidence in himself and his ability.

But in 1907 he decided to try to measure the charge of the electron, discovered by J. J. Thomson ten years before. The electron charge, one of the most fundamental constants of physics, was then known only in a very rough approximation, and efforts to improve the accuracy of the measurement had not succeeded. Millikan experimented with the method used by others, namely to produce a cloud of water droplets in an electric field, and to watch the slowest–moving part of this, which presumably consisted of droplets carrying only one unit of charge. He realized that the only way of doing better was to observe individual droplets, but these evaporated too quickly to be kept under observation for a long enough time. He then hit on the solution of using drops of oil instead of water. There are more steps to the experiment than can be described here, but the result was a very accurate determination of the charge. This experiment was one of the two for which he was to receive the Nobel Prize, the second American physicist to be so honored.

The other had to do with the photoelectric effect, and here the story is more complicated. It was known that light falling on a metal surface can eject electrons from it. It was found that the speed of the electrons did not depend on the intensity of the light, but on its color. Einstein explained this by making use of Planck's idea of light quanta. If light consists of individual quanta, each carrying an amount of energy dependent on the color of the light, then the result is understandable. But the experiments were not accurate, and therefore Einstein's theory was in some doubt.

Millikan could not accept this explanation. If light was made up of little particles, one would have to give up regarding light as waves, and there was plenty of evidence for the wave nature of light. In fact, the apparent contradiction between the wave and the particle pictures was one of the deep mysteries that got cleared up only by the new quantum mechanics in the late 1920s. So Millikan set out to disprove Einstein's explanation; but he was a good experimenter, and his very accurate measurements confirmed the explanation exactly. He now accepted the Einstein formula, but still maintained that the underlying theory could not be right. In any case, he had obtained a very accurate determination of Planck's constant, which enters into Einstein's formula, since the energy of each light quantum depends on it.

This, too, was the basis of his Nobel award. At first, although he had verified the Einstein formula, he could not believe the theory behind it. But, Kargon tells us, he claimed in his autobiography that he immediately saw that the only possible interpretation of the data was provided by Einstein's theory. It seems he did not like having been wrong, and conveniently forgot his incorrect pronouncements.

A similar situation arose again much later in connection with cosmic radiation. The dust cover of the book describes Millikan as the explorer of cosmic radiation, but the honor of discovering it belongs to the German Victor Hess. What Millikan did was important enough, for it was at first not clear that this was a radiation from outer space, and that it did not originate in some radioactive contamination of the atmosphere. Millikan studied carefully the variation with altitude, and proved that the radiation did indeed come from outer space. He gave it the name "cosmic radiation."

He believed, as did most people initially, that this was electromagnetic radiation, like X–rays or gamma rays, and from its penetrating power, which for such radiation would have provided information about its energy, he convinced himself that it was radiation emitted in the process of building up atoms, a possibility in which he had always been greatly interested. However, in 1927 the Dutch physicist Jakob Clay found that the cosmic rays depended strongly on latitude, and this clearly had to be due to the magnetic field of the earth, which can deflect charged particles but not electromagnetic radiation. Thus the rays consisted of charged particles.

Millikan did not want to accept this, and set out to disprove the latitude effect. He sent one of his collaborators on an expedition to test it, and at first the results seemed to be negative. He therefore insisted in a public debate with Arthur Compton that there was no latitude effect, but he had to retract afterward. Later he was to claim that he and his group had been the first to discover the latitude effect.

His first two important experiments had brought him fame long before the Nobel award in 1923, and he became involved in the organization of science. He played an important part in the early years of the National Research Council. During the First World War he served in the Army Signal Corps to help apply scientific techniques. After the war he returned to Chicago, but was soon persuaded to take on the presidency of the new California Institute of Technology, and the directorship of its physics department. His energy and enthusiasm helped to develop Cal Tech to its present position. At the same time he encouraged and guided a team of first–class physicists.

His conservatism in physics was matched by conservative views on other matters. He was against scientific research being financed by government, against too much government in general, against the New Deal; and he would have been strongly against the welfare state if the term had been in use in his time. The book leaves us in no doubt about his ability, but does not gloss over his occasional obstinacy or his wishful thinking about past errors, matters on which some histories tend to be silent. Millikan was not a revolutionary who started new ideas, but the author stresses—rightly—the importance of men like him for the progress of science.

Russell McCormmach's *Night Thoughts of a Classical Physicist* is a very different kind of book. Its subject, Professor Victor Jakob, is invented, but invented with great labor and erudition. The author is a distinguished historian of science, familiar with the realities of the period, and you can almost feel the strain of forcing the imaginary professor into the Procrustean bed of the real history. For example, one of the illustrations is von Menzel's oil painting *Room with a Balcony*, which exists, and which shows in the background six portraits, too blurred to be recognizable. The caption refers to "Jakob's portrait gallery," and another illustration purports to give this gallery in close-up. Arranged to match the layout of the painting, this now consists of contemporary portraits of the great physicists whom Jakob admired. But the result is confusing, because from the text the portrait gallery is supposed to be in Jakob's study, while the balcony room in the painting can by no stretch of imagination resemble a study of the period.

The scene is a small unnamed German university town in 1918. Jakob is an elderly, unsuccessful physicist. I suppose he has to be unsuccessful, for if he had made any real contributions, these would have been hard to fit in with the real history of physics. He is very conscious of not having risen to be an "Ordinarius," or ordinary professor. He is just an honorary ordinary professor, something between an extraordinary (associate) and ordinary (full) professor. In fact he has to teach theoretical physics for apparently no better reason than that he was never given a chair of experimental physics.

In his younger days he wrote some papers, but we are not told their subject, or with what kind of problems he was struggling, except that he believed in the importance of a "world ether." Of the changes in physics, he seems to have been aware only of the new theory of relativity, which was too mathematical for him, and the developing quantum theory of the atom. I wonder whether his creator has not here endowed him with too much insight; the new atomic theory did indeed exist and had considerable success, but would an old–fashioned physicist really have seen it as the new trend in the subject, which he was reluctant to follow? There is no sign of his reaction to the many contemporary discoveries that were not matters of opinion, such as those of the electron, of X–rays, of radioactivity, all of which gave Millikan the exhilarating feeling of participating in a rapidly developing subject.

Thus Jakob's "night thoughts" sum up the state of physics only in a rather vague way. He reflects on his lack of success, and he recalls many instances of the lack of good will toward him, from the head of his institute down to the custodian. In his dreams—at least I assume that the passages in italics, with their dreamlike logic, are dreams—the Ministry of Education official, who had the last word about academic appointments in Prussia, appears as a rather intimidating dictator. He also recalls his teachers and friends—the subjects of his portrait gallery—and what he learned from them. Particularly vivid is his memory of Paul Drude, another real physicist, whose suicide came as a great shock to him.

It is wartime, and he thinks about his country. He fought and distinguished himself in the Franco–Prussian war of 1870. He is still a good patriot, but he has become skeptical about the progress of the war, he sees hope for his country in the character and idealism of its people rather than in the control of territory. In fact, he expresses these thoughts in a speech at a patriotic meeting at the opening of the story, and during this speech he suffers a slight stroke. It is during the enforced rest following this that his "night thoughts" come to him.

At the end he dies in a bizarre accident in the mountains near his university town. The notes inform us that this happened to a real physicist. There are forty–five pages of notes, which justify the attribution of opinions to real physicists in their letters or conversations with the invented Jakob, and sometimes explain how Jakob's opinions and experiences are taken from those of real people. I find that I learn more from these notes than from the rather contrived narrative about an invented person, though he comes very much alive, but that is evidently a matter of personal taste.

The primary preoccupations of the physicists of the time were, of course, the perplexity created by the new facts and by the new insights into old facts, the advent of relativity, more or less complete at the time, and the quantum theory, still in its initial stages and quite incomprehensible to most scientists. In wartime also the need to make new techniques available to the military added to their problems. This last concern was international and can be recognized in the story of Millikan as well as of Jakob. Perhaps the German physicists were more deeply worried about the theoretical difficulties of the new discoveries, partly because of their proximity to the innovators, partly because of the greater number of theoreticians among them.

At the same time the personal squabbles and complaints among the German physicists were more intense than they were elsewhere. There are some of course in any profession, and more among the less able and less successful practitioners. That such tensions are so much more prominent in Jakob's story than in Millikan's biography is in part accounted for by Millikan's being a success story, and Jakob's one of failure. But from the notes it is clear that such situations were very widespread in the German universities of the period. They are coupled with complaints about appointments, which in Germany were handled by the state ministries. No doubt there are grievances about lack of promotion everywhere, but in America or England there is no centralized authority inviting a centralized feeling of grievance. Of course the question of anti–Semitism came up in connection with appointments and Jakob mentions it, although in spite of the sound of his name, he is not Jewish. There is mild reference to it also in the Millikan story. (Cal Tech could not be expected to appoint *two* Jews.) One feels that the weight of anti–Semitism in the universities in the two countries at that time was about the same.

In the notes, and reflected in Jakob's musings, are the thoughts of many eminent physicists, including Helmholtz, Planck, Einstein, Wien, and others, on many different subjects. Following Jakob's rambling reflections, they naturally appear in a rambling order. Few of these concern physics and those that do are mostly about semi–philosophical points such as the merits of the old mechanistic world picture, which many of the older physicists were reluctant to abandon, the possibility of replacing it by an electromagnetic world picture, or the difficulty in understanding Bohr's postulates about the emission of radiation. One misses arguments about more concrete problems, such as how to explain some experiment or how to verify some theory. These must have been discussed and their omission presumably reflects the tastes of Jakob or rather of his creator. The rest of the comments are about personal problems, about squabbles with colleagues and staff, about promotions and appointments, and about physicists' attitudes to the war; these add up to a picture of a life of physicists of the time.

The book represents much serious research and extensive knowledge of the period. It could have been written only by an eminent historian of science. Whether its unusual presentation makes it easier to absorb the impressions of the history of science that the author wants to convey is a matter of taste that must be left to each reader to judge.

What Einstein Did

A lbert Einstein was not only respected and admired by his fellow scientists as probably the greatest physicist of this century but he also achieved extraordinary fame among people who did not have the least idea of his work. What did he accomplish? Although he is most widely known as the discoverer of the theory of relativity, he made other vital contributions which are far less well known but which, by themselves, would have placed him in the front ranks of physicists. The most important of these was his part in creating the quantum theory. He gained the Nobel prize for this, not for his work on relativity.

Abraham Pais's book is thus particularly valuable as the first thorough study in one accessible volume of all of Einstein's major contributions to science. A physicist who knew Einstein when they were both at the Institute for Advanced Study at Princeton, Pais presents each phase of Einstein's work against the background of Einstein's previous ideas and those of his predecessors. In each case he tries, as far as possible, to reconstruct Einstein's train of thought. To do so, of course, he has to describe the technical aspects of each problem, and it will be difficult for any reader without some acquaintance with theoretical physics to follow his account. For those qualified to read it, however, his book is remarkably clear as well as authoritative.

The breadth of Einstein's vision and his originality were shown in his work of 1905, when, as a twenty–six–year–old technical expert, third class, in the Swiss patent office in Bern, he wrote five important papers, all carefully discussed in Pais's book. Two of these established what is now called the special theory of relativity. This starts from the long–established fact that a state of uniform motion cannot be distinguished from being at rest. It is a

This is a review of *'Subtle is the Lord...': The Science and the Life of Albert Einstein,* by Abraham Pais. New York: Oxford University Press, 1982. The book review is reprinted with permission from *The New York Review of Books.* Copyright © 1985 Nyrev, Inc. For the complete reference, please see the Acknowledgments section.

familiar fact that the passenger in a fast airplane experiences no sensation of being in motion, unless air turbulence or a change of direction causes the motion to be nonuniform. But it was then thought that by watching the speed of light one might be able to tell the difference. Physicists believed that light consisted of vibrations of a hypothetical, all–pervasive medium, the "ether," and that by observing the way light travels, we could discover whether we were at rest, or moving, relative to this ether. This seems reasonable if we think of sound waves. If we were sitting on the tail of a plane, our voices shouting to a friend on top of the cockpit would seem to be traveling slowly, because they had to go against the rush of air experienced by the plane; the sound of our friend's voice would reach us faster than normal sound. On a supersonic plane the sound of our voices would not reach the front at all.

We are all participating in the motion of the earth around its axis and in its orbit about the sun, and so, according to the ether theory, we should be exposed to a similar "ether wind." It was surprising enough that such a wind did not seem to affect us in any way. But at least it was expected that the motions of the earth should influence the progress of ether waves, i.e., of light, just as the flow of air past the plane affects the progress of sound. But in a delicate experiment to test this, the American physicists Albert A. Michelson and Edward Williams Morley demonstrated that there was no effect on the speed of light. The speed of light plus any added velocity was still equal to the speed of light.

This puzzle worried many, including the physicist H. A. Lorentz and the mathematician H. Poincaré, who came close to the right answer. In his 1905 papers Einstein showed that in view of the Michelson–Morley experiments and other evidence, we had to revise our ideas of space and time. The speed of light appears the same no matter at what speed we are traveling ourselves, and this seemingly absurd statement becomes possible because time looks different to observers moving with different speeds: two observers might disagree on which of two distant events was the earlier, and which the later one.

Many important consequences of this theory have by now been verified. They include the fact that bodies become heavier with increasing speed, thus requiring more force to accelerate them further; in the end an infinitely strong force would be required to make them reach the speed of light. Therefore no object can be accelerated to that speed, let alone beyond. This is of great practical importance for the physicists' particle accelerators, in which particles travel at a speed less than that of light by only one part in 10 billion.

Other important consequences included the famous equation $E=mc^2$, which expresses the fact that matter and energy are equivalent, and which is— wrongly—considered by some to be the basis of nuclear energy. (It is true that the uranium nucleus loses some of its mass, or weight, in the process of fission, but ordinary fuel and oxygen also lose some mass in the process of combustion. The difference is one of quantity; the loss in the first case is about one part in a thousand, in the second case one in 50 billion. In either case the loss of mass is incidental to the gain in energy, not its cause.)

These ideas were immediately accepted by the community of physicists, and from then on formed the basis of much further work. But they upset many philosophers, partly because they conflicted with beliefs that had been regarded by Kant and others as evident a priori. Many were confused by the name "relativity," which appeared to suggest philosophical relativity, the attitude that all knowledge is relative, whereas in fact the essence of the theory of relativity is to recognize the absolute physical reality which unites the different appearances that phenomena present to different observers.

Einstein's second important idea, more revolutionary than that of relativity, concerned the light quantum. Here again Pais explains very clearly Einstein's achievement. In 1900, Max Planck had described the light emitted by a hot body by a formula using, in an unconventional way, the principles of statistical mechanics—the description of heat in terms of the behavior of atoms. While Planck was trying to minimize the revolutionary nature of his step, Einstein saw that it required a new basic approach to the nature of light: one had to treat light as made up of separate quanta, small amounts of energy dependent on the "frequency" or "pitch" of the wave, which is about twice as large for blue as for red light. This led him to predict that in the photoelectric effect, i.e., the release of electrons from a metal surface under the influence of light, the energy of these electrons, which was known to be independent of the intensity of the light, should grow in strict proportion to its frequency. This prediction, for which Einstein was later awarded the Nobel prize, was soon confirmed by precise measurements.

But his basic concept of the light quantum took a long time to be accepted. He realized himself how basically this conflicted with the evidence that light consisted of waves. He therefore emphasized that in the description of light we must combine some wave with some particle features, an early indication of an idea that did not find its full expression until the quantum mechanics of the late 1920s.

The third important contribution of 1905 concerned Brownian motion, the irregular motion of small specks of dust, known to be caused by the irregular impact of the molecules of the surrounding gas or liquid, thus giving a vivid illustration of the fact that heat consists of the motion of molecules. Einstein's work opened a new way of "counting atoms," i.e., of determining how many atoms or molecules are contained in a given amount of matter. Another paper, which I shall not attempt to summarize, put forward other new ways of counting atoms, and these contributed much to the confidence of physicists in the existence of atoms.

Einstein was not satisfied with the special theory of relativity, because the restriction of his new conception of space and time to uniform motion seemed unnatural. At first sight there seems no way out of this dilemma, because nonuniform motion, whether the bumping of a car on a rough road, or the revolutions of a merry–go–round, causes sensations that make it evidently impossible to ignore that motion. But Einstein saw that the sensation caused by accelerated motion is of the same kind as that caused by another familiar phenomenon, namely gravity. He illustrated this by the picture of people in a freely falling elevator, for whom the acceleration would balance their gravity, so that they would feel weightless. The same idea is now more familiar from the experience of astronauts, who are weightless even while in the field of attraction of the earth, because the nonuniform motion of a coasting space-craft, as a result of the attraction, compensates for their weight. This "principle of equivalence" depends on the equality of weight and mass, which had already been established with a high degree of precision.

Pais quotes Einstein as describing the application of this idea to nonuniform motion as "the happiest thought of my life," but it took Einstein many years, many false starts, and much persistence to build on this principle a theory of gravitation. This involved interpreting gravity as a change in the structure of space, an alteration of the laws of geometry. Its predictions included a correction to the accepted orbit of the planet Mercury, and the bending of light rays passing close to the sun, observable during an eclipse. Observations during the eclipse of 1919 by expeditions organized on the initiative of the English astrophysicist Arthur Eddington confirmed this prediction, and thus led to the general acceptance of the "general" theory of relativity, while also catching the attention of the press. It was then that Einstein's fame began with the general public.

Besides these impressive pieces of work Einstein took an interest in many minor problems of physics, and even participated in experiments and in practical inventions. But after 1919 the main direction of the development of physics bypassed his central preoccupation. Why did this occur?

Having recognized that gravitation was part of the structure of space and time, Einstein wanted to link also electricity and magnetism with the spacetime structure, and he hoped that this would lead to an explanation of the existence of elementary particles, like the electron. His search for such a "unified field theory" occupied much of the rest of his life, but it ended in failure. Most of his colleagues did not believe that such a theory was possible, or even desirable. While this did not diminish their respect and affection for him, it left him isolated.

The nature of his disagreement with his colleagues did not turn simply on his search for unification. Much of the progress of our understanding of the laws of nature consists in recognizing how seemingly different phenomena are explained by common or related basic principles. The great strides made by fundamental physics during the last few years consist in a theory unifying electromagnetic phenomena with the "weak" force, which is responsible, for example, for some of the radiations from radioactive substances, and for the heat of the sun.

But the kind of unified theory Einstein was searching for was of a different nature. It was related to the fact that he was unhappy about the other great development in twentieth–century physics, which he himself had helped to initiate: the quantum theory and its finished form, quantum mechanics. This theory also requires that we adjust our intuitive prejudices, in particular that we abandon the principle of causality in its old form, and Einstein could never reconcile himself to that.

The difficulty is illustrated by the "wave–particle duality." Einstein himself had introduced the notion of a "light quantum" which makes a flash of light behave like a stream of small particles; at the same time experiments with diffraction of light showed light to consist of waves, and these experiments continued to remain valid. The theoretical ideas of de Broglie and Schrödinger, and the experimental discovery of electron diffraction by C. J. Davisson and L. H. Germer, and by G. P. Thomson, showed that electrons and other familiar objects known to us as particles also could behave like waves.

Einstein himself first mentioned the way out of this dilemma, which is now generally accepted. He spoke of *Gespensterwellen* or "ghost waves" that would regulate the *probability* with which particles were found at any particular place, but he mentioned this as a logical possibility that he found most unattractive. Later this idea was elaborated into a full theory by Max Born, Werner Heisenberg and Niels Bohr. In this theory Planck's "quantum of action," which he had introduced through the energy of a light quantum of a given color, sets a limit to the accuracy with which we can observe, or even define, details of the motion of any particle. As a result, we can no longer, even in principle, predict the motion of a particle exactly, but can give only the probability of it getting to a certain place. Einstein acknowledged the success of quantum mechanics in many fields, but tried to find inconsistencies in the theory; these were always disproved by Niels Bohr. Eventually Einstein accepted that the theory was not inconsistent, but felt that it was incomplete.

At the heart of his reaction was his reluctance to accept the part that chance played in the interpretation of quantum mechanics, and in the loss of determinacy. According to the quantum theory the position of particles is a matter of statistical probability, as Einstein himself had once suggested, but could not bring himself to accept. He did not believe, in his much cited remark, that "the Lord would play dice." He therefore hoped that the "unified field theory" for which he was searching would make it possible to maintain a deterministic description, in which the existence both of particles and of Planck's quantum of action would be shown by field equations. In this view he became isolated from the rest of the physics community.

It is hard to understand how this great man, with the flexibility of mind and courage to introduce into physics the most revolutionary ideas, could fail to accept conclusions that all his colleagues found convincing and confirmed by ample experimental evidence. Peter Kapitza, the Russian physicist, once observed that great physicists who, as they got older, lost contact with the latest developments in their subject, such as de Broglie or Schrödinger, were those who did not have many pupils, while others, like Rutherford or Niels Bohr, who stayed in the front line until their deaths, were surrounded by crowds of young collaborators. It is true that Einstein did not have many pupils, but he is unique among physicists and it may be dangerous to apply general rules to him.

Pais's book is based on Einstein's writings, on his letters, on the published recollections of many who had conversations with him, and on Pais's own talks with Einstein at the Institute for Advanced Study at Princeton, where Einstein used to chat with his young colleague about his work and his thoughts. Many of these conversations show Einstein's pungent use of language, mostly in German with which he felt at home.

Typical of his way of talking is the often–quoted phrase *"Raffiniert ist der Herrgott aber boshaft ist er nicht"* (Subtle is the Lord, but malicious He is not), which Pais uses as his title and as a kind of leitmotif to the book. This expresses in language typical of Einstein the confidence that there must be an ultimate simplicity in the natural law, and it also shows his playful but genuine feeling of respect for a higher order. The sentiment is similar to Niels Bohr's, in a different style: "Physics is the belief that a simple and consistent description of nature is possible."

Einstein is often called naive in matters outside physics, and with the pressure on him to make public statements on all conceivable subjects, he may occasionally have displayed ignorance of their subtleties. But Pais stresses that more often Einstein's seemingly naive utterances were the result of his absolute and uncompromising moral principles, and an unconventional approach to human problems. In a collection of essays from the centenary celebration in Jerusalem of Einstein's birth,[1] for example, Yehuda Elkana quotes from a letter to Chaim Weizmann: "Unless we find the way to honest cooperation and honest dealings with the Arabs, we have not learned anything on our way of two thousand years' suffering and deserve the fate that is in store

---

[1] *Albert Einstein, Historical and Cultural Perspectives: The Centennial Symposium in Jerusalem,* edited by Gerald Holton and Yehuda Elkana (Princeton, New Jersey: Princeton University Press, 1982), p. 237.

for us." While he had come to support Zionism, he was strongly opposed to chauvinism in any form.

Pais also raises the puzzling question why Einstein occupies such a unique position in the eyes of the general public. He points to the influence of the press, to the confusion of the term "relativity" with general philosophical skepticism, to the concern of the theory with the stars, which is always a field of scientific study capable of catching the imagination of the public.

Along with his account of Einstein's scientific work and his thoughts on other problems, Pais briefly gives some biographical details, including his unconventional education, his difficulties in finding a job until 1902, when the peaceful haven of a post in the Swiss patent office turned up, his two marriages, and his stays in several universities and teaching posts. In 1914 Einstein finally accepted an invitation to take up a full–time research appointment in Berlin; he stayed there until the rise of Nazism caused him to resign and move to Princeton, where he spent the rest of his life. But undoubtedly the strong point of the book is the story of Einstein's work. Pais is uniquely qualified to tell it, and after an impressively thorough study of all available sources, he has done so with admirable clarity.

Reminiscences of Cambridge
in the Thirties

My first glimpse of Cambridge was in the summer of 1928 when, as a graduate student in Heisenberg's department at Leipzig, I came to England for my summer holiday. I had enrolled in a summer course run by the Extra–Mural Department, which was on literary and cultural subjects. I was fascinated by England, so completely different from the Continental countries I knew, and particularly by Cambridge: the architecture, the student digs in which I stayed, grubby and primitive, with the ground-floor windows secured to make students keep the prescribed hours, but presided over with great warmth by the landlady, the unfamiliar bacon–and–egg breakfast, and the tea parties given by academic families for members of the summer course. All this was strange, but it appealed to me, and I felt this would be a good place to come back to.

Near the end of my holiday, when the physicists were back, I did come back to Cambridge, and I called on Dirac, whom I had first met in Leipzig. He was very charming and arranged for me to use the Cavendish Laboratory, and introduced me to R. H. Fowler. When Fowler heard that I came from Leipzig, he was delighted. "No doubt you can tell us about the new work by Bloch. We have no speaker for the next meeting of the Kapitza Club, so you must talk there." Neither my English nor my familiarity with the details of Bloch's paper were really adequate, but it never occurred to me to refuse. I do not remember much of the meeting, but at least there was no disaster.

Otherwise my main recollection is of trying to read in the unheated and bitterly cold laboratory library (it was by now late September). On one occasion I overheard one young man, who was studying a German paper, ask another whether the word "Blei" might perhaps mean lead, and I was able to confirm his conjecture, with the satisfaction of having made some very small contribution to the progress of physics at the Cavendish.

I came back to the Cavendish in very different circumstances in April 1933. I had a one-year fellowship from the Rockefeller Foundation, of which I had spent the first half with Fermi in Rome, and had arranged to be in Cambridge for the summer, in contact with Fowler and Dirac. This was the beginning of the Hitler era in Germany, and my wife and I were not intending to go back to Germany. After the end of the fellowship in October our plans were not clear, and our first child was expected the following August.

When we had found somewhere to live for a while, I set out to call on Fowler in the Cavendish. I went up to the first floor, where the offices were, and found a row of unmarked doors, and nobody about to ask. After pacing up and down for a while, I decided to try the most inconspicuous looking door to ask for instructions. I knocked and entered what turned out to be Rutherford's room, but only his secretary was there, who told me where to find Fowler.

Fowler was an old acquaintance from the previous visit, as was Dirac. I had met Blackett in Zurich, and the Blacketts took us under their wing and helped us settle in Cambridge. I met the rest of the physicists before long. My recollection of these encounters tends to fuse the summer of 1933 with the period of 1935 to 37, which I also spent in Cambridge, and even with brief visits during the intervening years, so I shall not attempt to distinguish.

Of course we soon got to know Rutherford, both at meetings in the Cavendish and at some of the parties held regularly at his house. It goes without saying that I found his command of physics and his energy as impressive as his reputation had led me to expect. What one had not been led to expect was his simple directness and his warm interest in people. He loved to tell stories, and I recollect, for example, one of his favorite stories about the new university library being opened by King George V and Queen Mary. An enthusiastic librarian showed the royal visitors the bookshelves, which were of a new design, and had certain advantages. To show interest the King asked what they were made of. "Of steel, your Majesty." "Will they last?" This earned a poke in the ribs from the Queen's umbrella: "Don't be silly, George; of course they will last!" Another of his reminiscences was of the earlier days in Cambridge, when the social life was strictly stratified, so that professors would tend to visit each other, and the heads of colleges would keep to themselves; the wife of a master was heard to remark "I wish I knew what they talk about in a professor's house."

It is said that Rutherford did not think much of theoreticians, but I never experienced any lack of consideration or courtesy. When I came back to Cambridge in 1935 to an appointment at the Mond Laboratory, of which Rutherford had taken charge when Kapitza was unable to return from Moscow, I met him in the courtyard a few days after my arrival. I was not sure he would remember who I was, but he greeted me and asked how I was settling down in Cambridge. "Have they paid you any money yet?" I said no, that would come at the end of the month. "Will you be able to last until then?" I know many a professor with less on his mind, who would not remember the practical problems of young people. This conversation with Rutherford set me an example which, I hope, I have always remembered when necessary.

I remember one occasion which, showed his capacity for introducing some badly needed informality. He was visiting Manchester to give a colloquium talk, and he was introduced by Lawrence Bragg, who spoke at length about his feeling of inadequacy in succeeding Rutherford as Professor at Manchester. People on the faculty board, he said, had been accustomed to turn for wisdom on difficult decisions to the professor of physics, and were disappointed to see it was only him sitting there. This left the audience with some feeling of embarrassment, which was happily defused when Rutherford got up and said, "Professor Bragg has expressed some doubts about his ability to fill my chair, but," and here he pointed to his own bulk and to the beginning of middle-age spread of Bragg, "I think he is well on the way to doing so."

Initially my closest contacts were with the theoreticians; with Fowler and Dirac, and the younger ones, Nevill Mott, H. R. Hulme, and H. M. Taylor.

Dirac had the reputation of being rather silent, but this is not an accurate description—he does talk, but not in idle conversation, only when he has something to say. What he does say is often unexpected, because his reaction to an idea is never trivial, and in this respect his style in conversation is like that in his physics.

He always answers direct questions, but does not feel the need to enlarge upon them. On questions about physics he is likely to answer yes or no, if he has thought out the problem, or "I don't know" otherwise. This does not make it easy to engage him in debate on unsolved questions. The same tendency can cause misunderstandings, as on one occasion when both Wigner and I (then in Manchester) were visiting Cambridge and separately called on Dirac to see the experiment on isotope separation in which he was then engaged. One could see gas from a compressor entering a little brass T-piece, with hoses coming out on both sides. Eventually light gas was to emerge on one side and heavier gas on the other. One could feel that the gas on one side was hot and that on the other was cold, so evidently something interesting was going on. Back in Manchester, Wigner complained that Dirac was secretive and refused to say what the principle of his device was. This had not been my experience, so I probed a little more, to discover that all Wigner had said was "It must have been very difficult to make the little brass piece" to which Dirac replied "No, that was quite easy." He had been asked a simple question and he had answered it, whereas Wigner, in Hungarian style, had asked for information, and had been refused. Wigner would not accept my interpretation so I bet him that, if he had asked Dirac for an explanation, it would have been given. When I later asked Dirac if he would have explained on demand, he said "I do not know." He never liked hypothetical questions.

Nineteen-thirty-three was a time of economic depression, and academic positions were in short supply. It would not have been surprising if the Cambridge physicists, and particularly the theoreticians, had resented the arrival of so many refugees, who would increase the competition for posts. Instead the new arrivals were received with great kindness, and physicists took an active part in organizing the Academic Assistance Council, which became later the Society for the Protection of Science and Learning, to give help and support to refugee scientists until they could find their feet. I remember one manifestation of this spirit: I had applied for an Assistant Lectureship at Manchester, and an informal reply from Bragg had been encouraging. Then one day a letter for me arrived from Manchester. Nevill Mott, who knew of my situation, saw the letter in the Cavendish, and cycled to my house in great excitement to bring it to me. (The letter turned out, in fact, to be a rejection, but later I was offered a grant from local Manchester funds similar to the Academic Assistance Council.)

The same year was also an exciting time in nuclear physics. The neutron was newly discovered. Cockcroft and Walton's accelerator was new and others, particularly Oliphant and Dee, were starting on the technique of using artificially accelerated particles. Chadwick, the discoverer of the neutron, was a particularly good person to learn from.

About that time Chadwick was engaged in observing the photodisintegration of the deuteron. He had not told anyone of this experiment yet but during one visit he teased Bethe and me by saying, "I bet you could not calculate the photodisintegration cross section of the deuteron!" We thought we could, and this started us off in nuclear physics. Chadwick's work on this was done in collaboration with Maurice Goldhaber, another refugee, then very young and not very experienced in the ways of the world. He caused some raised eyebrows by going around telling everybody what experiments they ought to be doing and how—what made this particularly irritating was that he was usually right.

When I came back to Cambridge in 1935 I had a research appointment in the Mond Laboratory. Since Kapitza was detained in Moscow and could not resume his Royal Society Professorship, it was decided to use the money thus saved to establish two Research Fellowships, to which J. D. Allen and I were appointed. The Mond Laboratory had been built fairly recently for Kapitza's magnetic and low–temperature work and was in the center of the courtyard containing the Cavendish and some other laboratories, with access through gates from Free School Lane and Downing Street, which were firmly locked at a certain hour in the evening.

After I had been there for a while, it was suggested that I might give some regular lectures. This was a little difficult because to lecture one had to have a Cambridge, MA. But the experts soon found a solution. It involved passing three resolutions, or "Graces" in the Senate House; one to create a new un-paid position of Assistant–in–Research at the Mond Laboratory, the second to add this post to the list of positions whose holders could be awarded an ex. officio MA, and the third to appoint me to that position and to give me the MA.

Another marginal contact with Cambridge teaching came in my second year when I was asked to supervise three final–year undergraduates for St. John's College. I had no idea what a supervisor was supposed to do, and nobody seemed to be able to give me a clear explanation, so I don't know what relation my conversations with the students had to the Cambridge tradi-tion. Still, it seems that no irreparable harm was done. One of the three was Charles Kittel, now at Berkeley, another, Rivlin, later became well–known in polymer research, and the third, Huck, went into schoolteaching to earn some money because his father died when he was about to graduate. He is now on the staff of the University of Surrey.

This teaching was done for St. John's College, and I was given limited dining rights ("up to once a week in term time at your own expense"). The Fellows I met there included Cockcroft, and this was a good opportunity to see him in a relaxed mood and willing to chat in a leisurely way. Otherwise he was very busy, looking after nuclear experiments and the building of the new High–Voltage Laboratory, as well as the Mond Laboratory in the ab-sence of a director. He also dealt with maintenance and restoration of college buildings. Driving around the country he would often notice a dilapidated wall of old bricks; he would then stop and buy the bricks, to replace modern machine–made bricks which had been used for repairs in the college. With all the activities he had to make very efficient use of his time, and could not spare any for idle conversation, so the encounters in the combination room after dinner were exceptional.

One episode which made a deep impression on me took place in the Mond Laboratory, where I worked in a room which also contained a drawing board. Cockcroft was busy with a drawing when the secretary came in to say the architect was on the phone and wanted to know whether the light switches in the New High–Tension Laboratory were to be black or brown. (In those days switches were either black or brown.) Anybody else would have taken some time for reflection and would have said, "It does not really matter, but, let me see, perhaps brown ... ." Cockcroft did not lift his eyes from the drawing, and without a moment's hesitation said "Brown" (or perhaps it was black, I don't remember).

One institution I remember from the early 1930s was the Physics Club. This was a colloquium held at irregular intervals (once or twice a term I think) started on the initiative of exiles from the Cavendish who felt isolated in London or elsewhere, and wanted to keep in touch. The meetings were usually held in the Cavendish or in London. These meetings were greeted with great enthusiasm by the German refugees, who were eager to participate. However, membership of the club was limited; many of the "old hands" like G. P. Thomson and Darwin did not want the meetings too large, and said that in a semi–public meeting they would be reluctant to ask stupid questions. The outsiders—including myself—could not understand this attitude, and thought it was unreasonable. I still thought so after I had been admitted to membership of the club. However, at first the rule was respected. But it was not, of course, possible to enforce it: one could not have a steward at the door asking for people's credentials, and gradually many non–members invaded the meetings. Perhaps this caused the decline of the club, but I do not remember for how long it continued—perhaps I just lost touch gradually after leaving Cambridge.

Finally, I would like to stress that this collection of anecdotes, most of them light–hearted, should not obscure the feeling of excitement with which I recall the period I was privileged to spend in one of the world's centers of physics, combining its ancient tradition of institutions with revolutionary new methods and ideas in physics.

Struggling with
Quantum Mechanics

John Gribbin has set himself the ambitious task of explaining to the non–scientific reader the nature and the interpretation of quantum mechanics. The task is made even more formidable by his covering in outline the whole of physics, from the basic laws to the transistor, the laser, and the structure of DNA.

Gribbin's approach is largely historical; he not only presents the ideas more or less in the sequence in which they were developed, but also characterizes the physicists to whom these ideas are due. This is informative and pleasant, but increases his difficulty in getting it all into 300 pages.

Gribbin has taken his title from a famous paradox concerning the interpretation of quantum mechanics. It indicates his intention to lead up to the problems of interpretation. This makes the task particularly difficult, because in learning how to interpret quantum mechanics physicists had to abandon some of their prejudices. Many ideas about space and time and motion seem intuitively obvious, but physicists had to learn that for events on the atomic scale we cannot trust our intuition, which derives from experience on the scale of everyday life.

Most physicists managed this adjustment, with some notable exceptions, such as Einstein and Schrödinger. Lately there has been a tendency to question the "Copenhagen interpretation" developed in the 1930s and to try to find a new interpretation. None has been found so far, and recent experiments have verified that some alternative ideas that had been suggested could not work, even if one were prepared to give up some of the predictions of quantum mechanics.

This is a review of *In Search of Schrödinger's Cat,* by John Gribbin. Colorado Springs, Co.: Wildwood, 1984. For the complete reference, please see the Acknowledgments section.

Gribbin makes it clear to readers that they have to accept the seemingly absurd statements of quantum mechanics, and that there is no way of avoiding them, except for the idea of "many worlds," to which I shall return.

He begins his task with a lively and readable sketch of the prequantum situation in physics. A few errors do not spoil the story. (The spectral lines of a compound are not the same as those of the constituent elements; and Einstein's conclusion that the photon has mass zero did not need a discussion of its statistics.)

Unfortunately, the exposition of the later parts is marred by many misunderstandings and errors. The point of the famous "clock in the box" paradox, raised by Einstein to disprove the uncertainty principle and resolved by Niels Bohr, is missed. The important paradox of Einstein, Podolsky and Rosen is garbled completely. "Phase space" is confused with configuration space. And some of the analogies used to explain difficult points do not work very well. The comparison of a matrix with a chess board is fine, but when an atomic transition is compared with a move on the board, the analogy becomes confused: the move is described by the two coordinates of the initial position and two of the final position, so the "matrix" now has four subscripts.

Historical details are sometimes inaccurate. Rutherford did not retire, but died in office. Yukawa did not suggest the name "meson" for the particle he postulated. Positrons were not mistaken for electrons "moving into an atomic nucleus." Some other statements may be matters of opinion, but I do not agree that the discovery of wave mechanics was unfortunate, nor that many physicists who apply wave mechanics today are under the illusion that the wave function represents real waves, nor that they have been taught to derive quantum mechanics from the uncertainty principle, which would be the opposite extreme.

Gribbin favors the "many worlds" interpretation of quantum mechanics. This is a dissident viewpoint, which avoids the usual step of the wave function "collapsing" on seeing the result of an observation. Instead there is the step of selecting, from a number of alternative worlds, the one we happen to be in. Since we could not, in principle, have any contact with the other "worlds", this seems to me only a difference in words, or in metaphysical beliefs.

I am sad, because I would have wished to see this enterprise succeed.

Kapitza Detained

Peter Kapitza was a young Russian physicist, who came in 1921 to the Cavendish Laboratory in Cambridge to work under Rutherford, though not on nuclear physics. His work was concerned with the effect of intense magnetic fields on matter at low temperatures. This work needed equipment which, for the time, was unusually heavy and expensive; it was the natural area for a man like Kapitza, who, besides being a first–rate physicist, had an engineer's training and outlook. Kapitza's approach differed from Rutherford's "string and sealing-wax" methods, but he gained Rutherford's support, which led eventually to a special laboratory, the Royal Society Mond Laboratory, being built for him.

Kapitza seemed to be a fixture in Cambridge, as a Royal Society professor, a fellow of Trinity College and later a Fellow of the Royal Society. He created the "Kapitza Club," for many years the physicists' main discussion forum in Cambridge.

All this time he remained a Soviet citizen, and intended ultimately to go home when his maturity as an individual and conditions in his country made this seem opportune. He returned most summers to Russia, and he seemed to have the unusual privilege of coming and going without formality, a privilege granted to few people other than diplomats and governments officials.

It was therefore a great shock to Kapitza and to his Cambridge colleagues when at the end of his 1934 summer visit to Moscow he was told that he could not leave because the Soviet Union needed him. The events that ensued are of absorbing interest to anyone concerned with the freedom of science, the fate of the individual scientist, or the conflict between international relations and national needs.

This is a review of *Kapitza, Rutherford and the Kremlin,* by Lawrence Badash. New Haven, Conn.: Yale University Press, 1985. For the complete reference, please see the Acknowledgments section.

Rutherford's first reaction was to try personal appeals to the Soviet authorities to change their mind, and only when this did not bring any results to make public protests. Finally, when all this failed, his efforts concentrated on getting reasonable working conditions for Kapitza in Moscow, including the transfer of his specialized equipment.

Kapitza's wife Anna had returned to Cambridge to her children, and there was an intense exchange of letters with her husband. Anna passed many of his letters to Rutherford, presumably in her translation. They make a fascinating story revealing at first–hand Kapitza's reaction to events.

He was in general a supporter of the Soviet regime and approved of its efforts to raise standards of education, science and industry. But he was indignant about the peremptory way in which he had been treated and about the bureaucracy. Kapitza did not see how he could continue his work in Moscow, not only because of the shortage of equipment and qualified help, but mainly because the authorities did not seem to trust him. He complained that, while a reorganization of the Academy of Science was being discussed, on which he was qualified to make constructive suggestions, his advice was not asked for, and when offered was ignored. He insisted for a time that he would do no more physics, but would learn physiology, for which no large equipment was required.

Eventually they decided to build a laboratory for him in Russia and his Cambridge equipment was transferred or copied; two of his Cambridge staff were sent to Moscow to help get the new laboratory started. He then settled down, and developed the laboratory into a center of excellence, and he did work there for which he was later awarded the Nobel Prize.

The letters are accompanied by full notes explaining the episodes and people mentioned in them. Other sections describe the situation as seen from Cambridge, and speculate about the motives of the Soviet authorities for their action, which was possibly based on an overestimate of what Kapitza's engineering genius could do for the country's industry. On these latter questions there must, of course, remain more uncertainty than about Kapitza's or Rutherford's views, even in the hands of a historian of the standard of Prof Badash. The book makes fascinating reading.

Does Physics Ever Come to an End?

The answer to the question I have been set, "Does physics ever come to an end?" depends, of course, on what we mean by physics, and on this the physics community will not be unanimous. Some will see physics as the search for the basic laws of nature, and regard it as the end of physics if this search ceases. This is a view often held by specialists in particle physics, who sometimes show a rather arrogant attitude to other branches of our subject. Others feel that a much broader range of problems are good physics and worth pursuing.

So we have really two questions, and I shall start by discussing the one relating to the search for the fundamental laws. Talk of an impending end is not new. For example, in the early 1930s, when atomic theory was complete, there remained only some problems with relativistic field theory, and there were two dimensionless parameters: the fine structure constant, and the electron–proton mass ratio. These two numbers seemed to be related. There was also the nucleus, but so little was known about the nucleus that there did not seem many problems. Besides, the nucleus evidently contained electrons, and we could not understand that, but it was evidently related to the problems with relativistic electron dynamics. So all remaining problems seemed to be related, and some expected that they would be resolved in a single step. Of course we know today how wrong was this view.

For a more general analysis I cannot do better than quote the talk on the future of physics which Richard Feynman gave at the M.I.T. Centenary. He reminded us that the development of physics has proceeded in steps, each of which solved the problems at one level by opening up a new and deeper level, with new problems, below. The properties of chemical compounds were illuminated by introducing molecules and atoms; the problem of the properties of the atom were explained in terms of nuclei and electrons; the nucleus was recognized as consisting of nucleons; the forces between these were seen by Yukawa to be based on new fields, in particular that of pions, and that and the problems of other new particles led us to the picture of quarks and gluons. Trying to understand the weak interactions led us to another new level of problems. In the light of this, Feynman said, there are three alternatives for future development.

The first possibility is that we might one day find the ultimate basic laws. That, he said, would mean the end of physics. It would be sad, he added, because it would leave us defenseless against the philosophers, who always try to prove that things must be the way they are—so far we have always been able to disconcert them by finding new and unexpected things.

His second possibility was that, before we reach the basic laws, we might run out of facilities. There may be a limit to the energy that can be reached in practice by accelerators, and to the intensities of their beams. Within these limits we may not have enough clues to piece together the fundamental story. Then physics would also end.

Finally, he said, it was possible that such limitations could be overcome by new and ingenious methods of accelerating particles, or of making discoveries without raising the particle energy. Yet we might never reach the ultimate basis. In that case we would go on discovering layer after layer of fundamental physics, and then we would get bored. We could not sustain our excitement in such an endless succession of levels. He therefore concluded that, whatever happens, we live in the golden age of physics. I think it is hard to quarrel with this analysis. Feynman's second alternative, the limits on energy, has acquired more plausibility since his talk. Considering the cost of the next generation of accelerators, and the effort needed to get them funded, one feels there may not be many generations ahead.

As regards to the search for the fundamental laws we must therefore answer the question in the affirmative. Physics will come to an end, unless you allow the rather gloomy kind of survival of Feynman's last alternative.

Feynman did not specify a time scale; in fact his talk was supposed to deal with physics for the next thousand years! There may well be some uncertainty at what stage we shall recognize which of Feynman's possibilities is realized. For example, the possibility of reaching the ultimate laws, so that fundamental physics would be complete, may well be controversial. Remember that, with the development of physics, also the range of questions which a physicist should ask, expanded. At the end of the last century, for example, Ohm's law was physics, but the value of the conductivity of copper was not. This was something you looked up in a table, or on which a chemist might have views.

Some will not be satisfied that physics is complete until many deep questions have been answered. Not only what are the ultimate constituents, what kind of particles or fields, or whatever new concepts we may meet, and their interactions, but also why just these particles and no others, why just those interactions, etc. Taking this questioning to its extreme would mean deriving all physics from a priori principles, and most of us would doubt that this is possible. So there may well be controversy whether physics is complete and whether the end has been reached.

Or consider the second possibility, that we run out of clues. Here of course there will always be theoreticians who refuse to give up, and who will try to construct theories from the available clues, reinforced by their ingenuity and imagination, much like the present fashions for strings and supersymmetry, though I leave you to decide whether this is physics or science fiction. Such approaches may succeed, so in time what appeared to be the second situation might become the first. This, of course, may also involve controversy. Since we assume there are no new experimental clues to be obtained, such theoretical laws may become a matter of faith. Dirac believed that a theory could become convincing because of its mathematical beauty, but estimates of beauty may not be unanimous.

In the third possibility, the infinite succession of levels, there would exist for a long time some physicists willing to plod on, and it is not clear when finally everybody would give up, or the community would discontinue the support of physics, but it would probably be within Feynman's thousand years. Eventually we are bound to find ourselves in one of the three situations outlined, and then the search for the basic laws of physics will come to an end.

Let us now turn to the other way of interpreting the question: including in physics all the beautiful work done in areas in which the laws of physics are well known. From a naive point of view, once the laws are known, all one has to do is to apply them to various situations, and activity that might be described as Applied Physics if it is not Engineering. But this is to misjudge the position completely.

There are indeed cases in which new and exciting devices are invented by making predictions from the laws of physics. Such an example is the laser. Its inventors saw the theoretical possibility of such a device, but this was no trivial matter; many people knew the laws of spectroscopy and had been applying them for many years. Yet nobody had seen the point. Of course seeing the possibility of the laser in principle was not the whole story. One had to know the spectra of various substances, to find combinations of levels and transitions suitable for the purpose, and then it needed experimental skill to implement it. Since then the development has continued, adding to the variety of lasers (including tunable ones) and also discovering many new ways in which lasers can be used. This is where the engineer comes in.

But this progress by theoretical prediction is by no means general. An almost completely opposite circumstance was the cause of superconductivity. It is not only that it was not predicted theoretically. After its surprise discovery we did not understand what was going on, and even after quantum mechanics had led to an understanding of most solid–state phenomena it remained a mystery, even though everybody was convinced it must be a consequence of the known laws. It was not until 1957 that the explanation was found by Bardeen, Cooper and Schrieffer. Their work built on many ideas that had been proposed to find an explanation, particularly by F. London and by Frohlich. The B.C.S. theory not only helped us to understand and therefore better to apply the phenomenon, but it also enlarged the scope of theory, since the techniques that were necessary for the explanation had important applications to other problems, including those of field theory.

Superconductivity has many important applications, and has in many instances been taken over by the engineers, but that is not always the case with exciting discoveries. The superfluiduity of liquid helium is another effect that came out of the blue, without being foreseen by theory, though Landau soon proposed a description which no doubt contains the essence of the phenomenon. The superfluidity of ^{3}He also involved a surprise. Before its discovery many suggested that ^{3}He would turn superfluid at low enough temperature; in fact many expected this would happen well before one reached millidegree temperatures. But the properties were unexpected. The analogy with superconductivity had suggested that the helium atoms are paired, and so they are, but not as expected in a singles-S state. Different pairing states make the phenomenology of superfluid ^{3}He very rich.

Superfluidity of either He isotope has not so far found practical applications and is not likely to do so, but it is exciting physics. The lack of applications ensures that the physicist can keep the field, without danger of being displaced by the engineer. This is not so with the current excitement over the high–temperature superconductivity, which also came as a complete surprise and has as yet no generally accepted explanation. Here the potential practical applications look so promising that there is a race between physicists and engineers. A colleague remarked that this is an ideal time to work on other problems in condensed–matter physics, because you have no competition— everybody is so busy working in high–temperature superconductivity. Perhaps this is a little exaggerated.

Another beautiful subject opened up in the post–war years is the study of phase transitions. For a long time it was taken for granted that in a second–order phase transition, such as the critical point of a gas–liquid system, or the

Curie point of a ferromagnet, the specific heat was discontinuous but finite. In fact, Onsager had given a rigorous solution of the two–dimensional Ising model, in which the specific heat at the Curie point is singular, but this was not noticed. Only when precise measurements showed that the specific heat at the critical point gets bigger as the accuracy of the measurement increases, was it realized that the correct behavior is singular, and from this the study of such transitions by means of the renormalization group has developed.

As one of the latest surprises I might mention the quantum Hall effect. A two–dimensional electron system (e.g., confined to the surface of a semiconductor) shows qualitatively the same Hall effect as a metal or a semiconductor, but there are periodic variations, which follow simple numerical relationships with incredible accuracy. This, too, has called for novel and sophisticated theoretical approaches.

Astrophysics in a sense straddles the boundary between my two categories: much of it is covered by the well–established laws of physics, but some of its aspects, such as the solar–neutrino problem and that of the "dark matter" in the universe, involve the frontier of known laws.

I could not possibly list all the recent exciting developments, I mention these few examples to illustrate, firstly, that there are nontrivial studies for which the skill, the experience and the imagination of the physicist are required. I would insist that they are certainly physics. Secondly, as the examples show, they arise unexpectedly, not from any systematic attempt to deduce consequences of the laws of physics. I see no reason why the flow of such discoveries should ever stop. I do not have the confidence of Feynman to make predictions for a thousand years, but if "physics" includes the type of phenomena I have described, there is certainly no end in sight.

So, depending on how you interpret the question, the answer is either "yes" or "no."

Microwave Cooking for "Foxes"

T he German title translates as "Microwave Cooking Course for Foxes." The readers whom the authors address as "foxes" are people who approach the art of cooking with cunning—in other words, intellectuals. The authors are indeed intellectuals: both are physicists. The senior author, a physicist of world renown, has books on more conventional cooking for foxes to his credit. His coauthor adds an element of feminine skill and of Viennese tradition.

The appeal to the "foxes" consists in an emphasis on understanding the principles of microwave cookery, the ways in which it differs from conventional methods, and the tricks needed to achieve the desired results. This includes, for example, a discussion of how the temperatures reached in the interior of the objects are influenced by their thickness, by being covered with cling film, or by varying the microwave power during the cooking process. The working of a browning dish is explained.

The many recipes given serve largely as illustrations of the principles, and often include, instead of a definite cooking time, instructions such as "continue until the liquid starts boiling." This facilitates adaptation to varying circumstances, but it also obliges the beginner to stand watching over the microwave oven instead of relying on the automatic timer. Where actual cooking times are given, they are, of course, a matter of taste, and in many cases I would prefer rather shorter times, but anyone who has understood the "course" will be able to make adjustments.

The authors are evidently firm supporters of the microwave method, and carefully explain its many advantages, while admitting the occasional disadvantages. They also mention the tasks which cannot be done in the micro-

This is a review of *Mikro-Wellen Kochbuch für Füchse,* by H. Maier-Leibnitz and T. Cless-Berrert. Munich: Piper, 1986. For the complete reference, please see the Acknowledgments section.

wave oven, beginning with boiling eggs (which would tend to explode), but they report success in many an ambitious tour–de–force, such as baking an *Apfelstrudel* by microwave.

There are examples of complete meals, including detailed time sequences. Many are designed for the single host, who wants to minimize the time spent in the kitchen during the presence of the guests. Here I find it surprising that they ignore the option of selecting dishes, such as casseroles, which do not lose by being kept warm, so that all the cooking can be done before the guests arrive.

The book is intended for users of microwave ovens who want to acquire a command over the techniques, and for those who want to make up their minds whether to buy such an oven. The seven "lessons" consist of abstract explanations, but discuss specific recipes as illustrations of the rules. However when the preface says, "You may—we hope—begin with any recipe without referring back," this must be understood to apply only after you have read and more or less understood the whole book, because there are many explicit or implicit references to what went before.

For any "fox" who wants to perfect his or her microwave technique (and who has a good working knowledge of German), this is a most valuable book.

First Encounter with
English Food

My first visit to England was a summer holiday in 1928, and I came again in 1933—as it turned out, to stay. The first experience of food was mainly in boarding–houses and rather average restaurants. The magnificent breakfast in these places fully justified its fame on the Continent, and the afternoon tea (known on the Continent as "five–o'clock–tea," although I have never met a place where it was served that late), was equally attractive. But the rest of the meals seemed strange. It took some time before I understood the system.

In a thoroughly democratic country it would not do for the cook to impose his or her taste on the guests, so everything was boiled until only a neutral matrix remained. Then at the table the guests could add salt, pepper, mustard, ketchup, Worcester sauce, etc., to generate any desired flavor. My theory seemed confirmed by the observation that many people would add salt, pepper and other condiments to their food before tasting it.

This custom led an English couple into trouble when at dinner in a famous restaurant in Paris they asked the waiter for salt and pepper. The patron appeared and threw them out. They had insulted his food by suggesting it was not perfectly seasoned.

I also had disappointing experiences caused by similarities with familiar foods: when first offered brown bread I expected rye, and was taken aback by Hovis. When I first met junket, I took it to be yogurt until I tasted it (or rather failed to detect any taste). Minted peas were unfamiliar and made my wife suspect that the cook had by mistake used tooth powder instead of salt.

In the 1930s coffee was, in the places we could afford, an overrated pastime, but this was not unfamiliar, because we were used to Germany. There in comparable places the coffee was similar, particularly in Saxony where the local coffee is often called *Blümchenkaffee* (flower coffee) because you could see through it the little flower painted at the bottom of the cup.

It took us some to discover the best of English food, then served mainly in private homes. Now, some fifty years later, excellent English food is much easier to come by, though the old style of cooking, conforming to my theory, can still be found occasionally.

One custom which seems to me strange is still met quite often. That is treating potatoes as a vegetable, to be served with a main dish which is itself based on potato (cottage pie) or on pasta or rice. Sometimes two different forms of potato (boiled and roast). Such a surfeit of carbohydrates seems peculiarly English.

Other puzzles did not relate to the nature of the food, but to terminology. Why did butcher shop windows distinguish between English and Canterbury lamb? Was not the cathedral city in England? (For that matter was "Botany Wool" some kind of cotton?)

One commodity which is even today not too widespread is fruit ice cream. Too often it is replaced by a concoction containing neither real fruit nor real cream. I might therefore end with an easy recipe, learned from our daughter, Mrs Joanna Hookway. Liquidize 8 ounces of fruit (e.g. strawberries or raspberries, if necessary defrosted; dried apricots, soaked, are excellent). Whip half a pint of cream, not too stiff, and add to the fruit, with the juice of half a lemon and sugar to taste. Put in the freezer and forget it. (The usual procedure of stirring the mixture during freezing to avoid crystallization is not necessary here, because crystallization is prevented by the air bubbles in the whipped cream, and the lemon juice stabilizes the whipped cream.)

Acknowledgments

Most of the essays in this volume were originally printed or presented elsewhere. The publisher and the author wish to express their grateful acknowledgment to the following publications and publishers for permission to reprint previously published articles. In each case, the material was re–worked for this publication. All photographs are reprinted here by permission from the Niels Bohr Library (with the exception of photograph on page 302) and were not part of the original publications.

I. A PHYSICIST'S PORTRAIT GALLERY

WOLFGANG ERNST PAULI, 1900–1958 was published in *Biographical Memoirs of Fellows of the Royal Society*, **5** (1960), pp. 175–187, and is reprinted here with kind permission from The Royal Society, 6 Carlton House Terrace, London SW1Y 5AG, United Kingdom.

PROFESSOR H. W. B. SKINNER, F.R.S., 1900–1960 was published in *Nuclear Physics*, **23** (1961), pp 1–4.

AN APPRECIATION OF NIELS BOHR was published in *Proc. Phys. Soc.*, **81** (1963), pp. 793–799.

TRUTH AND CLARITY originally appeared in *The Lesson of Quantum Theory* (eds. J. De Boer, E. Dal, and O. Ulfbeck, Elsevier 1986), pp. 379–380.

RUTHERFORD AND BOHR is the text of the Rutherford Memorial Lecture, delivered at the National Physical Laboratory of India, Delhi, on 10 November 1987. It was published in *Notes and Records of the Royal Society London*, **42** (1988), pp. 229–241, and is reprinted here with kind permission from The Royal Society, 6 Carlton House Terrace, London SW1Y 5AG, United Kingdom.

J. ROBERT OPPENHEIMER, 1904–1967 was published in the *Dictionary of Scientific Biography* (Sr. Ed. C. C. Gillespie, Scribner, New York, 1974), vol. **10**, pp. 213–218, and is reprinted here with kind permission from the American Council of Learned Societies, 228 East 45th Street, New York, NY 10017–3398, U.S.A.

THE GROWING PAINS OF ROBERT OPPENHEIMER is a review of *Robert Oppenheimer: Letters and Recollections*, edited by Alice Kimbal Smith and Charles Weiner. 250 pages. Cambridge, Mass.: Harvard University Press, 1980. It is reprinted here with permission from *Nature*, **286**, pp. 827–828, © 1980, Macmillan Magazines Limited.

HEISENBERG'S RECOLLECTIONS is a review of *Physics and Beyond: Encounters and Conversations*, by Werner Heisenberg, translated by Arnold J. Pomerans. 266 pages. New York: Harper & Row, 1971. The review was originally published under the title "Atomic Germans" and is reprinted with permission from *New York Review of Books*, 1 July 1971, pp. 23–24, © 1971, Nyrev, Inc.

WERNER HEISENBERG, 1901–1976, co-authored with Sir Nevill Mott, F.R.S., was published in *Biographical Memoirs of the Fellows of the Royal Society*, **23** (1977), pp. 213–251, and is reprinted here with kind permission from The Royal Society, 6 Carlton House Terrace, London SW1Y 5AG, United Kingdom.

A HEISENBERG BIOGRAPHY, is a review of *Uncertainty: The Life and Science of Werner Heisenberg*, by David C. Cassidy. 669 pages. New York: W. H. Freeman, 1991. The review was originally published under the title "The Uncertain Scientist" and is reprinted with permission from *New York Review of Books*, 23 April 1992, pp. 43–45, © 1992, Nyrev, Inc.

THE BOMB THAT NEVER WAS is a review of *Heisenberg's War: The Secret History of the German Bomb,* by Thomas Powers. 608 pages. New York: Knopf, 1993. The review is reprinted with permission from *New York Review of Books*, 22 April 1993, pp. 6–9, © 1993, Nyrev, Inc.

A PHYSICIST WHO ENJOYS IT is a review of *What Little I Remember*, by Otto R. Frisch. 219 pages. Cambridge: Cambridge University Press, 1979. The review is reprinted here with permission from *Nature*, **280**, pp. 257–258, © 1979, Macmillan Magazines Limited.

OTTO ROBERT FRISCH, 1904–1979 was published in *Biographical Memoirs of the Fellows of the Royal Society*, **27** (1981), pp. 283–306, and is reprinted here with kind permission from The Royal Society, 6 Carlton House Terrace, London SW1Y 5AG, United Kingdom.

PAUL ADRIEN MAURICE DIRAC, 1902–1984 originally appeared in *Physics Today*, January 1985, p. 111.

DIRAC was originally published in *Tributes to Paul Dirac* (ed. J. G. Taylor, IOP Publishing, 1986), pp. 35–37, and is reprinted here with the kind permission of IOP Publishing Ltd., Techno House, Redcliffe Way, Bristol BS1 6NX, United Kingdom.

DIRAC'S WAY was published in *Recollections of Dirac* (eds. B. Kursunoglu and E. P. Wigner). Cambridge, Mass.: Cambridge University Press, 1987, pp. 43–45.

PHYSICS AND HOMI BHABHA is a review of *Homi Jehangir Bhabha: Collected Scientific Papers*, edited by B. V. Sreekantan, Virendra Singh, and B. M. Udgaonkar. 1,023 pages. Bombay, India: Tata Institute, 1986. The review is reprinted with permission from *Nature*, **323**, pp. 212, © 1986, Macmillan Magazines Limited.

TWO MATHEMATICIANS is a review of *John von Neumann and Norbert Wiener: From Mathematics to the Technologies of Life and Death*, by Steve J. Heims. 547 pages. Cambridge, Mass.: MIT Press, 1980. The review was originally published under the title "Odd Couple" and is reprinted with permission from *New York Review of Books*, 18 February 1982, pp. 16–19, © 1982, Nyrev, Inc.

PHYSICIST EXTRAORDINARY is a review of *Oliphant: The Life and Times of Sir Mark Oliphant*, by Stewart Cockburn and David Ellyard. 369 pages. Axiom Books: Australia, 1981. The review is reprinted with permission from *Nature*, **296**, pp. 685, © 1982, Macmillan Magazines Limited.

LANDAU IN THE 1930S is reprinted from *Frontiers of Physics*, The Proceedings of the Landau Memorial Conference, 6–10 June 1988, (eds. E. Gotsman, Y. Ne'eman and A. Voronel, Pergamon, 1990), pp. 23–26.

WILLIAM GEORGE PENNEY, 1909–1991 was originally published in *Physics Today* (October 1991), pp. 138–142.

CONSERVATIVE REVOLUTIONARY is a review of *The Dilemmas of an Upright Man: Max Planck as Spokesman for German Science*, by J. L. Heilbron. 238 pages. Berkeley: University of California Press, 1986. The review is reprinted with permission from *New York Review of Books*, 20 November 1986, pp. 56–57, © 1986, Nyrev, Inc.

RECOLLECTIONS OF JAMES CHADWICK is based on a talk at the Chadwick Centenary celebration in Liverpool, 1991. It was originally published in *Notes and Records of the Royal Society London*, **48**(1) (1994), pp. 135–141, and is reprinted here with kind permission from The Royal Society, 6 Carlton House Terrace, London SW1Y 5AG, United Kingdom.

BELL'S EARLY WORK was published in *Europhysics News*, **22** (1991), pp. 69–70.

II. ATOMIC ENERGY AND ARMS CONTROL

THE FRISCH-PEIERLS MEMORANDUM was written by Rudolf E. Peierls and Otto R. Frisch in 1940 and submitted to the UK authorities.

DEFENCE AGAINST THE ATOM BOMB is reprinted with permission from *Nature*, **158**, p. 379, © 1946, Macmillan Magazines Limited.

ATOMIC ENERGY: THREAT AND PROMISE is reprinted from *Endeavour*, **6** (April 1947), © 1947, pp. 51–57, with kind permission from Elsevier Science Ltd., The Boulevard, Langford Lane, Kidlington OX5 1GB, United Kingdom.

BRITAIN IN THE ATOMIC AGE was published in the *Bulletin of the Atomic Scientists*, **26** (June 1970), pp. 40–46, © 1970 by the Educational Foundation for Nuclear Science, 6042 South Kimbark, Chicago, IL 60637, U.S.A.

LIMITED NUCLEAR WAR? was published in the *Bulletin of the Atomic Scientists*, **38** (May 1982), p. 2, © 1982 by the Educational Foundation for Nuclear Science, 6042 South Kimbark, Chicago, IL 60637, U.S.A.

AGONISING MISAPPRAISAL originally appeared in *Spectator*, 28 April 1961, pp. 592–593.

REFLECTIONS OF A BRITISH PARTICIPANT was published in the *Bulletin of the Atomic Scientists*, **41** (August 1985), pp. 27–29, © 1985 by the Educational Foundation for Nuclear Science, 6042 South Kimbark, Chicago, IL 60637, U.S.A.

ENERGY FROM HEAVEN AND EARTH is a review of *Energy from Heaven and Earth*, by Edward Teller. 346 pages. New York: W.H. Freeman & Son, 1979. The review was published in the *Bulletin of the Atomic Scientists*, **36** (May 1980), pp. 53–54, © 1980 by the Educational Foundation for Nuclear Science, 6042 South Kimbark, Chicago, IL 60637, U.S.A.

MEMORIES OF THE SECRET CITY is a review of *Reminiscences of Los Alamos, 1943–1945*, edited by Lawrence Badash, Joseph O. Hirschfelder, and Herbert P. Broida. 188 pages. Dordrecht, Netherlands: Reidel/Kluwer Academic, 1980. The review is reprinted with permission from *Nature*, **290**, p. 529, © 1980, Macmillan Magazines Limited.

COUNTING WEAPONS is a review of the books *Britain and Nuclear Weapons*, by Lawrence Freedman. 160 pages; *Countdown: Britain's Strategic*

Forces, by Stewart Menaul, 188 pages; *The War Machine*, by James Avery Joyce, 210 pages; and *Protest and Survive*, edited by E.P. Thompson and Dan Smith, 262 pages. The review originally appeared in *London Review of Books*, 5 March 1981, pp. 16–18.

FORTY YEARS INTO THE ATOMIC AGE is a review of *The Nuclear Chain Reaction Forty Years Later*, edited by Robert G. Sachs. 283 pages. Chicago: The University of Chicago, 1984. The review originally appeared in *Minerva*, **22**(2) (Summer 1984), pp. 285–291.

THE CASE FOR THE DEFENCE is a review of *Better a Shield Than a Sword: Perspectives on Defense and Technology*, by Edward Teller. 257 pages. New York: Free Press, 1987. The review is reprinted with permission from *Nature*, **328**, p. 583, © 1987, Macmillan Magazines Limited.

THE MAKING OF THE ATOMIC BOMB is a review of *The Making of the Atomic Bomb*, by Richard Rhodes. 886 pages. New York: Simon & Schuster, 1987. The review was originally published under the title "Making It" and is reprinted with permission from *New York Review of Books*, 5 November 1987, pp. 47–48, © 1987, Nyrev, Inc.

NUCLEAR WEAPONS: HOW DID WE GET THERE, AND WHERE ARE WE GOING? originally appeared in *Oxford Magazine*, No. 51, Fourth Week, Michaelmas Term, 1989, pp. 10–11.

ATOMIC HISTORY is a review of the books *British Scientists and the Manhattan Project: The Los Alamos Years*, by Ferenc Morton Szasz. 167 pages. New York: Macmillan/St. Martin's Press, 1992; *The Los Alamos Primer: The First Lectures on How to Build an Atom Bomb*, by Robert Serber. 98 pages. Berkeley: University of California Press, 1992. The review was originally published under the title "The Making of the Atomic Bomb" and is reprinted with permission from *Nature*, **357**, pp. 290–291, © 1992, Macmillan Magazines Limited.

"IN THE MATTER OF J. ROBERT OPPENHEIMER" was published in *Atomic Scientists Journal*, **4** (1955), pp. 145–154, and is reprinted here with kind permission from Taylor & Francis, Rankine Road, Basingstoke, Hants RG24 8PR, United Kingdom.

"SECURITY" TROUBLES was originally published as a chapter of *Bird of Passage: Recollections of a Physicist* by Rudolf E. Peierls, © 1985 by Princeton University Press. Reprinted by permission of Princeton University Press.

III. PHYSICS POLITICS AND PLEASURES

THE CONCEPT OF THE POSITRON is a review of *The Concept of the Positron. A Philosophical Analysis*, by Norwood Russell Hanson. 236 pages. Cambridge, Mass.: Cambridge University Press, 1963. The review was originally published in *History of Science*, **4** (1965), pp. 124–129, and is reprinted here with permission from The Welcome Institute for the History of Medicine, 183 Euston Road, London, NW1 2BP, United Kingdom.

THE SCIENTIST IN PUBLIC AFFAIRS: BETWEEN THE IVORY TOWER AND THE ARENA was published in the *Bulletin of the Atomic Scientists*, (November 1969), pp. 28–30, © 1969 by the Educational Foundation for Nuclear Science, 6042 South Kimbark, Chicago, IL 60637, U.S.A.

BORN—EINSTEIN CORRESPONDENCE is a review of *The Born—Einstein Letters*, commentaries by Max Born. 246 pages. New York: Macmillan. The review was originally published under the title "Strong Interactions" in the *New Scientist*, 6 May 1971, p. 339, and is reprinted with their kind permission.

IS THERE A CRISIS IN SCIENCE? 1971 Joseph Wunsch Lecture at the Technion (Israel Institute of Technology, Haifa, Israel, 1972) p. 16.

THE JEW IN TWENTIETH-CENTURY PHYSICS was originally published in *Next Year in Jerusalem*, by Douglas Villiers, editor. Copyright © 1975, 1976 by Douglas Villiers Publishing Limited. Used by permission of Viking Penguin, a division of Penguin Books USA Inc.

FROM WINCHESTER TO ORION is a review of *Disturbing the Universe*, by Freeman Dyson. 283 pages. New York: Harper and Row, 1979. The review originally appeared under the title "Nuclear Family" in *London Review of Books*, 19 June 1980, pp. 9–10.

THE PHYSICISTS is a review of *The Physicists*, by C. P. Snow. 192 pages. New York: Little, Brown, 1981. The review was originally published under the title "One Culture" and is reprinted with permission from *New York Review of Books*, 17 December 1981, pp. 44–45, © 1981, Nyrev, Inc.

FACT AND FANCY IN PHYSICS is a review of the books *The Rise of Robert Millikan: A Portrait of a Life in American Science*, by Robert H. Kargon. 205 pages. Ithaca, New York: Cornell University Press, 1982; *Night Thoughts of a Classical Physicist*, by Russell McCormmack. 217 pages. Cambridge, Mass.: Harvard University Press, 1982. The review was originally published under the title "Quantum Squabbles" and is reprinted with permission from *New York Review of Books*, 29 April 1982, pp. 34–35, © 1982, Nyrev, Inc.

WHAT EINSTEIN DID is a review of *'Subtle is the Lord...': The Science and Life of Albert Einstein*, by Abraham Pais. 552 pages. New York: Oxford University Press, 1982. The review is reprinted with permission from *New York Review of Books*, 28 April 1985, pp. 21–23, © 1985, Nyrev, Inc.

REMINISCENCES OF CAMBRIDGE IN THE THIRTIES originally appeared in *Cambridge Physics in the Thirties* (ed. John Hendry, Adam Hilger, Bristol, 1984), pp. 195–200.

STRUGGLING WITH QUANTUM MECHANICS is a review of *In Search of Schrödinger's Cat*, by John Gribbin. 302 pages. Colorado Springs, Co.: Wildwood, 1984. The review was originally published under the title "Prowling Around Quantum Mechanics" in the *New Scientist*, 13 December 1984, p. 45.

KAPITZA DETAINED is a review of *Kapitza, Rutherford and the Kremlin*, by Lawrence Badash. 127 pages. New Haven, Conn.: Yale University Press, 1985. The review was originally published under the title "Cambridge Russian who went home" in the *Financial Times*, 11 May 1985.

DOES PHYSICS EVER COME TO AN END? originally appeared in *Evolutionary Trends in the Physical Sciences* (eds. M. Suzuki and R. Kubo, Springer-Verlag, Berlin, 1991), © 1991, pp. 35–38.

MICROWAVE COOKING FOR "FOXES" is a review of *Mikro-Wellen Kochbuch für Füchse*, by H. Maier-Leibnitz and T. Cless-Berrert. 259 pages. Munich: Piper, 1986. The review was originally published in *Interdisciplinary Science Reviews*, **14** (1989), p. 187.

FIRST ENCOUNTER WITH ENGLISH FOOD originally appeared in *But the Crackling was Superb* (ed. N. Kurti, Adam Hilger, Bristol, 1993), pp. 203–205.

Subject Index

A

Abelson, Philip, 261
Accelerators, 95, 328
    cyclotrons, 19, 49, 120, 161
    European project, 220
    synchrotrons, 161
Acheson—Lilienthal report, 52, 231, 250
Advisory Council for Science, 91
Alpha-particles, 291
Alsos mission, 108, 329
Anderson, Carl, 49, 294
Arms Control and Disarmament Agency, 325
Arms race, 251–253, 264–268
Atomic bomb
    British program, 109, 111, 130, 167, 210–220, 229
    defense against, 195–197
    Frisch—Peierls Memorandum, 187–194, 228, 270
    German project, 49, 87–91, 102, 105–116, 228
    Manhattan Project, 50, 59, 88, 109, 113, 131, 154, 178, 229, 237–239, 269
Atomic energy
    academic research, 217–220
    control of, 231
    threat and promise, 198–209
Atomic Energy Authority (United Kingdom), 210–212
Atomic Energy Commission, 52, 54–55, 272–281, 284

Atomic Energy Research Establishment (Harwell), 134, 182, 210, 247
Atomic piles
    mobile, 206–208
    radioactive elements in, 208
Atomic power, production of, 200–203
Atomic Scientists Association, 213, 231
Atoms for Peace Conference, 92

B

Badash, Lawrence, 353
Baruch Plan, 52, 250
BCS theory, 95, 358
Becquerel, A.H., 258
Bell, John, 182–184
Bellamy, H., 137
Beta-decay, 12, 18, 183
Bethe, Hans, 113, 174, 253, 255, 324
Better a Shield Than a Sword: Perspectives on Defense and Technology (book review), 254–256
Bhabha, Homi, 149
Biological weapons, control of, 215
Blackett, P.M., 125, 294, 346
Bloch, Felix, 79, 94
Blount, Colonel B.K., 91
Bohr, Aage, 27
Bohr, Niels, 21–45, 58, 61, 90, 101, 106, 113, 129, 163, 176, 180, 259, 263–264, 315, 328
Bohr's Institute for Theoretical Physics, 119, 126, 316